Dynamical Astrochemistry

Dynamical Astrochemistry

David A. Williams
University College London, UK
Email: daw@star.ucl.ac.uk

Thomas W. Hartquist
University of Leeds, UK
Email: twh@ast.leeds.ac.uk

Jonathan M. C. Rawlings
University College London, UK
Email: jcr@star.ucl.ac.uk

Cesare Cecchi-Pestellini
INAF – Osservatorio Astronomico di Palermo, Italy
Email: cecchi-pestellini@astropa.unipa.it

and

Serena Viti
University College London, UK
Email: sv@star.ucl.ac.uk

ROYAL SOCIETY
OF CHEMISTRY

THE QUEEN'S AWARDS
FOR ENTERPRISE:
INTERNATIONAL TRADE
2013

Print ISBN: 978-1-78262-776-0
PDF eISBN: 978-1-78262-989-4
EPUB eISBN: 978-1-78801-270-6

A catalogue record for this book is available from the British Library

The Royal Society of Chemistry is a charity, registered in England and Wales, Number 207890, and a company incorporated in England by Royal Charter (Registered No. RC000524), registered office: Burlington House, Piccadilly, London W1J 0BA, UK, Telephone: +44 (0) 207 4378 6556.

Visit our website at www.rsc.org/books

Printed in the United Kingdom by CPI Group (UK) Ltd, Croydon, CR0 4YY, UK

Preface

Astrochemistry is now a well-established interdisciplinary subject. It originated in the astronomical identification of a wide variety of molecular species in interstellar space, a variety that now ranges from simple diatomics to cage molecules containing seventy carbon atoms, and—while confirmation is still lacking—there is strong evidence for much larger organic species. The total number of astronomically detected interstellar species is currently around two hundred and the detected isotopologues of these species bring that total to a much larger figure. The chemical networks that generate these species in different regions of interstellar and circumstellar space have been the subject of intense study for half a century. Models based on these networks imply that very many more molecular species must exist in interstellar space but are undetectable either because of low abundances or because those species have no suitable identified spectroscopic transitions by which they may be studied. The study of these networks has generated a huge demand for chemical reaction data in a variety of processes, including gas-phase reactions of many kinds, and at low and high temperatures, reactions occurring on the surfaces of interstellar dust grains of rather uncertain composition, solid-state reactions occurring within simple ices that can be deposited on these grains, and molecule formation by the erosion and destruction of grains. Astrochemistry has provided a significant stimulus to laboratory-based and theoretical studies of chemistry under physical conditions that are far from terrestrial.

At the same time, astrochemistry has revolutionized astronomy. It is obvious that radiation emitted by molecules can be a useful probe of density and temperature in the locations where the molecules are formed. For example, the rotational spectra of carbon monoxide and its isotopologues led to the discovery of huge amounts of cold gas within our own galaxy, the

Dynamical Astrochemistry
By David A. Williams, Thomas W. Hartquist, Jonathan M. C. Rawlings, Cesare Cecchi-Pestellini and Serena Viti

Published by the Royal Society of Chemistry, www.rsc.org

Milky Way, and in very many external galaxies. We now understand that these gas clouds, the so-called giant molecular clouds, provide the raw material for new stars and determine the star formation rates of galaxies. However, the information content in interstellar molecules is potentially much greater models of the chemical networks applied to a region of interstellar space where a variety of molecular species are observed depend on a number of fundamental parameters that categorize that region quite specifically, defining, for example, the local radiation field, the local flux of cosmic rays, the relative abundance of the elements, the mass ratio of gas to dust, and the degree of turbulence. Molecular observations interpreted in conjunction with sophisticated models of chemical networks can generate a remarkably complete description of a region of interstellar space. In this sense, molecules are diagnostics of interstellar space, particularly of those denser, darker, colder regions of space that before the advent of molecular astronomy were simply inaccessible to observations and therefore completely unknown to science.

The steady improvement in the amount and quality of chemical data, and in the sophistication of the models of the chemical networks, has led to a realisation that simple ideas that once seemed appropriate are no longer acceptable descriptions of the interstellar medium. Interstellar gas is now understood to be complex, always stirred into motion by the violent events such as novae, stellar winds, and supernovae explosions that occur within it; the gas dragged by gravity and ripped apart by galactic rotation. Shocks and turbulence are inherent within it, with turbulence operating on both large and small scales. Mixing of different types of gas occurs at turbulent interfaces. In particular, chemistry, in many circumstances, rarely has time to achieve steady-state in these environments. Most importantly, these rapidly evolving regions are some of the most interesting regions of interstellar space, and are the regions to which astrochemistry can make its most valuable contribution to astronomy.

This, then, is our motivation for writing this book. We argue that modern astrochemistry must embrace the evolving nature of the interstellar medium, a medium that is shocked and turbulent, and subject to large scale forces such as gravity or galactic rotation. It is now rarely adequate to describe an interstellar region as a spherical static cloud existing for an infinite time. The thrust of much of modern astronomy is the study of star and planet formation, and the impact of newly formed stars on their environments. We devote this book to studies of the dynamic interstellar medium through the chemical signatures that are developed in those regions. Our intended readership includes researchers in astrochemistry, researchers in astrophysics drawn into astrochemistry, and researchers in chemistry who have an interest in the many and intriguing problems arising in astrochemistry.

Chapter 1 gives a brief and simple introduction to astrochemistry, while Chapter 2 introduces formally the theory of shocks and turbulence. Chapter 3 describes the influence of turbulence on astrochemistry. Chapter 4 discusses the formation of stars, including both massive stars and of stars like the Sun. Stars have a dramatic impact on their environments, and some of these are discussed in Chapter 5, while the endpoints of stellar evolution, including stellar winds, novae and supernovae explosions, are discussed in Chapter 6. Chapter 7 makes some conclusions, lists areas where further work is required, and considers briefly the role of astrochemistry in the early Universe and in external galaxies.

David A. Williams, *University College London*
Thomas W. Hartquist, *University of Leeds,*
Jonathan M. C. Rawlings, *University College London,*
Cesare Cecchi-Pestellini, *Observatory of Palermo*
Serena Viti, *University College London*

Contents

Dynamical Astrochemistry
By David A. Williams, Thomas W. Hartquist, Jonathan M. C. Rawlings, Cesare Cecchi-Pestellini and Serena Viti

Published by the Royal Society of Chemistry, www.rsc.org

CHAPTER 1

Chemistry and Dynamics in the Interstellar Medium

1.1 Introduction

Eighty years ago, optical spectra of three simple molecular species, CH, CH^+, and CN, were observed in interstellar gas along lines of sight towards hot stars. These first detections stimulated detailed investigations of chemical reactions in interstellar space, a subject that has since become known as *astrochemistry*. The next detections, in 1963, were of maser emission at decimetric wavelengths from another simple hydride, interstellar OH, but the subject was totally transformed when millimetre and sub-millimetre wavebands became available for astronomy from the late 1960s. A flood of discoveries of interstellar molecular species was made in the following years and decades, and a high rate of discovery continues today. Observations confirm that by far the most abundant species of the interstellar molecules is molecular hydrogen; however, that species is difficult to detect except in particular circumstances. The next most abundant species is carbon monoxide, and it is readily detectable in its rotational spectrum. It is widely used to trace interstellar molecular gas in the Milky Way and other galaxies. Such observations led to the discovery of a new gaseous and largely molecular component in the Milky Way galaxy: the giant molecular clouds. These mainly molecular clouds have masses ranging up to one million solar masses ($M_{\odot}$), and are the main reservoirs of mass for the formation of new stars in the Milky Way.

The number of molecular species detected in interstellar and circumstellar regions of the Milky Way galaxy is currently (in 2017) about 200; about 60 of these species have also been detected in gas in external galaxies (a complete and up-to-date list is maintained by staff at the Universität zu

Dynamical Astrochemistry
By David A. Williams, Thomas W. Hartquist, Jonathan M. C. Rawlings, Cesare Cecchi-Pestellini and Serena Viti

Published by the Royal Society of Chemistry, www.rsc.org

Köln at https://www.astro.uni-koeln.de/cdms/molecules). These totals ignore the very large number of discoveries of *isotopologues*, *i.e.*, species that differ only in the substitution of isotopes, such as D for H, ^{13}C for ^{12}C, or ^{17}O for ^{16}O. The detected species include a large number of diatomics (more than 40) and triatomics (also more than 40), with numbers of detected species thereafter tending to decline with size. For example, there are currently just four detected molecular species containing 10 atoms, and the same is also true for species containing 11 and 12 atoms. Three large "cage" molecules (the fullerenes C_{60}, $C_{60}{}^{+}$, and C_{70}) have been detected in circumstellar regions, and—although no precise detections have yet been made—there is good reason to consider that the interstellar medium is populated by polycyclic aromatic hydrocarbons, each molecule containing several benzene-type rings and including perhaps up to 100 atoms. Evidently, the large masses of molecular gas in the Milky Way and other galaxies are chemically quite complex.

These astronomical observations stimulated intense activity in laboratory studies of gas-phase, surface and solid-state chemistry, and in computational modelling of chemistry in various types of astronomical regions. All these studies helped to establish astrochemistry, *i.e.*, the study of the chemical networks that are believed to lead to the formation of interstellar molecules in interstellar and circumstellar regions, and of the consequent harvest of astronomical understanding that such study made possible. In general terms, these networks are now reasonably well understood for most of the smaller molecular species (say, up to five atoms), although some outstanding problems remain. The chemistry governing the larger species is currently an active area of research. We shall summarize our present understanding of interstellar chemistry in Section 1.2.

From an astronomical perspective, observations of molecular spectra in interstellar and circumstellar regions are immediately useful: the basic observational data on spectral lines, their positions in the sky, relative strengths, and linewidths provide essential insight into the physical conditions of the regions in which the spectra originate by helping to constrain the gas number density and temperature in those regions. Shifts in spectral line positions enable the determination of the relative velocities of discrete structures in the interstellar gas, and linewidths may help to establish the presence and extent of some kind of turbulence in the gas. However, one can extract a much greater range of information from these observational data by using astrochemical models based on an understanding of the appropriate chemistry. As we'll see in the next section, the chemistry is determined not only by the density and temperature in the gas, but by other important astronomical parameters specific to the region in which the molecules are found. These include the local radiation field in various wavebands from UV to X-ray, the local flux of cosmic rays with energies in the MeV range, the local relative elemental abundances, and the physical and chemical nature of the local dust and the dust to gas ratio. Further, since interstellar chemistry is inherently time-dependent, observations may give insight into the timescales of relevant processes of the observed region and may help to describe its physical evolution. Thus, astrochemical models can provide very powerful diagnostics of physical conditions within a molecular region. The physics and chemistry of the interstellar medium are described in important texts by Draine (2011)[1] and by Tielens (2005).[2] An observational perspective of astronomical molecules is given in the textbook by Williams and Viti (2013),[3] and a study of the chemistry of cosmic dust may be found in the monograph by Williams and Cecchi-Pestellini (2016).[4]

Molecular observations and subsequent modelling have revealed a rich variety of molecular regions in the Milky Way and other galaxies. Some of these are quiescent; *i.e.*, essentially motionless, so that any chemistry occurring within them has ample time to reach steady state. Many diffuse clouds in the interstellar medium, in which chemistry is dominated by rapid photoionization and photodissociation by stellar radiation, may be in this condition. Also, some regions change, but so slowly that the chemistry is able to reach steady state in the conditions at any stage of evolution of that region. This may be the case, for example, in the very early stages of gravitational collapse of a diffuse gas cloud. However, such quiescence or quasi-quiescence certainly does not apply everywhere. Some physical changes are rapid. Regions may be affected by forces that cause changes to density, temperature and other important physical parameters on timescales that are rapid compared to the time required for the chemistry to attain steady state. For example, interstellar clouds may collide, and such collisions may lead to the generation of shock waves causing transient heating and compression of the gas, with consequent chemical changes. The later stages of gravitational collapse may accelerate gas to high velocity, generating turbulence that modifies the chemistry. Stellar outflows may impinge at high velocity on nearby gas causing turbulent mixing of gases at different temperatures and pressures. Stellar explosions such as novae and supernovae may accelerate gas to very high velocities with consequent changes in physical conditions and chemistry. In such cases, the chemistry in the gas is unlikely to attain steady state and so the molecules that can be observed are not necessarily those that arise in steady state. We shall briefly summarize the physical properties of various interstellar regions in Section 1.3 and Table 1.3 below.

Regions subject to such dynamical effects are some of the most interesting regions to study through their molecular spectra. Such studies reveal the evolution of intergalactic and interstellar gas during the formation of galaxies, stars, and planets, and the effects of stars on their environments. Studies of this kind cannot be approached through steady state models. The chemical networks that we shall discuss are highly nonlinear, and the approach to steady state is not a straightforward process.

This book is devoted to the astrochemistry of astronomical regions subject to rapid dynamical change and to the methods that may be useful in those studies. It is aimed primarily at astrophysicists who wish to investigate the dynamical interstellar and circumstellar media through their molecular emissions, and at astrochemists who wish to understand the sensitivity of interstellar and circumstellar chemistry to the physical environment.

The astronomical objects offered as examples in this book all belong to the Milky Way galaxy. However, the book's relevance is not limited to our galaxy. The physical and chemical processes that we shall describe are of general application to objects in all galaxies. Indeed, the value of astrochemical modelling is that it shows how chemical processes occur in regions of widely differing physical conditions.

1.2 Interstellar and Circumstellar Chemistry—A Brief Summary

There are three main types of chemical process occurring in interstellar and circumstellar regions, as physical conditions permit. These are:

(i) gas phase processes between atoms, ions and molecules, promoted by photoionization, photodissociation, and cosmic ray ionization;
(ii) reactions on the surfaces of bare grains, with either prompt or delayed ejection of products into the gas phase;
(iii) accumulation of molecular ices on the surfaces of dust inside dark clouds, with activation of these ices by cosmic rays or photons and subsequent reactions between the radicals so generated to form species of greater complexity.

An enormous amount of attention has been given to gas-phase schemes, *i.e.*, type (i), although one could properly argue that the truly fundamental process of interstellar chemistry in the local Universe (where dust is abundant) is of type (ii) in the formation of H_2. As we shall see in Section 1.2.1, this molecule is essential in all gas phase schemes based on type (i) chemistries, apart from those operating in the ejecta of novae and supernovae (see Chapter 6). Systems of type (ii) depend sensitively on the physical and chemical nature of the surface of the grains. The specific case of H_2 formation has recently received considerable attention in the laboratory and through *ab initio* calculations. It appears from recent research that the more realistic the physical description, the more favourable the H_2 formation process is predicted to be, and that the reaction is likely to occur over a significant temperature range. Type (iii) chemistries are now benefitting from detailed experimental and theoretical attention, and while there is a general consensus that schemes of this type can generate chemical complexity from simple initial compositions, there is as yet no agreement on the details.

The presence of dust grains involved in types (ii) and (iii) chemistries is inferred primarily from observational studies of interstellar extinction. These show that a range of sizes and compositions of dust grains must be present, and in particular that there are many more small grains than large.

A commonly adopted size distribution for assumed spherical grains of radius a is that $dn(a)/da$ varies as $a^{-3.5}$, within adopted minimum and maximum radii. An early model of this type for grains of silicates and graphite is called the Mathis–Rumpl–Nordsieck (or MRN) model. It was shown to be capable of fitting extinction data from the infrared to UV, and it has since been substantially modified and extended by Draine,[1] (see also ref. 4).

1.2.1 Gas-phase Chemistries

The basic chemical problem is to understand how the available elements O, C, N, S, *etc.*, as trace atoms in a gas of H and H_2, can combine to form molecules (see Table 1.1 for relative elemental abundances in the Sun, often used as a standard, and taken from Asplund *et al.* (2009);[5] slightly different values taken from studies of nearby B stars are also used—see Nieva and Przybilla 2012).[6] Most of the interstellar gas is at a sufficiently low temperature that energy barriers in, for example, the O–H_2 potential energy surface inhibit direct reactions between O and H_2 to form OH, which could undergo further reaction with H_2 to form H_2O. In localized regions of higher temperature, however, such reactions may proceed efficiently at nearly the total collisional rates. Direct atom-exchange reactions of this type are important in regions of elevated temperature that may arise in circumstellar regions, and may account for the abundances of hydrides in such regions.

Circumstellar regions may also be very dense compared to interstellar densities. At sufficiently high densities, above $\sim 10^{10}$ molecules cm^{-3}, three body reactions become important. In a system of three species, A + B + C, if a third body, A, collides with an already colliding pair, BC*, it may remove energy (indicated by *) from the pair, leaving it in a bound state, BC. Such reactions are generally efficient, and were particularly important for generating molecular hydrogen in high density, dust-poor conditions in the Early Universe.

1.2.1.1 Basic Oxygen and Carbon Chemistry

In the bulk of the interstellar gas, however, such reactions are inefficient, and the gas needs to be energized in some way. The principal energy

Table 1.1 Total relative elemental solar abundances (extracted from Asplund *et al.* 2009[5]) for elements X with an abundance greater than one part in 10^7 relative to hydrogen H.

X	[X]/[H]	X	[X]/[H]	X	[X]/[H]
H	1	Mg	4.37×10^{-5}	K	1.32×10^{-7}
He	9.55×10^{-2}	Al	2.95×10^{-6}	Ca	2.14×10^{-6}
C	2.95×10^{-4}	Si	3.55×10^{-5}	Cr	4.79×10^{-7}
N	7.41×10^{-5}	P	3.23×10^{-7}	Mn	3.31×10^{-7}
O	5.37×10^{-4}	S	1.45×10^{-5}	Fe	3.47×10^{-5}
Ne	9.33×10^{-5}	Cl	1.86×10^{-7}	Ni	1.74×10^{-6}
Na	2.04×10^{-6}	A	2.75×10^{-6}	Co	0.98×10^{-7}

sources available in interstellar clouds are cosmic rays and stellar UV radiation. X-rays may also be very important in localized circumstellar regions. Cosmic rays penetrate most interstellar regions, whereas stellar UV radiation is confined to diffuse clouds, where the visual extinction caused by interstellar dust is less than a few magnitudes, and to the peripheries of dark clouds. Cosmic rays ionize hydrogen atoms and molecules to provide H^+ and (mainly) $H_2{}^+$. This hydrogen molecular ion extracts a hydrogen atom from H_2 to create $H_3{}^+$, a stable triangular ion—protonated dihydrogen—which happens to have low proton affinity so it is capable of donating a proton to many other species, for example:

$$H_3{}^+ + O \rightarrow OH^+ + H_2.$$

This OH^+ ion can react directly and efficiently in successive ion–molecule reactions with H_2 to form H_2O^+ and H_3O^+:

$$OH^+ + H_2 \rightarrow H_2O^+ + H;$$

$$H_2O^+ + H_2 \rightarrow H_3O^+ + H.$$

Exothermic ion–molecule reactions of this type are efficient, occurring in almost every collision of the reacting pair. The ions created in these ion–molecule reactions rapidly react with electrons in dissociative recombination reactions to form O, H, OH, (hydroxyl) and H_2O (water). Subsequent reactions may begin to generate chemical complexity; for example, reaction of C^+ with H_2O generates HCO^+ (the formyl ion), which dissociatively recombines with electrons to give CO (carbon monoxide). Both HCO^+ and CO are very important observed interstellar species.

$H_3{}^+$ ions also provide an entry into carbon chemistry; the reaction of $H_3{}^+$ with a carbon atom forms CH^+, which (similarly to OH^+) can successively abstract H atoms from hydrogen molecules to form $CH_2{}^+$ and $CH_3{}^+$, while a slow radiative association of $CH_3{}^+$ with H_2 forms $CH_5{}^+$. Recombination of these ions with electrons may lead to C, CH (methyladyne), CH_2 (methylene), CH_3 (methyl), and CH_4 (methane), and reactions of these species may lead to further chemical complexity; for example, reaction of CH_3 with N atoms can generate hydrogen cyanide, HCN.

1.2.1.2 Basic Nitrogen and Sulfur Chemistry

These straightforward schemes do not operate for all species. Nitrogen atoms have a low proton affinity and cannot accept a proton from $H_3{}^+$, although N_2 molecules will do so, creating N_2H^+ (protonated nitrogen), which—after dissociatively recombining with electrons—can provide nitrogen hydride, NH. However, in a separate channel, nitrogen ions, N^+, created by cosmic ray ionization of N atoms, do react directly with H_2, creating NH^+ and—after successive H-atom abstractions—other hydride ions, leading to the formation, after recombination, of the hydrides NH, NH_2 and NH_3 (nitrogen hydride, aminyl, and ammonia).

Like nitrogen, sulfur has a lower proton affinity than H_2, so cannot accept a proton from H_3^+. The main entry to the chemistry of sulfur-bearing interstellar species seems to be in the reaction of S atoms with radicals such as OH and CH. Simple networks of atom exchange reactions generate species such as SO, SO_2, CS, HCS, and H_2CS (sulfur monoxide and dioxide, carbon monosulfide, thioformyl, and thioformaldehyde, respectively). Reactions of O atoms with thioformyl give carbonyl sulphide, OCS. All of these sulfur-bearing molecules are detected interstellar species.

The entry schemes indicated above are applicable in the bulk of the interstellar gas, which is cool, relatively dense, dark (*i.e.*, mostly opaque to stellar radiation), and mainly molecular. Diffuse clouds are warmer but still fairly cool, less dense, fairly transparent to stellar radiation, and of varying H/H_2 balance. Where H_2 is less abundant, the role of H_3^+ is consequently restricted. In that case, cosmic rays drive the chemistry by creating H^+ ions from H atoms. For example, accidental resonance charge exchange between H^+ and O atoms generates O^+ ions, which drives the oxygen hydride chemistry as described earlier. If unshielded by dust, the mean interstellar radiation field gives a high rate of photodissociation and photoionization, so that relative molecular abundances tend to be low in such regions.

1.2.1.3 More Complex Molecules

These entry routes show how a range of species involving a particular element may be established through gas-phase reactions activated by cosmic rays or stellar radiation. Chemical complexity may then grow rapidly as products from different entry routes react. For example, CH_3^+ and H_2O, being products from oxygen and carbon schemes, combine with the emission of a photon, $h\nu$, to form protonated methanol:

$$CH_3^+ + H_2O \rightarrow CH_3OH_2^+ + h\nu,$$

which dissociatively recombines with electrons to form methanol:

$$CH_3OH_2^+ + e \rightarrow CH_3OH + H,$$

an important species found in many interstellar and circumstellar locations. Similarly, CH_3^+ may also radiatively associate with methanol, ultimately to form dimethyl ether, CH_3OCH_3:

$$CH_3^+ + CH_3OH \rightarrow CH_3OCH_4^+ + h\nu$$

$$CH_3OCH_4^+ + e \rightarrow CH_3OCH_3 + H.$$

Evidently, chemical complexity may readily be attained in schemes such as these, which are initiated by ion–molecule reactions with hydrogen molecules, followed by dissociative recombinations and neutral–neutral exchanges. However, the number of possible reactions is enormous. It is now routine for computational models of interstellar or circumstellar chemistry to include several thousands of reactions. The requirement for accurate determinations of rate coefficients over appropriate temperature ranges is intense. Fortunately, databases of reactions of interest in interstellar and circumstellar chemistries, together with critically assessed values of their rate coefficients, are available to enable modellers to construct appropriate chemical networks.

1.2.1.4 Databases

The *UMIST Database for Astrochemistry* (udfa.ajmarkwick.net) is a long established database. The first release was in 1991, and the fifth release, named RATE12, was made in 2012 with some recent updates in 2014. A paper describing RATE2012 is McElroy *et al.* (2013).[7] The current database includes information on 6173 reactions, involving 467 species of possible relevance to interstellar and circumstellar chemistry. The reaction types include cosmic ray ionizations, ion–neutral and neutral–neutral exchanges, and dissociative recombinations, as discussed above, and a variety of other types including photoprocesses by the local stellar radiation field and also by the radiation field created in recombination following the cosmic ray ionization of molecular hydrogen. Reactions involving negative ions (radiative attachment, mutual neutralization) are also included. Searches in Udfa are made by species, and the database supplies lists of formation and destruction reactions of that species.

Rate coefficients k (cm^3 s^{-1}) for two-body reactions are fitted to a conventional Arrhenius-type expression of the form

$$k = \alpha\ (T/300)^{\beta} \exp\ (-\gamma/T)$$

where T is the gas temperature in Kelvin. For photoprocesses caused by the mean interstellar radiation field the rates (s^{-1}) are given in the form

$$k = \alpha \exp(-\gamma A_V)$$

where α for these processes represents the rate coefficient in the region unshielded by extinction caused by dust, A_V is the visual extinction due to dust at 550 nm at the relevant point in the region, and γ is a factor that computes the additional extinction at the wavelengths at which the photoprocess occurs. The visual extinction, A_V, is 1.086$\times$ the optical depth at the relevant point. It is important to note that these expressions for rate coefficients and rates are approximations, since the form of the spectral energy distribution (SED) of the local radiation field has been assumed, and the expression implies that the photoprocess occurs essentially at a single wavelength (not usually the case) and for a one-dimensional slab (almost certainly not the case). In many circumstellar situations, the SED of the local radiation field differs significantly from that of the mean interstellar radiation field, and the rates of photoprocesses may therefore need to be recalculated for specific purposes.

McElroy *et al.* (2013)[7] evaluate the gas-phase chemistry for a conventional model of a dark interstellar cloud and of a carbon-rich circumstellar envelope. They also compare their results for the dark cloud model with those arising from another database, the *Kinetic Database for Astrochemistry* (KIDA). The results are fairly similar.

KIDA (kida.obs.u-bordeaux1.fr) covers similar ground to Udfa, but also provides data on reactions of relevance to planetary atmospheres as well as the interstellar and circumstellar media. It also provides some information on surface reactions. A full description of KIDA can be found in Wakelam *et al.* (2012).[8] Reaction rate data are presented in a similar format to that in Udfa, and users may search for species (which may be either reactants or products) or for a particular reaction. An independent expert assessment of the data is also provided. Chemical networks for the study of dense interstellar clouds are revised annually, taking account of frequent updates. Codes for studying the time-dependence of the chemistry in zero- and one-dimensional regions are provided.

Both databases now include anion chemistry. For example, the associative detachment reaction

$$C^- + H_2O \rightarrow H_2CO + e$$

is a possible route to form formaldehyde. Associative detachment reactions are found to improve the match of predicted long chain hydrocarbon abundances with observations, so it appears that negative ions (anions) may play a significant role in gas phase chemistry alongside the chemistry driven by the reactions of positive ions (cations). However, anions are formed in radiative electron attachments to a neutral species, M,

$$M + e \rightarrow M^- + h\nu,$$

which may be rather slow.

1.2.2 Surface Processes on Bare Interstellar Grains

We saw in Section 1.2.1 that molecular hydrogen is an essential partner in the gas-phase ion–molecule chemistry that is important in interstellar and circumstellar regions. While some gas-phase routes to form H_2 exist, and may be important in some circumstances, it is clear these routes are not sufficiently rapid to account for the H_2 formation rates that are necessary in the diffuse interstellar medium to compete with the fast destruction of these molecules by starlight. By default, it became accepted that H_2 was formed in reactions on the surfaces of the bare dust grains of silicate or carbon that were evidently present in diffuse interstellar clouds. These rates of formation implied that most of the H atoms arriving at the surfaces of dust grains in the diffuse interstellar clouds must leave as part of an H_2 molecule. This stringent requirement was not able to be tested until the 1990s, when advances in laboratory techniques and in computational facilities enabled accurate and detailed investigations to be made into the interaction of H atoms with surfaces and the consequent formation of H_2 molecules.

Most of the theoretical and experimental treatments (see ref. 4) have considered the interaction of H atoms with graphitic surfaces, because these are well defined and are also probably appropriate in the interstellar medium. These studies confirm that the initial interaction is of a physisorption bond in which the atom is held at 0.3 nm from the surface. The experimentally determined bond energy on a single graphite layer is 39.2 ± 0.5 meV; the theoretical treatment shows this to be slightly site-dependent.

A physisorbed H atom may become chemisorbed on a graphitic surface with a bond energy of about 0.67 eV. On a perfect surface, the transition to chemisorption requires a surface distortion that implies an energy barrier of about 0.2 eV. This is sufficient to preclude H chemisorption on perfect graphite in the cool interstellar medium. However, deviations from surface perfection seem to permit barrier-free chemisorption. These deviations include the presence of a nearby H atom, or atomic substitutions in the lattice, or other lattice defects. A second H atom, arriving directly from the gas or from the surface physisorbed state is predicted by theoretical studies to react exothermically, providing an H_2 molecule that may be ejected promptly (with internal and kinetic excitation) or after a delay (with excitation corresponding to the grain temperature).

Experiments have been performed to study H_2 formation on a variety of surfaces, including graphitic material, silicates, and ices. Results show that both prompt and delayed ejection of H_2 may occur. Studies of H_2 formation on silicates (amorphous olivine films) show that H_2 formation from physisorbed H atoms is restricted to very low surface temperatures, formation *via* chemisorbed atoms may take place at elevated surface temperatures, depending rather sensitively on the chemisorption energy. It appears, therefore, that H_2 formation is possible not only in cold regions of interstellar space, but also occurs in regions that are warm (such as photon-dominated regions, PDRs) and in which the grains themselves can be heated to temperatures as great as a few hundred Kelvin.

Taken together, the theoretical and experimental studies present a consistent picture of the formation of H_2 on the surfaces of interstellar dust. H atoms collide with and stick to the surfaces of grains. Subsequent collisions of the bound H atoms with incident H atoms produce H_2 molecules with high efficiency, and at least some of the products are released promptly with internal and kinetic excitation. It is not possible to determine from theory or laboratory experiment *ab initio* an H_2 formation rate, as this depends on the model of dust that is adopted. However, using conventional models of dust with the results of the theoretical and experimental studies, the H_2 formation rate is consistent with that inferred from H_2 measurements of optically thin H_2 along lines of sight to bright stars, $\sim 10^{-17} n_H n(H) T^{1/2}\,cm^{-3}\,s^{-1}$ where $n_H = n(H) + 2n(H_2)$, and $n(H)$ and $n(H_2)$ are the number densities of H and H_2, respectively.

The possible contribution of surface reactions forming hydrides other than H_2 or other species remains controversial. There have been laboratory studies of sequential hydrogenation of nitrogen atoms in various matrices and ice analogues. These confirm that hydrogenation of N atoms is efficient and probably barrier-free. Modelling studies show that if this is the case, then the observed abundances of NH, NH_2, and NH_3 along low density diffuse paths can be accounted for, while gas-phase reactions significantly fail to provide all three nitrogen hydrides. Similar studies of oxygen hydride formation in surface reactions have also been made, but it appears that these do not affect the predicted oxygen hydride abundances significantly because the gas phase routes forming them are efficient, whereas gas-phase entry routes to nitrogen hydrides are slow. Nevertheless, the existence of interstellar water ice (see Section 1.2.3) is clear proof that the hydrogenation of O atoms does occur efficiently on grains.

1.2.3 Chemistry in Interstellar Ices

Taken together, gas-phase reactions (type (i) chemistry) with molecular hydrogen (and possibly other hydrides) provided by surface reactions (type (ii) chemistry) are very successful in providing viable formation routes and reasonable matches (say, within a factor of about three) to the observed abundances of many of the molecular species detected in interstellar clouds. However, this fair agreement tends to disappear for larger species (up to about 10 atoms), so that there may be a discrepancy of one or more orders of magnitude between observational and computed abundances. Indeed, in some cases, there may be no viable chemical pathway to form a particular molecular species. For example, no viable pure gas-phase route to the detected and fairly abundant species propylene (CH_3CHCH_2) appears to exist.

There is one remaining type of chemistry that demonstrably generates chemical complexity from simple species. This is type (iii): the processing of interstellar ices. In darker regions of the interstellar medium, interstellar dust grains accumulate mantles of dirty ice containing large amounts of simple molecular species that might be considered as building blocks of larger molecules (see Table 1.2 for relative abundances of ice species, as measured in different types of region). For example, under irradiation water might be a source of OH, methane a source of CH_3 and these two radicals

Table 1.2 Typical relative abundances and (lower, upper) quartiles of molecular components in interstellar ices, as measured in different interstellar regions (extracted from Öberg *et al.* 2011[11]). *X*CN is believed to be OCN^-.

Component molecule	Low mass stars	High mass stars
H_2O	100	100
CO	38 (20, 61)	13 (7, 19)
CO_2	29 (22, 35)	13 (12, 22)
CH_3OH	7 (5,12)	8 (8, 16)
NH_3	5 (4, 6)	16 (10, 17)
CH_4	5 (4, 7)	4 (2, 4)
*X*CN	0.6 (0.2, 0.8)	0.8 (0.4, 1.4)

may combine in appropriate circumstances to form methanol, CH_3OH. Methanol could be a source of the CH_2OH radical, which might react with a CH_3 radical to form ethanol, CH_3CH_2OH. Methanol might also be a source of the radical CH_3O, which could in suitable circumstances combine with CH_3 to form CH_3OCH_3, dimethyl ether. All these product molecules have been detected in the interstellar medium. In principal, quite complex species would be built if these radicals could be made available under appropriate conditions. Hence, type (iii) chemistry, *i.e.*, the chemical processing of ices, has become and remains an important topic of theoretical[9] and experimental[10] study.

Since the simple molecules present in the ice do not react with each other at the temperatures at which ice exists in interstellar clouds ($\lesssim$ 100 K), the ices must be activated in some way, to create the reactive radicals. Recent experiments have explored the products of low temperature ices exposed to UV, X-rays, energetic electrons, fast protons and heavy nuclei.[10,12–14] All methods of excitation produce similar results, although some experiments produce a wider range of product species than others. Essentially, the important step is to create radicals within the ice. Studies of methanol ice and methanol ice mixed with CO or CH_4 and irradiated by UV have identified products CO, CO_2, CH_4, HCO, H_2CO, CH_2OH (hydroxymethyl), CH_3CHO (methyl formate), CH_3OCH_3 (dimethyl ether), CH_3CH_2OH (ethanol), $(CH_2OH)_2$ (ethylene glycol) and other species. Evidently, chemistry of type (iii) is very effective in the laboratory, and the range of products is very suggestive of radical–radical association chemistry even though the details of the processing remain to be firmly established. Laboratory work is moving on from the qualitative phase, and is now providing quantitative results on the formation of particular species by specific irradiation of ices (*e.g.*, Öberg *et al.* 2009)[10] and enabling a coherent picture of ice evolution, based on observational data, to be established.[11]

Theoretical approaches also support the view that radicals generated and contained within ices can contribute to the formation of a very wide range of complex species. The conventional approach is to assume that UV photons from a weak cosmic ray generated radiation field are responsible for photodissociation, leading to the presence of radicals within the ice. If these radicals are mobile in the ice, then a vast variety of organic species of the kind listed above can be created. Computational methods involving both rate equations[15,16] and Monte Carlo approaches[17] have been developed. It is difficult to treat this problem with sufficient accuracy, because the binding energies and mobilities of the many species involved are very poorly known.

Molecules that are the products of reactions in the ices may be retained in the ices until the grains pass near to a star and are heated so that thermal desorption and ice evaporation occurs. In addition, non-thermal desorption may occur, even far from a star.[18] This desorption may be driven by the weak cosmic ray generated radiation field that initiates the solid-state chemistry, as discussed in the previous paragraph. In addition, a fast cosmic ray particle may, when passing through the ice, deposit enough heat to cause local warming. Another source of energy is the energy released when reactions occur in or on the ice. The most promising exothermic reaction appears to be the formation of H_2 from H atoms. Alternatively, intermittent heating events may generate an extremely dense evaporating gas in which three-body reactions occur, so that the radical associations occur very rapidly. Radical associations may lead directly to saturated molecules, which take no further part in the chemistry (*e.g.*, $2CH_3 \rightarrow C_2H_6$ (ethane)), or they may generate radicals (*e.g.*, $CH_3 + CH_2 \rightarrow C_2H_5$), which may proceed to a second stage of reaction (*e.g.*, $C_2H_5 + OH \rightarrow C_2H_5OH$). There are well over one hundred such possible products, many of which are observed species in the interstellar medium.[19]

There is a severe lack of fundamental data on these proposed mechanisms. Nevertheless, it is encouraging that the envisaged processes should be able to generate efficiently a great variety of the larger organic species detected in the interstellar medium and which seem to present problems for gas-phase schemes. The results of laboratory studies of ice processing stimulated by a variety of mechanisms are unequivocal: ice processing is effective in generating chemical complexity from simple reactants.

It should be noted, however, that in using molecules as diagnostics of physical conditions in various interstellar and circumstellar regions, it is currently the rather simple and small molecules that are the most effective. These molecules are likely to have their main chemistry determined by gas-phase processes (type (i) chemistry). It is possible that when the chemistry of the larger species (type (iii) chemistry) is better constrained, the larger molecules will also become important diagnostics of astrophysical conditions.

1.3 Interstellar and Circumstellar Dynamics

On current estimates, the total mass of the Milky Way galaxy is about 10^{11} $M_\odot$ of which about 7% is gas. There are many phases of this gas. Two ionized phases are observed; these include very diffuse ionized gas at a temperature $\sim 10^6$ K, and somewhat denser ionized gas near bright stars with a temperature of $\sim 10^4$ K. Together, these hot ionized phases probably occupy more than half the interstellar volume, but contain only about 0.25% of the total mass of the interstellar medium of the Milky Way galaxy.

The mainly neutral gas contains most of the mass of gas, and also occurs in several phases. Warm neutral gas (typically, $n_H \approx 0.6$ cm^{-3}, $T_K \approx 6000$ K) occupies much of the remainder of the volume and includes about a quarter of the total mass of gas. Several other phases are identified: cool neutral gas ($n_H \approx 30$ cm^{-3}, $T_K \approx 100$ K), diffuse gas ($n_H \approx 100$ cm^{-3}, $T_K \approx 80$ K), and dense molecular gas ($n_H \approx 10^3 - 10^7$ cm^{-3}, $T_K \approx 10 - 300$ K). The three densest phases contain about $\frac{3}{4}$ of the total interstellar mass but occupy only about 1% of the interstellar volume. It is in this 1% of volume (containing most of the interstellar mass) that many dramatic astronomical events occur. Fortunately, these regions are also where molecules are abundant and can be useful diagnostic probes of the physical conditions. Such work leads to a

Table 1.3 Types of molecular regions in interstellar and circumstellar space, with approximate typical values of various physical parameters (data taken from van Dishoeck *et al.* 2013[20] and modified and extended).

Type of region	n_H(cm^{-3})	*T*(K)	A_V(mag)	Hydrogen state	Signature molecule	Ice?	Dynamical age (My)
Diffuse	10^2	30–100	≤ 1	H, H_2	H_2 (UV)	N	10
Translucent	10^3	15–50	1–5	H_2, H	CO	N	3
Dense PDR[a]	10^4–10^5	50–500	≤ 10	H_2, H	H_2 (IR)	N	1
Cold dense (IRDC[b])	10^4–10^5	10–20	≥ 10	H_2	CO	Y	1
Pre-stellar core	$\geq 10^5$	8–15	10–100	H_2	NH_3	Y	0.3
Protostellar envelope							
—Cold outer	10^4–10^7	8–100	10–100	H_2	H_2CO	Y	0.1
—Warm inner[c]	10^7–10^9	≥ 100	100–1000	H_2	$HCOOCH_3$	N	0.1
Strong shock	10^4–10^5	≥ 1000	≤ 3	H_2, H	SiO	N	0.01
Protoplanetary disc							
—Outer	10^6–10^{10}	10–500	≤ 100	H_2	H_2CO	Y, N	0.01
—Inner	10^9–10^{15}	100–3000	≤ 1000	H_2	N_2H^+	N	0.01
Evolved stellar envelope							
—Outer	$\leq 10^8$	10–100	≤ 50	H_2	HC_3N	N	0.01
—Inner	10^{10}–10^{13}	100–2000	≤ 2000	H_2	HCN	N	0.01

[a]PDR = photon dominated region.
[b]IRDC = infrared dark cloud (birthplace of massive stars).
[c]The warm inner regions are also known as *hot cores*.

more detailed description of neutral interstellar gas, and Table 1.3 lists the classifications obtained from observations of the molecular regions in the interstellar medium, together with indicative values of various physical and chemical parameters associated with each class.

The interstellar medium is a dynamic environment. On the large scale, galactic rotation stirs all the gas. On a local scale, gravitational collapse of gas clouds leads to infall and possibly to star formation. The newly formed stars generate jets, outflows and winds that impinge on the progenitor clouds, causing shocks and turbulence that have characteristic chemical signatures. Fast stellar outflows may also accelerate clumps of neutral gas to exceptionally large velocities, generating high speed dissociative shocks. At later stages of star formation, stellar winds may impinge on circumstellar planet-forming discs, eroding them on relatively short timescales and controlling planet formation. Depending on their masses, evolved stars may generate huge explosions as novae and supernovae and mix their ejecta with the surrounding interstellar gas. This process enriches the interstellar medium with newly formed dust grains and with atomic isotopes, some of which may previously have been absent from the gas. The mixing process is turbulent and violent. Supernovae explosions are responsible for heating the very tenuous million-degree gas through high speed shocks, and the supernova rate is adequate to maintain gas at this high temperature on a galactic scale. Supernovae explosions also impinge more locally on cooler denser gas and may be responsible for creating structure within neutral gas clouds and initiating further (sequential) star formation. On a larger scale, powerful winds from active nuclei at the centres of galaxies may cause galaxy-wide phenomena.

A few images illustrating interstellar situations in which dynamics plays an important role are shown in Figures 1.1–1.4. In Figure 1.1, a false colour infrared image, the 20 $M_{\odot}$ runaway star ζ Ophiuchi is moving to the left in this image at about 24 km s^{-1}. Its wind compresses the ambient interstellar material, shocking and heating it. Emission from the shocked region is shaped by the relative motion into the characteristic bow structure of shocks (shock theory is discussed in Chapter 2). In Figure 1.2, the image is of the Herbig–Haro region HH24 and shows tightly confined jets emerging from an obscured newly forming star (HH objects will be discussed in Chapter 5). Variations in the velocity of the jets generate shocks within them, and the leading fronts of the jets generate bow shocks. At a later stage, these jets will become less confined and will open out to become wider outflows, eventually becoming more general stellar winds (these topics will be discussed in Chapters 4 and 5). In Figure 1.3 the Hubble image is of the Veil Nebula, an edge-on view of a portion of the interaction of a supernova blast wave with interstellar gas (interstellar shocks of various kinds will be met in Chapter 3 and chemistry in the ejecta from stellar explosions will be described in Chapter 6). The image shows emission from the very detailed structure of the shock. In this image, the emission is from atoms: red is emission from H atoms, green from S atoms, and blue from O atoms. Figure 1.4 shows

Figure 1.1 The 20 $M_\odot$ star ζ Ophiuchi is moving towards the left in this infrared false-colour image, with a velocity of about 24 km s^{-1} relative to the local interstellar gas. The stellar wind impacts on the interstellar gas and creates a characteristic bow shock, heating the gas so that it emits in infrared lines.
© NASA/JPL-Caltech.

dynamics on a galactic scale; the image is of M82, a massive starburst galaxy. The intense starburst activity in the galaxy's central regions ejects hydrogen from interstellar clouds out of the galactic plane.

1.4 Structure of This Book

The interstellar medium is a fluid that may be acted on by various forces, arising from pressure gradients, gravity, and (since the gas is partially ionized) electrical and magnetic constraints. In extreme cases, the response to these forces sets up a shock, in which a pressure-driven disturbance propagates faster than the local signal speed. These shocks can be the most dramatic evidence of dynamics in the interstellar medium, and shocked regions have characteristic spectral signatures, as the figures shown in the previous section illustrate. We discuss in Chapter 2 the equations of fluid mechanics for neutral and partially ionized fluids and the "jump" conditions that relate the preshock and postshock states of the fluid. Applications of theory presented in Chapter 2 are given in Chapter 3, in which the chemistry in shocked and turbulent regions of various types is explored.

The collapse of gas clouds under gravity may lead to star formation. Astrochemistry provides a diagnostic tool for following that collapse to very late stages. We describe the physics and chemistry of star-forming regions in Chapter 4. Newly formed stars are seen to exhibit fast jets, confined outflows,

Figure 1.2 An image of the Herbig–Haro 24 region in the Orion B molecular cloud. A protostar is accreting material from a rotating accretion disk, but the protostar itself is optically obscured by dusty infalling gas. The star generates highly collimated jets that lie along the rotation axis of the system, and shocks within the jets make the jets visible.
© ESA/Hubble & NASA, D Padgett/GSFC, T Megeath/University of Toledo and B Reipurth/University of Hawaii.

and winds. These may impact on nearby gas at high speed, so that mixing layers of disparate gases are set up. These layers may generate characteristic signatures. The stellar winds may also impact on circumstellar protoplanetary disks, if present. Other types of interface may also be generated in circumstellar regions. These circumstellar topics are covered in Chapter 5.

Mass loss from evolved stars leads to the formation of nebulae called planetary nebulae. It's a misleading title; these objects have nothing to do with planets. It is simply that early images showed extended objects, like planets as opposed to point-like stars. Stars towards the ends of their lives may take part in explosions, as novae (in the case of low mass stars) or as supernovae (for high mass stars). These events are important in the cycle of material that enriches the interstellar medium with heavy elements and dust. Astrochemistry can be used to study the mixing of stellar and circumstellar gas. These topics are described in Chapter 6. Some other topics are dealt with in Chapter 7 in which the state of this research area is described and further work is indicated.

Figure 1.3 An image of a portion of the Veil Nebula, a supernova remnant. A 20 $M_\odot$ star in the Cygnus constellation exploded some 8000 years ago to create a bubble of very hot gas expanding very rapidly into the surrounding cool interstellar gas. This image shows a small portion of the shocked gas in the impact zone, with filaments of red emission from atomic hydrogen, green from sulfur and blue from oxygen.
© NASA, ESA, and the Hubble Heritage Team (STScI/AURA).

Figure 1.4 An image of the starburst galaxy M82, showing masses of atomic hydrogen (in red) being ejected from the centre of the galaxy at high speed.
© NASA. ESA, and the Hubble Heritage Team (STScI/AURA), J. Gallagher (university of Wisconsin), M. Mountain (STScI), and P. Puxley (National Science Foundation).

In this book we follow the current (rather idiosyncratic) conventions of astrochemistry. Number densities are normally given per cubic centimetre (cm^{-3}) so that bimolecular rate coefficients have units of $cm^3\ s^{-1}$. Energies may be given in a variety of units, as appropriate; for example, as equivalent temperatures, or in electron volts (eV), or in CGS or SI units. All equations using electromagnetic terms are written in SI units, so that charge is measured in Coulombs and magnetic field is measured in Tesla. Relatively small (often near-stellar) astronomical distances may be given in astronomical units (au). The au is defined as the mean Earth–Sun distance (1.496×10^{11} m). Larger distances, such as that between near-neighbour stars, or the size of molecular clouds, may be given in parsecs. One parsec is the distance at which the diameter of the Earth's orbit subtends one second of arc, and is 3.08×10^{16} m. Distances within galaxies are more conveniently measured in kiloparsecs (kpc) and for distances between galaxies megaparsecs (Mpc) may be a more useful unit.

The amount of dust along a particular path is measured by the extinction along that path. Extinction A_λ in magnitudes at a wavelength λ is a logarithmic measure of the reduction in brightness of an object, defined by $A_\lambda = 2.5\ \log_{10}(I_0/I)$, of an object whose measured brightness is I but whose brightness in the absence of dust along the line of sight would have been I_0. On a sufficiently large scale in the Milky Way galaxy, dust and gas are well mixed, and dust causes about 1.8 magnitudes of extinction per kiloparsec at a visual wavelength ($\lambda = V$) of 550 nm. Of course, more dust means more gas, and the column density of hydrogen atoms (counting H in both H atoms and H_2 molecules) along a path with visual extinction A_V is $N_H = 1.9\times10^{21} A_V\,\mathrm{cm}^{-2}$. Thus, extinction can be used as a measure of the amount of gas that is present along a line of sight.

References

1. B. T. Draine, *Physics of the Interstellar and Intergalactic Medium*, Princeton Series in Astrophysics, 2011.
2. A. G. G. M. Tielens, *The Physics and Chemistry of the Interstellar Medium*, Cambridge University Press, 2005.
3. D. A. Williams and S. Viti, *Observational Molecular Astrophysics*, Cambridge University Press, 2013.
4. D. A. Williams and C. Cecchi-Pestellini, *The Chemistry of Cosmic Dust*, Royal Society of Chemistry, 2016.
5. M. Asplund, N. Grevesse, A. Sauval and P. Scott, *Annu. Rev. Astron. Astrophys.*, 2009, **47**, 481.
6. M.-F. Nieva and N. Przybilla, *Astron. Astrophys.*, 2012, **539**, 143.
7. D. McElroy, C. Walsh, A. J. Markwick, M. A. Cordiner, K. Smith and T. J. Millar, *Astron. Astrophys.*, 2013, **550**, 36.
8. V. Wakelam, E. Herbst, J.-C. Loison, I. W. M. Smith and 30 co-authors, *Astrophys. J. Suppl.*, 2012, **199**, 21.
9. R. T. Garrod and S. Widicus Weaver, *Chem. Rev.*, 2013, **113**, 8939.
10. K. I. Öberg, R. T. Garrod, E. F. van Dishoeck and H. Linnartz, *Astron. Astrophys.*, 2009, **504**, 891.

11. K. I. Öberg, A. C. A. Boogert, K. M. Pontoppidan, S. van den Broek, E. F. van Dishoeck, S. Bottinelli, G. A. Blake and N. J. Evans II, *Astrophys. J.*, 2011, **740**, 109.
12. Y.-J. Chen, K.-J. Chuang, G. M. Muñoz-Caro, M. Nuevo, C.-C. Chu, T. S. Yih, W.-H. Ip and C.-Y. R. Wu, *Astrophys. J.*, 2014, **781**, 15.
13. C. J. Bennett, S.-H. Chen, B.-J. Sun, A. H. H. Chang and R. I. Kaiser, *Astrophys. J.*, 2007, **660**, 1588.
14. A. L. F. de Barros, A. Domaracka, D. P. P. Andrade, P. Boduch, H. Rothard and E. F. da Silveira, *Mon. Not. R. Astron. Soc.*, 2011, **418**, 1363.
15. R. T. Garrod, S. Widicus Weaver and E. Herbst, *Astrophys. J.*, 2008, **682**, 283.
16. R. T. Garrod, A. I. Vasyunin, D. A. Semenov, D. S. Weibe and Th. Henning, *Astrophys. J.*, 2009, **700**, L43.
17. A. I. Vasyunin, D. A. Semenov, D. S. Wiebe and T. Henning, *Astrophys. J.*, 2009, **691**, 1459.
18. J. F. Roberts, J. M. C. Rawlings, S. Viti and D. A. Williams, *Mon. Not. R. Astron. Soc.*, 2007, **382**, 733.
19. J. M. C. Rawlings, D. A. Williams, S. Viti, C. Cecchi-Pestellini and W. W. Duley, *Mon. Not. R. Astron. Soc.*, 2013, **430**, 264.
20. E. F. van Dishoeck, E. Herbst and D. A. Neufeld, *Chem. Rev.*, 2013, **113**, 9043.

CHAPTER 2

Shocks and Turbulence and Their Effects on Chemistry

2.1 Introduction

The eminent Dutch astronomers J. H. Oort and H. C. van de Hulst suggested that collisions between interstellar clouds affect the composition of interstellar matter just as they and others were beginning to understand the distribution of diffuse interstellar clouds.[1] Since such collisions are supersonic, they generate shocks.

As mentioned in Chapter 1, the first detections of spectral line features formed due to the absorption by interstellar molecules of optical light from background stars occurred in the late 1930s. Though J. Hartmann observed analogous calcium atomic ion features as early as 1904,[2] the interstellar, rather than circumstellar, origin of the atomic ion features was first firmly established in 1923 by J. S. Plaskett.[3,4] This advance led A. S. Eddington to devote his influential 1926 Royal Society Bakerian Lecture[5] to the then very young field of "diffuse matter in interstellar space"; he correctly inferred that the average density of interstellar matter is about 10^{-24} g cm^{-3}. The recognition of the cloudy nature of the distribution of the interstellar gas began in 1936 when C. S. Beals noted that the line profiles that he analysed suggested that there might be "some irregularity" in the distribution of the interstellar gas.[6] Subsequent work reported by W. S. Adams in a paper published in 1943 showed clearly that the calcium is concentrated in separate clouds.[7]

Studies of continuum, as well as line, absorption of stellar radiation contributed to the understanding of the nature of interstellar clouds. In his 1968 book *The Bowl of Night*, F. P. Dickson[8] described how in 1744 J. P. L. de Chesaux, who assumed that the Universe is infinite and Euclidean,

Dynamical Astrochemistry
By David A. Williams, Thomas W. Hartquist, Jonathan M. C. Rawlings, Cesare Cecchi-Pestellini and Serena Viti

Published by the Royal Society of Chemistry, www.rsc.org

suggested that continuum interstellar absorption causes the night sky to be dark. In 1826 H. W. M. Olbers independently advanced a similar argument. In fact, such absorption would give rise to the re-emission of radiation at longer wavelengths, which would prevent the sky from being dark at those longer wavelengths. The resolution of the so-called Olbers's Paradox concerning the darkness of the night sky is now known to lie in the non-Euclidean nature of the Universe. However, continuum interstellar absorption does occur. G. Alter and later A. Li reviewed the subsequent history of the study of interstellar continuum absorption and the dust that causes it.[9,10] Although W. Herschel did not mention absorption, in 1785 he famously wrote "... for instance, in the body of the Scorpion is an opening, or hole, ...", which is now known to be due to absorption by interstellar dust.[11] In 1847 W. Struve attributed the apparent reduction of the number density of stars with distance from the Sun to absorption and estimated that visible light is attenuated by a factor of about 2.5 over a distance of 1 kpc, which is roughly 3×10^{19} m; Struve's value of the factor is about one half the currently accepted value. Li[10] has credited R. J. Trumpler with providing in 1930 "the first unambiguous evidence for interstellar absorption and reddening, which led to the establishment of the existence of interstellar dust". Li's comment is relevant to continuum, rather than line, absorption, and reddening refers to the attenuation of higher frequency light exceeding that of red light. S. Chandrasekhar attributed "the commonly accepted notion that interstellar matter occurs in the form of clouds" to V. Ambartsumian, who performed statistical analyses of the continuum absorption of starlight.[12] In Chandrasekhar's own influential paper, co-authored with G. Münch and reporting results for the mean and mean square deviation of the optical depth of interstellar clouds,[13] he and Münch referred to key papers of Ambartsumian and collaborators published in 1938, 1940 and 1946.

In an article published in 1946 Oort and van de Hulst argued that the distributions of interstellar gas and dust are identical.[1] They noted that elements more massive than hydrogen and helium can be removed from the gas phase and incorporated in the dust. They also suggested that some mechanisms act to remove material from the dust and inject it back into the gas phase. They considered the possibility that supersonic collisions between interstellar clouds drive grain–grain collisions that lead to the injection. In 1965 N. C. Wickramsinghe showed that sputtering of grains by gas-phase material heated in cloud–cloud collisions also contributes significantly to the destruction of grains.[14] In the following year T. P. Stecher and D. A. Williams proposed that the elevated gas-phase temperatures induced by interstellar cloud – cloud collisions result in the reaction barriers of chemical exchange reactions on grain surfaces being overcome and significant gas-phase molecular production.[15] At nearly the same time T. O. Carroll and E. E. Salpeter considered the effects of the elevated temperatures produced in cloud–cloud collisions on gas-phase chemistry affecting the abundance of OH and suggested that CH^+ might be produced in such environments by the endothermic reaction between C^+ and H_2.[16]

In 1973 P. Aannestad[17] included sputtering, gas-phase chemical reactions and cooling by molecular line emission in more detailed calculations of the thermal and density structures and molecular abundance profiles behind shocks propagating with speeds of 5–18 km s^{-1} into non-magnetised clouds with atomic pre-shock hydrogen number densities of close to 50 cm^{-2} and fractional H_2 abundances as high as 2×10^{-2}. Five years later E. R. Iglesias and J. Silk reported results for molecular abundances produced behind 10 km s^{-1} shocks propagating through higher density, primarily molecular, non-magnetised clouds.[18]

A number of molecular observations made in the early part and middle of the 1970s stimulated work on molecular level populations and astrochemistry in shocks in molecular clouds. For example, P. A. Aannestad and G. B. Field suggested that the interstellar ultraviolet absorption features of rotationally excited H_2 detected using the Copernicus satellite[19] originate in shocked interstellar clouds.[20] Other important observations that stimulated advances in shock modelling include some of those referenced by B. Zuckerman and P. Palmer who reviewed results for the Orion infrared nebula/molecular cloud showing the existence of a 4 km s^{-1} wide spike superposed with a roughly 30 km s^{-1} wide plateau in each of a number of emission lines;[21] for some molecular lines, *e.g.* those of H_2S and SiO, the ratios of the strengths of the plateaus to the spikes are markedly higher than for others due to chemistry occurring in shocked gas in the region.[22] The detection by T. N. Gautier and his collaborators of H_2 infrared emission produced by the radiative decay of vibrationally excited levels in the Orion region[23] stimulated many of the developments in the modelling of shocks in the magnetised molecular regions described below.

Cloud–cloud collisions do not provide the only mechanism for producing shocks in molecular astronomical sources. Shocks can be driven by the propagation of ionisation and photodissociation fronts formed when stars with sufficient ultraviolet luminosities are born, by the winds and outflows of young and evolved stars, by pulsations in stars, by supernovae and by galactic superwinds produced when multiple supernovae occur in starburst regions. They can also be driven by the nonlinear steepening of finite amplitude waves and by closely related nonlinear processes operating in turbulent regions.

In this chapter we shall provide introductions to hydrodynamics, which is also called fluid mechanics, and magnetohydrodynamics (MHD) and their applications to shock structures. We shall also introduce some concepts concerning turbulence and outline how it can affect the chemistry in, and the structure of, some astronomical sources. The next section concerns the equations of hydrodynamics and their use in studying sound wave propagation and the change of flow parameters caused by a shock. Section 2.3 contains a description of chemistry occurring at temperatures of about 10^3 K and its importance for the cooling that affects postshock dynamics. Section 2.4 is an introduction to the equations of single-fluid MHD and their application to wave propagation and the change of parameters caused by a shock. The topics of the subsequent section are multi-fluid MHD shock models and some of their applications in astrochemistry. A more detailed exposition of the dynamical effects, and destruction, of grains in such shock models is found in Section 2.6. Section 2.7 concerns some of the basic concepts of hydrodynamic and MHD turbulence. It also contains a summary of some proposed ways in which turbulence affects the structures of astronomical sources.

2.2 Basics of Single-fluid Hydrodynamics

Most media consist of multiple components. For example, one of the components of the Earth's atmosphere is composed of neutral nitrogen molecules, while neutral oxygen molecules comprise another and trace ions make up a third. Each component is a collection of particles.

A necessary condition for a fluid description of the dynamics of one of the components to be valid requires that the collision mean free path between the particles in that component is short compared to the length scale over which the mean flow parameters (*e.g.* density) vary. When this condition is satisfied, the physical properties of the flow are specified well by the position-dependent and time-dependent values of the density, mean velocity and temperature of the particles.

When the dynamics of all of the components of the medium can be treated adequately with fluid descriptions, and the mean velocities and temperatures of all of the fluids equal one another, the medium can be well modelled as a single fluid.

2.2.1 The Single-fluid Hydrodynamic Equations

By making appropriate assumptions one can derive the single-fluid equations by calculating velocity moments of the Boltzmann equation, which in turn can be obtained through the integration of the Liouville equation over the coordinates and velocities of all but one particle. However, the single-fluid equations can also be derived from a consideration of conservation laws and Newton's law of motion.

2.2.1.1 Mass Conservation

Consider the mass M in a volume V bound by a closed surface S. The time rate of change of the mass in that volume is

$$\partial M/\partial t = \int_V \partial\rho/\partial t \; \mathrm{d}V, \tag{2.1}$$

where ρ, t and $\mathrm{d}V$ are the mass density, time and infinitesimal volume element, respectively. The integral is over the specified volume. Mass will be gained or lost through the surface enclosing it at a rate given by

$$\partial M/\partial t = -\int_S \rho(\boldsymbol{v}\cdot\boldsymbol{n}) \; \mathrm{d}S \tag{2.2}$$

where $\boldsymbol{v}$, $\boldsymbol{n}$ and $\mathrm{d}S$ are the velocity, unit vector normal to the surface and infinitesimal surface element. The integral is over the closed surface bounding the volume and with the use of Gauss's law can be re-written to give

$$\partial M/\partial t = -\int_V \nabla \cdot (\rho \boldsymbol{v})\, \mathrm{d}V. \tag{2.3}$$

Eqn (2.1) and (2.3) yield the continuity equation

$$\partial \rho/\partial t + \nabla \cdot (\rho \boldsymbol{v}) = 0, \tag{2.4}$$

one of the three key equations of single fluid hydrodynamics.

2.2.1.2 Momentum

In the absence of forces, momentum is conserved, and if a Cartesian system with coordinates x_1, x_2 and x_3 is adopted, the conservation of the ith component of the momentum can be considered in the same manner as the conservation of mass above. Take p_i to be the ith component of the momentum contained in the volume bounded by a closed surface. Then

$$\partial p_i/\partial t = \int_V \partial(\rho v_i)/\partial t\, \mathrm{d}V, \tag{2.5}$$

and

$$\partial p_i/\partial t = -\int_S \rho v_i(\boldsymbol{v} \cdot \boldsymbol{n})\, \mathrm{d}S. \tag{2.6}$$

Use of Gauss's law and eqn (2.5) and (2.6) yields

$$\partial(\rho v_i)/\partial t + \sum_j \partial(\rho v_i v_j)/\partial x_j = 0. \tag{2.7}$$

More generally

$$\partial(\rho v_i)/\partial t + \sum_j \partial(\rho v_i v_j)/\partial x_j = f_{\mathrm{T}i} + f_{\mathrm{o}i} \tag{2.8}$$

where $f_{\mathrm{T}i}$ is the ith component of the force per unit volume due to the thermal pressure P, and $f_{\mathrm{o}i}$ is the ith component of the net force per unit volume due to other forces.

One can consider a simple plane-parallel flow depending on only x_1 and t. Focus on a thin region that, as shown in Figure 2.1, is bound by identical circles perpendicular to and centred on the x_1 axis and the cylinder containing them. Each circle has area A and one circle is in the $x_1 = x_\mathrm{A}$ plane and one is in the $x_1 = x_\mathrm{A} + \delta x$ plane. The force due to thermal pressure at $x_1 = x_\mathrm{A}$ on the region is $P(x_\mathrm{A})A\boldsymbol{e}_1$ and the force at $x_1 = x_\mathrm{A} + \delta x$ on the region is $-P(x_\mathrm{A} + \delta x)A\boldsymbol{e}_1$, where $\boldsymbol{e}_1$ is the unit vector in the x_1 direction. Using a first order Taylor expansion one can show that the net thermal force on the region is

$$f_{\mathrm{T}i}\, A\delta x = -\partial P/\partial x_i\, A\delta x. \tag{2.9}$$

This result can be generalised, and one concludes that

$$\partial(\rho v_i)/\partial t + \sum_j \partial(\rho v_i v_j + \delta_{ij} P)/\partial x_j = f_{\mathrm{o}i} \tag{2.10}$$

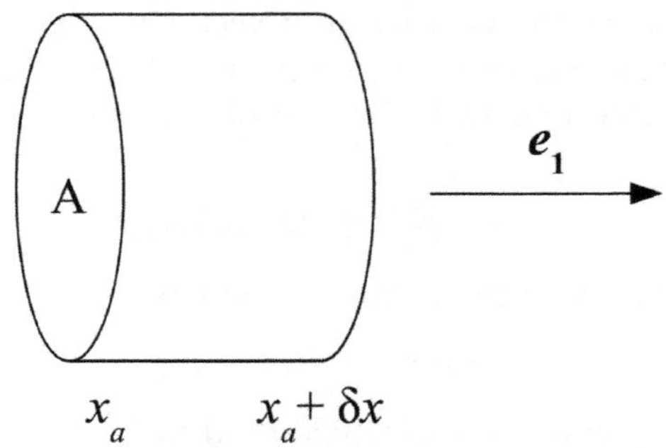

Figure 2.1 The small volume considered in the argument that the force per unit volume due to thermal pressure is minus the gradient of the thermal pressure. The volume is bound on two sides by parallel circles. Each is of area A and is perpendicular to, and centred on, the x_1 axis.

where δ_{ij} is the Kronecker delta function. Eqn (2.10) is another of the three key equations of single-fluid hydrodynamics.

2.2.1.3 Energy

The energy density is

$$\rho_E = \tfrac{1}{2}\rho v^2 + U \tag{2.11}$$

where U is the internal energy per unit volume. For a gas with a ratio of heat capacities γ

$$U = P/(\gamma - 1). \tag{2.12}$$

Take E to be the total energy in the volume bound by a surface. Following the same approach that led to eqn (2.4) and (2.8) one starts with

$$\partial E/\partial t = \int_V \partial(\tfrac{1}{2}\rho v^2 + U)/\partial t \; dV \tag{2.13}$$

and in the absence of work done by forces

$$\partial E/\partial t = -\int_S (\tfrac{1}{2}\rho v^2 + U)\,(\boldsymbol{v}\cdot\boldsymbol{n})\; dS. \tag{2.14}$$

One then obtains

$$\partial(\tfrac{1}{2}\rho v^2 + U)/\partial t + \nabla\cdot[(\tfrac{1}{2}\rho v^2 + U)\,\boldsymbol{v}] = \varepsilon_T + \varepsilon_0 \tag{2.15}$$

where ε_T is the energy change per unit time per unit volume due to work done by the thermal pressure force, and ε_0 is the energy change per unit time per unit volume due to other mechanisms. ε_T can be obtained by following an approach similar to that used to get f_{Ti}. The work done per unit time at $x_1 = x_A$ is $P(x_A)v_1(x_A)\,A$ and that done at $x_1 = x_q + \delta x$ is $-P(x_A + \delta x)v_1(x_A + \delta x)A$. One concludes that in general

$$\partial(\tfrac{1}{2}\rho v^2 + U)/\partial t + \nabla\cdot[(\tfrac{1}{2}\rho v^2 + U + P)\,\boldsymbol{v}] = \varepsilon_0, \tag{2.16}$$

the final of the three key equations of single-fluid hydrodynamics.

For a gas with a constant value of γ, eqn (2.12) and (2.16) yield

$$\partial\{\tfrac{1}{2}\rho v^2 + P/(\gamma - 1)\}/\partial t + \nabla \cdot [\{\tfrac{1}{2}\rho v^2 + \gamma P/(\gamma - 1)\}\, \boldsymbol{v}] = \varepsilon_0. \tag{2.17}$$

2.2.2 Single-fluid Hydrodynamic Sound Waves

Viscosity and thermal conduction contribute to f_{oi} and ε_0 and the damping of sound waves. However, we shall focus now on sound waves in media in which $f_{oi} = \varepsilon_0 = 0$ and also assume that γ is a constant. For these assumptions, eqn (2.4), (2.10) and (2.17) imply that

$$P = K\rho^{\gamma} \tag{2.18}$$

where K is a constant.

Consider a small perturbation superposed on an otherwise motionless, uniform medium. Assume that the perturbation is plane-parallel, the velocity is in the x_1 direction and ρ, v_1 and P depend on only x_1 and t. Take

$$\rho = \rho_0 + \rho_1, \tag{2.19}$$

$$v_1 = v_{11}, \tag{2.20}$$

and

$$P = P_0 + P_1 \tag{2.21}$$

where ρ_1, v_{11} and P_1 are small quantities and ρ_0 and P_0 are constants. Use of eqn (2.19), (2.20) and (2.21) with eqn (2.4), (2.10) and (2.18) and the retention of only those terms that are first order in small quantities yield

$$\partial\rho_1/\partial t + \rho_0\, \partial v_{11}/\partial x_1 = 0, \tag{2.22}$$

$$\rho_0\, \partial v_{11}/\partial t + \partial P_1/\partial x_1 = 0, \tag{2.23}$$

and

$$P_1 = \gamma P_0 \rho_1/\rho_0. \tag{2.24}$$

Now take

$$\rho_1 \propto v_{11} \propto P_1 \propto \exp(\mathrm{i}\omega t - \mathrm{i}kx_1) \tag{2.25}$$

where ω is a real constant and k is a positive real constant. Eqn (2.22), (2.23) and (2.25) give

$$\mathrm{i}\omega\rho_1 - \mathrm{i}k\rho_0 v_{11} = 0 \tag{2.26}$$

and

$$\mathrm{i}\omega\rho_0 v_{11} - \mathrm{i}kP_1 = 0. \tag{2.27}$$

Eqn (2.24), (2.26) and (2.27) yield

$$\omega^2 - c_s^2 k^2 = 0 \tag{2.28}$$

where

$$c_s = (\gamma P_0/\rho_0)^{1/2} \tag{2.29}$$

is the adiabatic sound speed. Eqn (2.28) is the dispersion relationship from which ω can be calculated for a specified k and has two roots

$$\omega = \pm kc_s. \tag{2.30}$$

2.2.3 Jump Conditions for Plane-parallel Hydrodynamic Shocks

A non-magnetised single-fluid medium transmits information at the sound speed. If an object moves through a medium at a speed larger than the sound speed, the medium cannot adjust far upstream and flow out of the path of the oncoming object. The shock is a sharp dissipation layer behind which the speed of the object relative to the medium, which has been accelerated, is subsonic.

The flow time through a shock front is sufficiently short that larger-scale time variations of the flow are negligible when the flow very near the shock is considered. Thus, when studying the relationship between immediate postshock physical parameters and their preshock counterparts one can assume that the flow is steady (*i.e.* neglect time variations). Consequently, the time derivatives in eqn (2.4), (2.8) and (2.17) are set to zero. f_{i0} and ε_0 are also set to zero. If, as depicted in Figure 2.2, the flow is plane-parallel, in the x_1 direction and depends on only x_1, then eqn (2.4), (2.10) and (2.17) imply that

$$\rho_a v_{1a} = \rho_b v_{1b}, \tag{2.31}$$

$$\rho_a v_{1a}^2 + P_a = \rho_b v_{1b}^2 + P_b, \tag{2.32}$$

and

$$\tfrac{1}{2}\rho_a v_{1a}^3 + \{\gamma/(\gamma-1)\}\, P_a v_{1a} = \tfrac{1}{2}\rho_b v_{1b}^3 + \{\gamma/(\gamma-1)\}\, P_b v_{1b}. \tag{2.33}$$

v_{1a} and v_{1b} are the speeds of the preshock flow and postshock flow as measured in the frame of the shock. ρ_a and P_a are the preshock density and pressure, and ρ_b and P_b are the postshock density and pressure.

The Mach number of the shock is defined as

$$M_a \equiv v_{1a}/c_{sa}, \tag{2.34}$$

where c_{sa} is the preshock adiabatic sound speed. Eqn (2.31)–(2.34) can be used to obtain a quadratic equation for the ratio of ρ_b to ρ_a. The solutions of it are 1 and

$$\rho_b/\rho_a = v_{1a}/v_{1b} = \{(\gamma+1)\, M_a^2\}/\{(\gamma-1)\, M_a^2 + 2\}. \tag{2.35}$$

Further algebra gives $P_b = P_a$ or

$$P_b/P_a = \{2\gamma M_a^2/(\gamma+1)\} - (\gamma-1)/(\gamma+1). \tag{2.36}$$

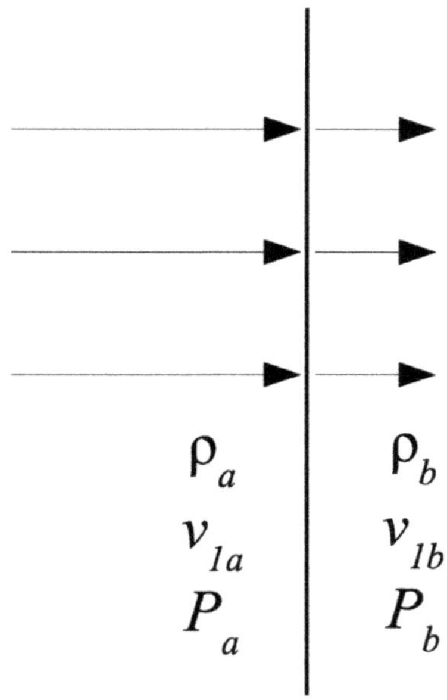

Figure 2.2 The media on either side of a plane-parallel hydrodynamic shock as viewed in a frame that is co-moving with the shock. Region a is upstream and material in it flows towards the shock. Region b is downstream and material in it flows away from the shock. Simple algebraic expressions relate the density, velocity and pressure of the postshock material to the corresponding parameters characterising the upstream material.

Eqn (2.35) and (2.36) are called the Rankine–Hugoniot relations or conditions.

For a strong shock $M_a \gg 1$. For a strong shock and $\gamma = 5/3$ the Rankine–Hugoniot jump conditions give

$$\rho_b/\rho_a = v_{1a}/v_{1b} \cong 4, \tag{2.37}$$

and

$$P_b \cong \tfrac{3}{4}\,\rho_a v_{1a}^2. \tag{2.38}$$

The postshock temperature is

$$T_b \cong (3/16)\,\mu m_H\, v_{1a}^2/k_B \tag{2.39}$$

where μ, m_H and k_B are the mean mass per particle in atomic units, the atomic mass unit and the Boltzmann constant, respectively.

2.3 Postshock Molecular Processes

Rotational and vibrational excitation of H_2 can contribute significantly to the internal energy per unit volume, which is seen from eqn (2.16) to be important for the dynamics. The radiative decay of the excited levels leads to postshock cooling, which also affects the dynamics. However, H_2 is not the only molecule that is a significant coolant behind shocks. In part, due to the temperature dependence of some reaction rates, the postshock chemistry significantly enhances the abundances of a number of molecules including some, *e.g.* H_2O, that can be important coolants. The chemistry also controls the fractional ionisation, which in a weakly ionised region influences the effect of the magnetic field on the dynamics. In sufficiently fast shocks excitation can lead to the dissociation of molecules.

2.3.1 H_2 Level Populations

For a medium consisting of identical hard spheres the internal energy per unit volume U is simply $3\rho k_B T/2\mu m_H$. However, molecules have internal degrees of freedom. For a medium consisting of only neutral atomic hydrogen, helium and molecular hydrogen the internal energy per unit volume is

$$U = (\rho/\mu m_H)[(3/2)\,k_B T + \sum_{v,J} f(H_2, v, J) E(v,J)\; x(H_2)/\{x(H) + x(He) + x(H_2)\}] \quad (2.40)$$

where $x(j)$ is the ratio of $n(j)$ the number density of particles of species j to n_H, the number density of hydrogen nuclei. $f(H_2, v, J)$ is the fraction of the H_2 that is in the level with vibrational and rotational quantum numbers v and J. $E(v,J)$ is the energy difference between the v,J level and the 0,0 level.

The excitation of the H_2 rovibrational levels occurs more slowly than the thermalisation of the translational degrees of freedom. Hence, in a detailed treatment of postshock structure $\gamma = 5/3$ is assumed when eqn (2.35) and (2.36) are used to get the initial postshock conditions. In general, in order to obtain conditions further downstream the integration of the hydrodynamics equations is then performed simultaneously with the integration of the rate equations governing the populations of the v, J levels, which are used to calculate U. However, except in a relatively thin layer just behind the shock front, the steady state equilibrium solution to the rate equations can be employed as the hydrodynamics equations are integrated. The rate equations contain terms arising due to collisionally induced excitation and de-excitation and radiative decay. Sometimes the absorption of radiation from external sources or collisionally induced dissociation is included.

The homonuclear nature of H_2 has the consequence that collisions with neutral species and radiative processes do not couple levels with even values of J to levels with odd values of J. Collisions of H^+ with H_2 do achieve such coupling,[24] but often on a timescale that is long compared to the dynamical time. Consequently, normally H_2 can be treated as two separate species in the calculation of U.

Sometimes the detailed treatment of the hydrogen level populations is not required to evaluate U. For example, for gas with $n_H = 1000\ \text{cm}^{-3}$ and $T = 1500$ K, one can reasonably approximate the molecular hydrogen contribution to U by $n(H_2)k_B T$. These values of n_H and T are sufficiently high that the populations of the $v = 0$ levels with even J are nearly thermalised with respect to one another and the populations of the $v = 0$ levels with odd J are nearly thermalised with respect to one another. However, these values are sufficiently low that H_2 remains primarily vibrationally unexcited. Collisionally induced dissociation is important over a wide part of the region of temperature-density parameter space for which the vibrational level populations are thermalised.[25]

The radiative decay of rotationally and vibrationally excited H_2 contributes to ε_o in eqn (2.16) and (2.17). The subject of radiative losses and cooling receives further attention in Section 2.3.3.

2.3.2 Chemistry in Shocked Material

The properties of preshock media vary considerably and the shocks propagating into them have a distribution of speeds. Thus, the range of postshock temperatures is wide. We shall focus on regions with temperatures of hundreds to thousands of degrees, which are sufficient for many reactions that either possess reaction barriers or are endothermic to proceed at higher rates than in lower temperature media. The effects of such reactions on gas-phase abundances have consequences for the postshock dynamics and for molecular line observations.

We shall consider postshock chemistry in two different types of clouds. Clouds of one type include those studied with the Copernicus satellite,[19] which transmit at least about a third or half of the optical light impinging on them and are often referred to as being diffuse. Clouds of the other type are often referred to as being dark and transmit at most of the order of a percent of the optical light incident on them.

2.3.2.1 Ionisation Structure

In unshocked diffuse clouds most of the gas-phase carbon exists as C^+, which is usually the most abundant ion in such clouds. The reaction

$$C^+ + H_2 \rightarrow CH^+ + H \tag{2.41}$$

is endothermic by about 0.4 eV and is sufficiently rapid at $T = 1000$ K that it removes most of the C^+ in diffuse cloud material that passes through a wide variety of shocks. In a primarily molecular cloud the formation of CH^+ initiates a sequence of hydrogen abstraction reactions with H_2 leading to the formation of other molecular ions such as ${CH_2}^+$ and ${CH_3}^+$. Such ions are removed by dissociative recombination reactions such as

$${CH_3}^+ + e \rightarrow CH + H_2 \tag{2.42}$$

and

$${CH_3}^+ + e \rightarrow CH_2 + H, \tag{2.43}$$

which are much more rapid than the radiative recombination of C^+. Consequently, the fractional ionisation drops by roughly an order of magnitude in diffuse postshock material before it cools much.[26,27] Photodissociation and photoionisation contribute to maintaining some C^+ and CH^+ in the shocked gas.

In dark clouds the ionisation is induced by cosmic ray protons, each of which is moving at a minimum speed of the order of 0.1*c*, where *c* is the speed of light. The ionisation of H_2 leads primarily to the production of H_2^+, which reacts with H_2 to form H_3^+. Such ions transfer protons to abundant neutral molecular species such as CO and H_2O to produce molecular ions such as HCO^+ and H_3O^+. The abundant molecular ions are removed by dissociative recombination. They also charge exchange with gas-phase neutral metallic atoms leading to metallic atomic ions, such as Mg^+, being the most abundant gas-phase ions.[28] Due to the slowness of radiative recombination relative to dissociative recombination ions like Mg^+ are removed primarily by recombination on grain surfaces,[29] which in some shocks also contributes significantly to the removal of gas-phase molecular ions.

Due to the thermal speeds of electrons being much greater than those of ions, in dark clouds grains become sufficiently negatively charged that the magnitude of the electric potential on the surface of a grain becomes comparable to several times k_BT/e if the gas-phase number density of electrons greatly exceeds the number density of grains times the average number of electrons carried by a grain.[30] Here e is the magnitude of the charge of a proton or electron and the temperature of ions and electrons are equal. The elevation of T in postshock material contributes to grains becoming the dominant carriers of negative charge in some shocked regions in dark clouds.[31]

2.3.2.2 Some Neutral–Neutral Chemistry

The key neutral–neutral reactions that are too slow in quiescent molecular regions to be of importance but play major roles in the chemistry of shocked regions[17,18,22] include:

$$O + H_2 \rightarrow OH + H, \tag{2.44}$$

$$OH + H_2 \rightarrow H_2O + H, \tag{2.45}$$

$$S + H_2 \rightarrow SH + H, \tag{2.46}$$

and

$$SH + H_2 \rightarrow H_2S + H. \tag{2.47}$$

OH and H_2O production is efficient if T is at least many hundreds of degrees, but significant production of SH and H_2S requires T to exceed 1000 K. In shocked diffuse molecular material photodissociation can ensure that OH and SH remain abundant where H_2O and H_2S are also plentiful. In shocked regions where H and H_2 are both abundant, the reverse reactions corresponding to reactions (2.45) and (2.47) also contribute to OH and H_2O, as well as SH and H_2S, being abundant simultaneously.

Reactions of OH and SH with other species help lead to the chemistry in shocked regions being rich. Examples of such reactions include

$$Si + OH \rightarrow SiO + H, \tag{2.48}$$

and

$$C + SH \rightarrow CS + H. \tag{2.49}$$

In sufficiently strong shocks the chemistry of all elements is influenced by the sputtering of grains and the evaporation of grains due to grain–grain collisions. Grain destruction has a particularly important influence on the abundances of gas-phase silicon-bearing species.[32]

In quiescent regions CO is generally the most abundant molecule other than H_2. CO is sufficiently stable that shocks, except those that trigger the dissociation of H_2, have little effect on its abundance.

2.3.3 Molecular Line Radiative Losses

As mentioned in Section 2.3.1 the radiative decay of excited levels of molecular hydrogen contributes to ε_0 in eqn (2.16) and (2.17) and affects the dynamics. The radiative losses due to transitions from excited levels of some other molecular species are also included in calculations of ε_0.[33–35] The effects of radiative transfer must be incorporated into the calculation of the energy losses due to important coolants other than H_2.

The high abundance of CO relative to other molecules except H_2, and in some cases OH and H_2O, and the small energy separations of its rotational levels compared to those of H_2, OH and H_2O, result in the line emission of CO leading to significant radiative losses at temperatures below several hundred K. Due to the high abundance of H_2 relative to other molecules, it is the most important coolant at T above several hundred degrees unless $n_H > 10^4\ \text{cm}^{-3}$, in which case H_2O is an important coolant.[33] If the fractional abundances of atomic hydrogen and molecular hydrogen are simultaneously substantial OH cooling can be important at the same time as H_2O cooling.

2.3.4 Dissociative Shocks

In sufficiently dense molecular postshock gas cooling due to the collisionally induced dissociation of H_2 dominates over molecular line radiative cooling at $T > 4000$ K. Behind shocks propagating through high density non-magnetised media at speeds of at least 24 km s^{-1} H_2 is nearly completely dissociated.[36] The dissociation rate decreases with decreasing n_H, if $n_H < 10^5\ \text{cm}^{-3}$, due to the increasing importance of the radiative decay of excited levels, and at low densities the rate is several orders of magnitude smaller than at high densities.[37,38] Consequently, the minimum speed of a shock behind which most molecules are dissociated increases with decreasing preshock density.[39] The inclusion of the effects of a magnetic field also increases this minimum speed, and calculations made for reasonable assumptions indicate that it is about 45 $\mathrm{km\,s^{-1}}$ for a high density and magnetised preshock medium.[33]

Of course, molecular reformation can begin as the postshock gas cools.[40] In regions in which some collisional or radiative induced ionisation has occurred H_2 is formed in the gas-phase by

$$\mathrm{H} + e \rightarrow \mathrm{H}^{-} + \mathrm{h}\nu, \tag{2.50}$$

followed by

$$\mathrm{H}^{-} + \mathrm{H} \rightarrow \mathrm{H}_2 + e. \tag{2.51}$$

The formation rate of H_2 on grains at $T > 100$ K is uncertain and models of postshock H_2 reformation based on a range of assumptions exist.[40]

2.4 Basics of Single-fluid Ideal MHD

Magnetic fields often significantly affect the dynamics of astrophysical media. The equations of single-fluid ideal MHD are obtained through the addition of appropriate contributions to $\boldsymbol{f}_0$ in eqn (2.10) and ε_0 in eqn (2.16) or (2.17). Boyd and Sanderson's book *The Physics of Plasmas* contains a good introduction to the single-fluid ideal MHD equations and their applications to linear waves and shocks.[41]

2.4.1 The Single-fluid Ideal MHD Equations

The equation for mass conservation is unaffected by the presence of a magnetic field and is eqn (2.4). We shall restrict our consideration of contributions to $\boldsymbol{f}_0$ and ε_0 to those made by electric and magnetic fields in an electrically perfectly conducting, non-dissipative medium and designate the relevant contributions $\boldsymbol{f}_B$ and ε_B.

The medium is assumed to be charge neutral implying that the electric force per unit volume is zero. The magnetic force per unit volume is

$$\boldsymbol{f}_B = \boldsymbol{J} \times \boldsymbol{B} \tag{2.52}$$

where $\boldsymbol{B}$ and $\boldsymbol{J}$ are the magnetic field and the current density. Ampère's Law is

$$\nabla \times \boldsymbol{B} = \mu_0 \boldsymbol{J} + \mu_0 \varepsilon_0\, \partial \boldsymbol{E}/\partial t, \tag{2.53}$$

where $\boldsymbol{E}$ is the electric field and μ_0 and ε_0 are the vacuum permeability and vacuum permittivity, respectively. Typically in MHD, sufficiently low frequency phenomena are considered so that the displacement current term can be neglected and

$$\boldsymbol{J} = (1/\mu_0)\, \nabla \times \boldsymbol{B} \tag{2.54}$$

is used. Eqn (2.52) and (2.54) imply that

$$\boldsymbol{f}_B = (1/\mu_0)\, (\nabla \times \boldsymbol{B}) \times \boldsymbol{B}. \tag{2.55}$$

The use of a vector identity yields

$$\boldsymbol{f}_{\mathrm{B}} = -\nabla(B^2/2\mu_0) + (1/\mu_0)(\boldsymbol{B}\cdot\nabla)\boldsymbol{B}. \tag{2.56}$$

Thus, eqn (2.10) becomes

$$\partial(\rho v_i)/\partial t + \sum_j \partial(\rho v_i v_j + \delta_{ij}P)/\partial x_j = (1/\mu_0)[(\nabla\times\boldsymbol{B})\times\boldsymbol{B}]_i. \tag{2.57}$$

The magnetic field does no work, but the work done per unit volume per unit time by the electric field is

$$\varepsilon_{\mathrm{E}} = \boldsymbol{J}\cdot\boldsymbol{E}. \tag{2.58}$$

Eqn (2.54) and (2.58) give

$$\varepsilon_{\mathrm{E}} = (1/\mu_0)(\nabla\times\boldsymbol{B})\cdot\boldsymbol{E}. \tag{2.59}$$

Eqn (2.59) and a vector identity imply that

$$\varepsilon_{\mathrm{E}} = -(1/\mu_0)\nabla\cdot(\boldsymbol{E}\times\boldsymbol{B}) + (1/\mu_0)\boldsymbol{B}\cdot(\nabla\times\boldsymbol{E}). \tag{2.60}$$

Faraday's Law is

$$\nabla\times\boldsymbol{E} = -\partial\boldsymbol{B}/\partial t. \tag{2.61}$$

The current density is assumed to obey Ohm's Law

$$\boldsymbol{J} = \sigma\,(\boldsymbol{E} + \boldsymbol{v}\times\boldsymbol{B}). \tag{2.62}$$

In ideal MHD the conductivity σ is assumed to be infinite, and

$$\boldsymbol{E} + \boldsymbol{v}\times\boldsymbol{B} = 0. \tag{2.63}$$

Eqn (2.60), (2.61) and (2.62) yield

$$\varepsilon_{\mathrm{E}} = -\,\partial(B^2/2\mu_0)/\partial t - (1/\mu_0)\,\nabla\cdot[\boldsymbol{B}\times(\boldsymbol{v}\times\boldsymbol{B})]. \tag{2.64}$$

Eqn (2.17) and (2.64), and the assumption that $\varepsilon_0 = \varepsilon_{\mathrm{E}}$, imply that

$$\begin{aligned}&\partial\{\tfrac{1}{2}\rho v^2 + P/(\gamma-1) + B^2/2\mu_0\}/\partial t\\ &\quad + \nabla\cdot[\{\tfrac{1}{2}\rho v^2 + P\gamma/(\gamma-1)\}\boldsymbol{v} + (1/\mu_0)\boldsymbol{B}\times(\boldsymbol{v}\times\boldsymbol{B})] = 0.\end{aligned} \tag{2.65}$$

For a given time-dependent and spatially varying magnetic field, one can use eqn (2.4), (2.57) and (2.65) to calculate ρ, $\boldsymbol{v}$ and P. Together they can be used to obtain eqn (2.18). An additional equation is required for $\boldsymbol{B}$ to be calculated simultaneously with these other parameters. Taking the curl of eqn (2.62), using eqn (2.54) and (2.61), a vector identity and

$$\nabla\cdot\mathbf{B} = 0, \tag{2.66}$$

one finds that

$$\partial\boldsymbol{B}/\partial t = \nabla\times(\boldsymbol{v}\times\boldsymbol{B}) + (1/\sigma\mu_0)\nabla^2\boldsymbol{B}. \tag{2.67}$$

In ideal MHD this reduces to

$$\partial\boldsymbol{B}/\partial t = \nabla\times(\boldsymbol{v}\times\boldsymbol{B}). \tag{2.68}$$

Eqn (2.4), (2.57), (2.65) and (2.68) form a complete set.

2.4.2 MHD Waves

To study linear MHD waves one can adopt an approach similar to that employed in Section 2.2.2. Begin by assuming that

$$\boldsymbol{v} = \boldsymbol{v}_1, \tag{2.69}$$

and

$$\boldsymbol{B} = \boldsymbol{B}_0 + \boldsymbol{B}_1 \tag{2.70}$$

where $\boldsymbol{v}_1$ and $\boldsymbol{B}_1$ are small and $\boldsymbol{B}_0$ is a constant vector. Also assume that

$$\rho_1 \propto \boldsymbol{v}_1 \propto P_1 \propto \boldsymbol{B}_1 \propto \exp(\mathrm{i}\omega t - \mathrm{i}\boldsymbol{k} \cdot \boldsymbol{x}) \tag{2.71}$$

where $\boldsymbol{k}$ is a constant vector. Without loss of generality one may take

$$\boldsymbol{B}_0 = B_0 \boldsymbol{e}_3 \tag{2.72}$$

and

$$\boldsymbol{k} = k_2 \boldsymbol{e}_2 + k_3 \boldsymbol{e}_3 \tag{2.73}$$

where B_0, k_2 and k_3 are real constants and $\boldsymbol{e}_2$ and $\boldsymbol{e}_3$ are unit vectors in the x_2 and x_3 directions, respectively.

Use of eqn (2.19), (2.21), (2.69), (2.70) and (2.71), eqn (2.4), (2.24), (2.29) and (2.57), eqn (2.68), and eqn (2.72) and (2.73) allows the derivation of the dispersion relation

$$[\omega^2 - k_3^2 v_A^2][\omega^4 - k^2 \, (c_s^2 + v_A^2) \, \omega^2 + k^2 k_3^2 \, c_s^2 v_A^2] = 0 \tag{2.74}$$

where

$$v_A = B_0/(\mu_0 \rho_0)^{\frac{1}{2}} \tag{2.75}$$

is the Alfvén speed.

Eqn (2.74) may be considered to be a cubic equation for ω^2. Its three roots are

$$\omega^2 = k_3^2 \, v_A^2 \tag{2.76}$$

and

$$\omega^2 = \tfrac{1}{2} k^2 [c_s^2 + v_A^2] \, [1 \pm (1 - \delta)^{1/2}] \tag{2.77}$$

where

$$\delta = (4k_3^2/k^2) \, c_s^2 v_A^2/(c_s^2 + v_A^2)^2. \tag{2.78}$$

In the process of arriving at eqn (2.74) one uses eqn (2.24) and derives other equations, similar to eqn (2.26) and (2.27), relating ρ_1, $\boldsymbol{v}_1$, P_1 and $\boldsymbol{B}_1$ to one another. Once the value of ω has been obtained for a specified $\boldsymbol{k}$ these relations can be used to explore the nature of the waves. The nature depends upon which of the roots given by eqn (2.76) and (2.77) is selected.

The waves corresponding to the root given by eqn (2.76) are called Alfvén waves or intermediate waves.[42] They are transverse waves with $\boldsymbol{v}_1$ and $\boldsymbol{B}_1$

parallel to one another but perpendicular to $\boldsymbol{B}_0$ and $\boldsymbol{k}$. They propagate parallel to $\boldsymbol{B}_0$ and are to first order non-compressive with $\rho_1 = P_1 = 0$. Some exact solutions to the single-fluid ideal MHD equations correspond to finite amplitude circularly polarised Alfvén waves that are non-compressive, but even these waves couple to compressive waves and can lose energy to them.[43]

The waves corresponding to the root given when one uses the plus sign in eqn (2.77) are called fast magnetosonic or fast magnetoacoustic waves. Those corresponding to the root given when one uses the minus sign in eqn (2.77) are called slow magnetosonic or slow magnetoacoustic waves. The magnetosonic waves are compressive and given the assumed geometry for such a wave the component of $\boldsymbol{v}_1$ in the x_1 direction is zero. If $c_\mathrm{s} \ll v_\mathrm{A}$, for a slow wave the magnitude of the component of $\boldsymbol{v}_1$ parallel to $\boldsymbol{B}_0$ is much larger than the magnitude of the perpendicular component, while for a fast wave the opposite is true. If ω_f and ω_s are the frequencies of a fast wave and a slow wave, respectively, with wavevector $\boldsymbol{k}$

$$\omega_\mathrm{f} \geq k_3 v_\mathrm{A} \geq \omega_\mathrm{s}. \tag{2.79}$$

For a fast wave for which $\boldsymbol{k}$ is perpendicular to $\boldsymbol{B}_0$

$$\omega^2 = k^2(c_\mathrm{s}^2 + v_\mathrm{A}^2). \tag{2.80}$$

If $c_\mathrm{s} < v_\mathrm{A}$ for a slow wave for which $\boldsymbol{k}$ is parallel to $\boldsymbol{B}_0$

$$\omega^2 = k^2 c_\mathrm{s}^2. \tag{2.81}$$

2.4.3 Single-fluid MHD Shocks

The approach introduced in Section 2.2.3 to derive the jump conditions for hydrodynamic shocks is applicable to MHD shocks. First we shall apply such an approach to a shock that propagates perpendicular to an upstream magnetic field and then consider a shock that propagates obliquely to an upstream magnetic field.

2.4.3.1 Perpendicular Shocks

We assume that in the shock frame the upstream material moves with velocity $\boldsymbol{v} = v_{1\mathrm{a}}\ \boldsymbol{e}_1$ and has a uniform magnetic field given by $\boldsymbol{B}_\mathrm{a} = B_\mathrm{a}\ \boldsymbol{e}_2$. Eqn (2.31) remains true, and the equations analogous to eqn (2.32) and (2.33) are obtained from eqn (2.57) and (2.65), respectively. They are

$$\rho_\mathrm{a} v_{1\mathrm{a}}^2 + P_\mathrm{a} + B_\mathrm{a}^2/2\mu_0 = \rho_\mathrm{b} v_{1\mathrm{b}}^2 + P_\mathrm{b} + B_\mathrm{b}^2/2\mu_0 \tag{2.82}$$

and

$$\tfrac{1}{2}\ \rho_\mathrm{A} v_{1\mathrm{a}}^3 + \{\gamma/(\gamma-1)\}\ P_\mathrm{a} v_{1\mathrm{a}} + B_\mathrm{A}^2 v_{1\mathrm{a}}/\mu_0 = \tfrac{1}{2}\ \rho_\mathrm{b} v_{1\mathrm{b}}^3 + \{\gamma/(\gamma-1)\}\ P_\mathrm{b} v_{1\mathrm{b}} + B_\mathrm{b}^2 v_{1\mathrm{b}}/\mu_0. \tag{2.83}$$

In addition from eqn (2.68) one has

$$v_{1\mathrm{a}}\ B_\mathrm{a} = v_{1\mathrm{b}}\ B_\mathrm{b}. \tag{2.84}$$

Eqn (2.31), (2.82), (2.83) and (2.84) can be used to derive a cubic equation for ρ_b/ρ_a. One solution is $\rho_b/\rho_a = 1$. Division of the cubic equation by $(\rho_b/\rho_a - 1)$ gives the following quadratic equation for ρ_b/ρ_a

$$2(2-\gamma)(\rho_b/\rho_a)^2 + [2\gamma(\beta_a + 1) + \beta_a\gamma(\gamma - 1)M_a^2](\rho_b/\rho_a) - \beta_a\gamma(\gamma + 1)\, M_a^2 = 0, \tag{2.85}$$

where

$$\beta_a = 2\mu_0 P_a/B_a^2. \tag{2.86}$$

For $\gamma < 2$ only one root of eqn (2.85) is positive. For that root to be greater than unity and relevant to a shock

$$\gamma M_a^2 > \gamma + (2/\beta_a). \tag{2.87}$$

Use of eqn (2.34), (2.75) and (2.86) shows that eqn (2.87) is equivalent to

$$v_{1a}^2 > c_{sa}^2 + v_{Aa}^2. \tag{2.88}$$

Eqn (2.80) and (2.88) imply that for a perpendicular shock to occur the shock speed must exceed the speed of those fast magnetosonic waves that propagate perpendicular to the magnetic field.

The Alfvénic Mach number of a shock is defined as

$$M_{Aa} = v_{1a}\,(\mu_0\,\rho_a)^{\frac{1}{2}}/B_a. \tag{2.89}$$

Eqn (2.31) and (2.84) imply that

$$B_b/B_a = \rho_b/\rho_a. \tag{2.90}$$

Hence, before radiative losses become important the immediate postshock magnetic pressure is limited by the moderate nature of the jump in density that occurs in a non-radiative shock no matter how strong it is. In contrast, eqn (2.36) implies that the postshock thermal pressure is large behind a very strong hydrodynamic shock. Consequently, immediately behind very strong perpendicular MHD shocks (*i.e.* those for which $M_a \gg 1$ and $M_{Aa} \gg 1$) the thermal pressure dominates the magnetic pressure. As a result, in the limit that the shock is very strong, the immediate postshock conditions are very similar to those given by eqn (2.35), (2.36) and (2.90). However, the onset of radiative cooling of the shocked material will decrease the thermal pressure and increase the density and magnetic pressure. In postshock gas that cools sufficiently the magnetic pressure becomes larger than the thermal pressure and limits further compression of the gas.

Of course, if β_a is not a large quantity, and the inequality given in eqn (2.88) is only weakly satisfied, the jump conditions differ markedly from those given by eqn (2.35), (2.36) and (2.90). In fact, the increases in ρ_b, P_b and B_b are reduced.

2.4.3.2 Oblique Shocks

Single-fluid oblique shocks are addressed in different levels of detail in the books by Boyd and Sanderson,[41] Jeffrey and Taniuti[44] and Somov.[45] Such a shock propagates at an angle with respect to the magnetic field not equal to 0 or any other integer multiple of $\pi/2$.

The algebra required to obtain an equation for oblique shocks analogous to eqn (2.85) from eqn (2.31), (2.57), (2.65) and (2.68) is rather more involved than that used in the derivation of eqn (2.85). The standard approach is to work in the de Hoffmann–Teller frame, which is the one in which the upstream velocity with respect to the shock front is parallel to the upstream magnetic field. The transformation to this frame is achieved by the addition of a uniform velocity parallel to the shock front.

For oblique shocks, the equation analogous to eqn (2.85) has a number of solutions. To classify them one identifies four velocity zones: I: $v_1 > v_f$; II: $v_f > v_1 > v_{A1}$; III: $v_{A1} > v_1 > v_s$; IV: $v_s > v_1$. Here, as before, v_1 is the component of the flow velocity parallel to axis 1 as measured in the frame of the shock, and v_{A1} is the component of the Alfvén velocity parallel to axis 1. The spatial variation of the flow is assumed to depend on only x_1. Six algebraic solutions to the equation analogous to eqn (2.85) correspond to shocks for which v_1 undergoes I to II, I to III, I to IV, II to III, II to IV and III to IV transitions. The I to II shocks and III to IV shocks are called fast-mode shocks and slow-mode shocks, respectively. The other shocks are called intermediate shocks or trans-Alfvénic shock waves.

A fast-mode shock or a slow-mode shock is compressive and leads to the magnetic field remaining in the same plane and the component of the magnetic field parallel to the shock front maintaining its original sign. The perpendicular component of the magnetic field is larger behind a fast-mode shock than ahead of it. In contrast, a slow-mode shock reduces the perpendicular component. Most studies of MHD shocks conducted by astrophysicists have concerned perpendicular shocks, which are a limiting type of fast-mode shock. However, slow-mode shocks are present in astronomical molecular regions. For example, in simulations of interstellar molecular cloud formation triggered by shocks driven into diffuse gas, slow-mode shocks followed fast-mode shocks and significantly enhanced the thermal pressure and density.[46]

The magnetic field downstream of an intermediate shock is in the same plane as the magnetic field upstream of the shock. However, the component of the magnetic field parallel to the shock front changes sign through the shock. Intermediate shocks have been described as being non-evolutionary. A discontinuity is evolutionary if no small perturbations imposed on the discontinuity surface cause any *instantaneous large* changes in the discontinuity.[45] Arguments, based on the comparison of the numbers of boundary conditions and outgoing waves at MHD discontinuities, imply that intermediate shocks are non-evolutionary. They led to the common view that intermediate shocks are unphysical. However, starting in the late 1980s and early 1990s time-dependent, plane-parallel numerical solutions of single-fluid MHD equations, including dissipation, indicated that under some circumstances structures similar to intermediate shocks do not experience instantaneous large changes.[47] Though some higher dimensional simulations have been interpreted as containing intermediate shocks,[48] Falle and Komissarov suggested that the apparent presence of such shocks in simulations of this type may be due to numerical shock structures not being sufficiently thin compared to the length scales over which the global flows vary.[49]

2.5 Basics of Multi-fluid Models of MHD Shocks

Much of the molecular material in galaxies is in regions where the fractional ionisation is about 10^{-4} or less. A magnetic field acts directly on only the charged species. Consequently, a non-uniform field induces a velocity of any charged species relative to the neutral material. The relative motion is called ambipolar diffusion, and the relative speed is limited by collisions between the charged species and the neutrals.

In some cases, all charged species are assumed to have the same average velocity, which in general varies with time and position, and a two-fluid description is adopted. In other cases, the velocities of the gas-phase ions, the electrons and dust particles of different sizes are assumed to differ from each other and a multi-fluid model is used.

Dissipation in two-fluid and multi-fluid shock models often occurs over regions that are very broad compared to all of the relevant collision mean free paths, but sometimes some of the dissipation occurs in a region having a thickness that is comparable to the mean free path for neutral–neutral reactions. The fluid equations for each species are similar to those of a single-fluid model but include source terms accounting for the coupling between the fluids. Two-fluid and multi-fluid models have been developed for diffuse clouds and for dark clouds and used in attempts to explain observations, including those of CH^+ absorption lines and H_2 emission lines. Some shocks in weakly ionised, magnetised media are subject to the Wardle instability, which has some similarities to the Rayleigh–Taylor instability, but does not occur in single-fluid shocks.

2.5.1 J-type and C-type Shocks

As mentioned above, collisions limit the relative speeds between charged species and neutrals. Consider a medium consisting of ions and electrons that can be treated as a single charged fluid and of neutrals that can be treated as a single neutral fluid, which is coupled primarily by ion–neutral collisions to the charged fluid. The timescale for a neutral experiencing no forces other than those due to collisions with ions to slow is given by $1/\alpha_{ni}n_i$ where α_{ni} is the relevant momentum transfer collision rate coefficient and n_i

is the number density of ions. The value of α_{ni} is about 2×10^{-9} cm^3 s^{-1} if the neutral species is H_2 and the mass of H_2 is significantly less than that of an ion.[50] The analogous slowing down time for ions is $1/\alpha_{in}n_n$ where α_{in} and n_n are the appropriate rate coefficient and the number density of neutrals, respectively. The ratio α_{in}/α_{ni} is equal to the ratio of the mean masses of the neutrals and ions.

If the angular frequency ω of a wave is much larger than $\alpha_{in}n_n$ the speed of the wave is not significantly affected by the presence of the neutrals and can be well estimated from only the magnetic field strength and direction, the propagation direction, the ion density and the pressure of the ion–electron fluid. We shall refer to such a wave as a decoupled wave. If ω is much smaller than $\alpha_{ni}n_i$ the speed of the wave is significantly affected by the presence of the neutrals and the density and pressure of the neutrals must be included in the calculation of the propagation speed. We shall refer to such a wave as a coupled wave. Clearly the speed, v_{fc}, of a coupled fast-mode wave is much less than that, v_{fd}, of a decoupled fast-mode wave with the same propagation direction.

As earlier, consider the upstream flow to have velocity $v_{1A}\boldsymbol{e}_1$ in the frame of the shock. If $v_{1A}>v_{fd}$, no information can propagate upstream from the shock and it will be a sharp discontinuity. If $v_{fd}>v_{1A}>v_{fc}$ the decoupled waves can carry some information upstream. This will cause at least some of the dissipation to occur in a region in which the material decelerates continuously. In fact, if radiative cooling in the dissipation zone keeps the temperature in neutral material sufficiently low that the neutral fluid does not undergo a supersonic to subsonic transition in the shock frame, all dissipation will occur in a region of continuous deceleration. Draine called such shocks C-type shocks, and shocks in which the flow of at least one fluid undergoes a jump-like, discontinuous transition through which compression occurs J-type shocks.[51] The flow structures in J-type and C-type shocks are shown schematically in Figure 2.3.

One estimate of the thickness of the dissipation region in a perpendicular C-type shock is based on the assumption that throughout it the magnitude of the gradient of the magnetic pressure is comparable to the force per unit volume on ions due to their collisions with neutrals.[51] This gives a thickness of the order of

$$\Delta_1 = v_{Ac}^2/\alpha_{ni}n_i v_1, \tag{2.91}$$

where the subscript c indicates that the speed is that of coupled waves, and usually all of the quantities on the right hand side are evaluated for the upstream medium. However, Chen and Ostriker gave a different estimate,[52] which was

$$\Delta_2 = 2^{7/4}\, v_1{}^{1/2} v_{Ac}{}^{1/2}/\alpha_{ni}n_i, \tag{2.92}$$

where usually all quantities on the right hand side are evaluated for the upstream medium.

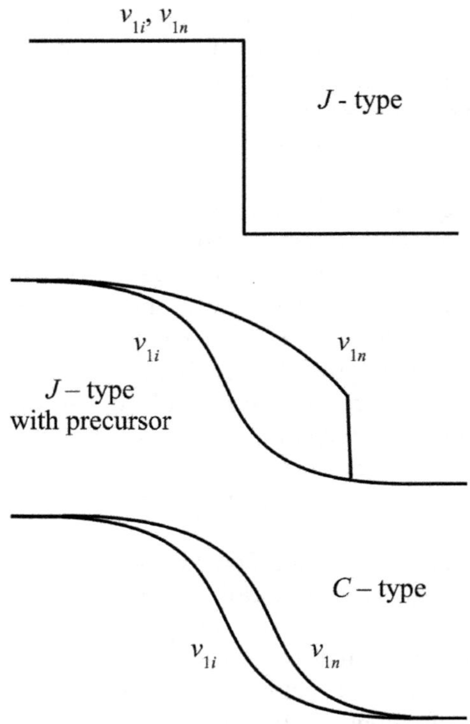

Figure 2.3 The velocity structures of J-type and C-type shocks. The panels show the velocities of the neutral and charged components as measured in frames that are co-moving with the corresponding shocks. In each panel the upstream medium is to the left and material in it is flowing towards the right. The first panel represents the structure of a J-type shock in which all components of the medium are decelerated simultaneously at the same surface. The second panel shows the structure of a J-type shock in which the charged component is decelerated smoothly and continuously but the neutral fluid undergoes some impulsive, as well as some continuous, deceleration. The third panel shows the structure of a C-type shock in which the decelerations of the charged and neutral components are continuous but not equal.

Section 2.1 contains a mention of the observations of H_2 infrared emission arising near the Orion Nebula, and in Section 2.3.4 mentions the fact that in sufficiently dense non-magnetised regions shocks with speeds of greater than 24 km s^{-1} will dissociate nearly all of the H_2. The Orion infrared emission features have velocity widths that exceed this speed limit, which stimulated interest in the speed limit for non-dissociating shocks that the two-fluid and multi-fluid models of shocks in magnetised media would give. As noted in Section 2.3.4, it is higher than 24 km s^{-1}, which provided evidence that C-type shocks are of relevance to understanding some observations.

2.5.2 The Equations Governing Time-dependent Plane-parallel Multi-fluid Flow

Falle presented the equations governing plane-parallel, multi-fluid flow in which the fluid and electromagnetic field variables depend on only t and x_1.[53] There are three main equations for each fluid. For fluid k they are:

$$\partial\rho_k/\partial t + \partial(\rho_k v_{1k})/\partial x_1 = \sum_{l \neq k} S_{kl}; \tag{2.93}$$

$$\partial(\rho_k v_{ik})/\partial t + \partial(\rho_k v_{1k} v_{ik} + P_k)/\partial x_1 = n_k q_k\Big(E_i + \sum_{jm} \varepsilon_{ijm} v_{jk} B_m\Big) + \sum_{l \neq k} F_{ikl}; \tag{2.94}$$

$$\begin{aligned}&\partial(\rho_k v_k^2/2 + U_k)/\partial t + \partial[(\rho_k v_k^2/2 + U_k + P_k)v_{1k}]/\partial x_1\\ &= H_k + \sum_{l \neq k} G_{kl} + n_k q_k \sum_i v_i\Big(E_i + \sum_{jm} \varepsilon_{ijm} v_{jk} B_m\Big).\end{aligned} \tag{2.95}$$

q_k is the charge on a single particle in fluid k and ρ_k, n_k, v_{ik}, v_k^2, P_k and U_k are the mass density, particle number density, ith component of the velocity, square of the speed, pressure and internal energy per unit volume of that fluid. E_i and B_m are the ith component of the electric field and mth component of the magnetic field, respectively, and ε_{ijm} is a Levi–Civita symbol. S_{kl} and G_{kl} are the rates per unit volume at which mass and energy, respectively, are transferred from fluid l to fluid k. F_{ikl} is the rate at which the ith component of the momentum per unit volume of the kth fluid changes due to transfer from the lth fluid. H_k is the rate per unit volume at which the energy density of fluid k changes due to external sources and losses, including those due to radiative cooling. Draine presented a detailed review of the evaluation of the various source terms.[54]

Due to the neglect of the displacement current density, Maxwell's equations give:

$$\partial B_1/\partial t = 0; \ \partial B_2/\partial t = \partial E_3/\partial x_1; \ \partial B_3/\partial t = -\partial E_2/\partial x_1; \tag{2.96}$$

$$\partial B_2/\partial x_1 = \mu_0 J_3; \ \partial B_3/\partial x_1 = -\mu_0 J_2. \tag{2.97}$$

The ith component of the current density is

$$J_i = \sum_k n_k q_k v_{ik}, \tag{2.98}$$

and charge neutrality

$$\sum_k n_k q_k = 0, \tag{2.99}$$

is assumed.

In many computational studies for charged fluids the left hand sides of eqn (2.94) and (2.95) have been assumed to be negligible due to the small fraction of the mass that the charged species carry in a variety of weakly ionised molecular regions.

2.5.3 Some Applications of Perpendicular Shock Models and an Instability

As mentioned in Section 2.1, in the 1970s the existence of a number of observations stimulated an increase in the effort to develop models of shocks in molecular regions. Consequently, a good deal of work on shocks in molecular regions has involved applications of models to understand data. Of course, research on the basic properties, such as the stability, of the shocks also occurred.

2.5.3.1 *The CH^+ Problem*

By the early 1970s the origin of interstellar CH^+ had been a puzzle for more than three decades. Gas-phase models of the chemistry in quiescent regions with temperature T of 100 K or less gave results for the CH^+ abundance that were too small. Therefore, by the late 1970s studies of CH^+ production in shocks were made with single-fluid hydrodynamic models.[55] Some of the key reactions affecting the CH^+ abundance in a shocked diffuse molecular cloud material are given in Section 2.3.2.1. Such models face substantial constraints including those provided by the observational measurements of the populations of some excited rotational levels of H_2 and of the OH abundance in some directions for which CH^+ data exist.

Subsequent studies of CH^+ formation in C-type shocks[27,56] were conducted in part because the significant breadths of the dissipation regions in such shocks give rise to lower temperatures for neutrals than found in hydrodynamic J-type shocks of the same speeds. The reduction in temperature leads to lower populations of more highly excited rotational levels and a smaller OH abundance, but it also retards the endothermic reaction (2.41), which produces CH^+. However, the streaming of ions relative to neutrals throughout the dissipation region provides energy, which can drive the reaction. This led to the hope that CH^+ could be sufficiently abundant in some dissipation regions that do not contain higher abundances of highly excited rotational levels or of OH other than indicated by observations. However, model results did not provide clear evidence that this hope was justified.[57] In addition, the observed line widths of CH^+ and velocity shifts of CH^+ features relative to the features of other species provided additional challenging constraints on the shock models.

As noted in Section 2.3.2.1, reaction (2.41) initiates chemistry that leads to a reduction in the fractional ionisation. This leads to a broader dissipation region in a C-type shock in a diffuse interstellar cloud than would occur if the flux of ions were constant through the shock.[26] For a model[27] with preshock atomic hydrogen and molecular hydrogen number densities of 0.4 cm^{-3} and 9.8 cm^{-3}, respectively, a shock speed of 12 km s^{-1}, a preshock magnetic field strength of 5×10^{-10} T and a preshock ion number density of about 7×10^{-3} cm^{-3}, the thickness of the dissipation region is about 1×10^{15} m. The maximum neutral temperature, maximum ion temperature, maximum electron temperature, maximum relative speed of ions and neutrals, minimum ion number density and maximum CH^+ number density in the dissipation region are about 1000 K, just under 4000 K, slightly over 4000 K, about 6 $km\,s^{-1}$, about $5\times10^{-3}\,cm^{-3}$ and about $2\times10^{-4}\,cm^{-3}$, respectively. For the assumed upstream conditions, the upstream Alfvén speed for coupled waves v_{Aca} is about 2 $km\,s^{-1}$, and eqn (2.92) gives a better estimate of the thickness of the dissipation region than eqn (2.91). The neutral density increases across the shock by a factor of about $2^{1/2}v_{1A}/v_{Aca}$.

2.5.3.2 *Shocked Material in the BN–KL Region in Orion*

The detection in the mid-1970s of infrared molecular line emission from the Orion molecular cloud was mentioned in Section 2.1. The results of a key effort to develop shock models for the origin of this emission were published in 1982.[58] Model results for a shock propagating at 36 $km\,s^{-1}$ into a medium with a hydrogen nuclei number density of $2\times10^5\,cm^{-3}$ and magnetic field strength of 4.5×10^{-8} T match the observationally measured strengths of H_2, CO and OH emission features. The corresponding upstream Alfvén speed for coupled waves is about 2 $km\,s^{-1}$. Since the region is dark and very dense, the fractional ionisation was taken to be less than 5×10^{-7}, which led to gas-phase ion–neutral collisions providing less drag on the neutral gas than grain–neutral collisions. As addressed more fully in Section 2.6, the grains are charged, causing the magnetic field, as well as collisions with neutrals, to influence their motion. In the evaluation of the drag on neutrals due to the collisions with grains, the grain to gas mass ratio and $\rho_{gr}r_{gr}$ were assumed to be 5×10^{-3} and 5×10^{-3} g cm^{-2}, respectively, where ρ_{gr} and r_{gr} are the density of the material of which the grains are composed and the radius of each grain, respectively. If the grains are well-coupled to the magnetic field, the use of eqn (2.92) and the replacement of $\alpha_{ni}n_i$ with $5\times10^{-3}\rho_a(3/4\rho_{gr}r_{gr})(v_{1a}/2)$ give an estimate of the thickness of the dissipation region of 3×10^{14} m.

The model does not fully account for the emission line profiles due to the presence of higher velocity wings. These were attributed to emission occurring as and after molecules reform behind a dissociating shock. Such a shock could occur in a wind from the central young stellar object. The wind would be decelerated by a shock, and the shocked wind material would be separated by a contact discontinuity from material that had passed through the 36 $km\,s^{-1}$ shock moving into the ambient material. Alternatively, parts of the surface of the shock driven into the ambient material could be traversing in-homogeneities having number densities below $2\times10^5\,cm^{-3}$, resulting in those parts of the shock surface to propagate fast enough to induce dissociation.

2.5.3.3 *A More Recent Example of the Application of Perpendicular Shock Models*

As observations of infrared emission from shocked regions have advanced over the last few decades, models of perpendicular C-type shocks have continued to have utility. For example, such models have been used to understand regions observed during the Herschel HIFI Water in Star Forming Regions programme.[59] One particular study concerned the water emission originating in an outflow in the high-mass star-forming region IRAS 17233-3606.[60] The results suggest that a shock model developed to explain SiO emission from the region[61] is probably consistent with the corresponding H_2O observational data. The model preshock hydrogen nuclei number density, speed and magnetic field strength are 10^6 cm^{-3}, 32 km s^{-1} and 10^{-8} T, and the upstream Alfvén speed for coupled waves is 0.2 km s^{-1}. The modelling led to the conclusion that the water fractional abundance in the observed region and total mass of water are $1.2\times10^{-5} - 2.5\times10^{-5}$ and 12.5 solar masses, respectively.

2.5.3.4 Dust Grain Destruction

In the application addressed immediately above, SiO emission was considered. In quiescent, dense, dark star forming regions, much of the elemental silicon is contained in grains. However, in sufficiently strong shocks the silicon can be released by grain–grain collisions or by sputtering of the grains due to collisions with neutrals streaming relative to the grains. Tielens, McKee, Seab and Hollenbach performed an important study[62] of the microphysics of grain mass loss due to grain–grain collisions and reviewed key results concerning the microphysics of grain sputtering that have been employed in investigations of the destruction of ice mantles on grains as well as the refractory cores of grains. The results for grain–grain collisions from this study have been used along with sputtering yields given by May and collaborators[63] in a number of perpendicular shock models incorporating grain destruction.[64–66] Important aspects of grain destruction are treated in more detail in Section 3.5.2.

Results[65] obtained for perpendicular shocks show: (a) that grain–grain collisions only contribute significantly to the destruction of the cores of silicate grains at preshock hydrogen nuclei number densities of above about 10^5 cm^{-3}; (b) that if sputtering dominates the destruction of silicate cores, shock speeds of at least about 27 km s^{-1} are required for much elemental silicon to enter the gas phase; (c) that shocks with speeds that are several km s^{-1} smaller can introduce non-negligible abundances of silicon-bearing species into the gas phase when the density is high enough for grain–grain collisions to dominate over sputtering. These conclusions are similar to those obtained earlier with an approximate treatment of oblique shock dynamics mentioned again below.[67] The SiO gas-phase abundance will depend on how the elemental silicon is injected into the gas phase. Sputtering may introduce more Si and less SiO directly into the gas phase than grain–grain collisions.[65] The gas-phase chemistry affecting the SiO abundance[32,66] includes reaction (2.48) and

$$\mathrm{Si} + \mathrm{O_2} \rightarrow \mathrm{SiO} + \mathrm{O}. \qquad (2.100)$$

Though the treatment in this section is mostly focused on silicate core destruction, the destruction of ice mantles and the destruction of carbonaceous cores are also considered in some of the cited work.

2.5.3.5 The Wardle Instability

If a shock is unstable a small perturbation of the flow grows over a finite time to modify the flow significantly. Wardle identified an instability occurring in some shocks in which the streaming of charged particles relative to neutrals is important.[68] In a frame co-moving with a C-type shock the charged species start to decelerate further upstream than the neutrals, though the collisions of the charged species with the neutrals limit the streaming. Suppose that there is a non-uniform perturbation in the density of the ions. Where that density is highest the collisional drag per unit volume is highest with the consequence that the relative streaming speed and upstream deceleration of the ions are smallest, which leads to a localised bend in the magnetic field lines. The bending tends to align the field lines somewhat with the flow of the neutrals. This increased alignment allows more ions to flow into the region of enhanced ion density and leads to even more frequent collisions. This further reduces the deceleration of the ions by the magnetic field, which causes an additional increase in the alignment. A linear analysis of this instability for a perpendicular shock propagating into a medium in which v_{fc} is about equal to v_{Ac} showed that it occurs when v_{1a}/v_{Aca} is greater than about 5.

2.6 Detailed Treatments of Grains in Multi-fluid Shock Models

If grains contain one percent of the mass, are composed of material having a density of 1 $g\,cm^{-3}$ and have radii of 0.1 μm, they have a fractional abundance of about 5×10^{-12} relative to hydrogen nuclei. As noted in Section 2.3.2.1, in dark regions grains become negatively charged, which causes them to be coupled to some degree to the magnetic field. As implied in Section 2.5.3.2, if the grains are well-coupled to the magnetic field, the drag on neutrals due to collisions with grains is significant for the neutral fluid dynamics when the fractional ionisation is less than of the order of $10^{-6}\,(v_{1a}/10\text{ km s}^{-1})(r_{gr}/0.1\ \mu\text{m})^{-1}$.

However, the grains are not always well coupled to the magnetic field. To determine whether they are one must calculate the charge on them. The calculation of the charges on grains in some shocks is complicated by the streaming of the grains relative to the other charged species, which necessitates the use of collision rates evaluated for velocity shifted Maxwellian distribution functions for the gas-phase ions.[69] In shocks in sufficiently dense regions grains carry a substantial fraction of the negative charge, which must be taken into account in a self-consistent fashion when one calculates the fractional ionisation and grain charge.[31] The Hall parameter is the product of the angular gyrofrequency and the slowing down time due to collisions. When the Hall parameter is less than or comparable to unity, the grains may not be well coupled to the magnetic field. However, when the grains are important carriers of charge and current, the value of the Hall parameter does not solely govern the degree to which the grains are coupled to the magnetic field.[31] Sometimes when grains are important carriers of charge and current in oblique shocks and have Hall parameters comparable to, or smaller than, unity, they can induce dynamical features that were not addressed before the mid-1990s.[53,70–73]

2.6.1 A Runaway Effect in a Perpendicular Shock Model

Pilipp, Hartquist and Havnes constructed plane-parallel, time-independent, perpendicular shock models in which the neutrals, ions, electrons and grains were treated as separate fluids having velocity components parallel to the shock front as well as along the shock propagation direction.[31] They also calculated the average charge carried by a grain and the gas-phase ion and electron number densities self-consistently. For one of their models they assumed a shock speed of 15 $km\,s^{-1}$ and a cosmic ray induced ionisation rate of H_2 of 3×10^{-17} s^{-1}, and that grains contain one percent of the upstream mass and have radii of 0.1 μm. They took the upstream values of the H_2 number density, elemental heavy metal (*e.g.* Mg) to H_2 abundance ratio and magnetic field strength to be $8\times10^6\,cm^{-3}$, 4×10^{-7} and 4.74×10^{-7} T, respectively. In part of the roughly 10^{12} m wide dissipation region, the fractional ionisation drops by about two orders of magnitude relative to its upstream value of about 3×10^{-9}, and the grains become the dominant carriers of negative charge with each carrying up to about 40 negative elementary charges.

The onset and incomplete development of a potential runaway effect emerge. The effect is due in part to the modest magnitude of the grain Hall parameter leading to the grains partially decoupling from the magnetic field, which results in charge separation of the grains and ions, and an associated electric field component in the shock propagation direction. An $\boldsymbol{E}\times\boldsymbol{B}$ drift velocity component parallel to the shock front arises, and the drift velocity of the grains is sufficient to increase the rate of ion recombination onto grains non-negligibly. This can lead to a reduction in both the fractional ionisation and the negative charge carried by the grains, and an associated increase in the component of the electric field along the shock propagation direction because the component must be stronger to prevent the less charged grains from separating even more from the ions. The increase in this component causes the drift velocity of the grains to rise even further.

2.6.2 A Particle Trajectory Approach to Grain Dynamics in Perpendicular Shocks

Guillet, Pineau des Forêts and Jones[74] adopted a particle trajectory approach[75] to follow the grain dynamics in time-independent, plane-parallel perpendicular shocks.[75] In this approach grains are assumed to have no velocity component relative to the neutrals in the direction parallel to the magnetic field and the two dimensional trajectories of many individual grain test particles are calculated. This allows the gyration of the grains, which in some cases occurs on non-negligible fractions of the dissipation regions' widths, to be followed. The evolution of the charge on each grain and the forces due to the electric and magnetic fields, and the drag due to collisions, are included. The effects of the grains on the ionisation structure and dynamics of the gas-phase fluids are treated self-consistently, and, if enough test particles are used, the charge distribution function for grains is obtained. A grain size distribution function is also assumed, and these authors have self-consistently included the evolution of the size distribution and its effects on the ionisation structure, grain charging and the dynamics of the gas-phase fluids and grains in studies, which are mentioned in Section 2.5.3.4, of grain destruction.[65]

These studies confirm that the inclusion of a population of grains large enough to be charged, small enough to have Hall parameters much greater than unity, and numerous enough to provide most of the grain surface area, can have significant effects on the dynamics. For example, in comparison to models such as those described in the previous section, the small grains can lead to a further reduction of the electron density, and fewer negative elementary charges on, and weaker coupling to the magnetic field of, larger grains. The small grains can also be responsible for enough drag on the neutrals to significantly limit the width of the dissipation region of a shock or even trigger the existence of a discontinuous jump in the flow parameters of the gas-phase neutrals.

2.6.3 Grains in Oblique Shocks

The approach described in Section 2.6.1 has also been applied to time-independent plane-parallel oblique shocks.[70] Results show that when grains, having Hall parameters comparable to or less than unity, carry a significant fraction of the negative charge but little of the positive charge, the magnetic field rotates around the shock propagation direction. Integration of the time-independent multi-fluid equations in the downstream direction enables only intermediate shock solutions to be obtained.

Wardle pointed out that this is due to the saddle-point nature of the downstream conditions.[71] He overcame this problem by integrating the equations in the upstream direction and found solutions for fast-mode shocks in which field rotation occurs. He showed that the inclusion of field and velocity components that do not lie in the plane of the shock propagation direction and the upstream magnetic field leads to the dissipation regions in the model shocks being narrower than they would be otherwise.[71] The approach does not allow the inclusion of non-equilibrium microphysics and chemistry. Wardle also made various simplifying assumptions, including that the various fluid temperatures are constant and that the grain charge and gas-phase ionisation structure do not need to be calculated self-consistently. However, the limitations of the approach and the assumptions did not prevent him from achieving significant insight into the general nature of fast-mode oblique shocks. He even included a distribution of grain sizes, and in later work an approximate treatment of the thermal structure was adopted.[72]

The integration of the time-dependent equations to find steady oblique fast-mode shock solutions[53,73] allows the inclusion of non-equilibrium microphysics and chemistry. Time-dependent oblique shock models including self-consistent treatments of the thermal structure, grain charge and ionisation have been used to investigate the interactions of shocks with density inhomogeneities.[76]

The first investigation of grain destruction in plane-parallel oblique shocks was based on a model for which a reasonable value of the width of the dissipation region was one of the input parameters and the properties of the dissipation region were taken to be uniform over that width.[67] The rest of the treatment, *e.g.* of the three-dimensional velocities of all species, the temperatures, the grain charge and ionisation structure, was self-consistent. This investigation was the first of C-type shocks to address destruction due to grain–grain collisions, which required the introduction of grains with different sizes, as well as sputtering. The only other investigation of grain destruction in oblique shocks included sputtering but not grain–grain collisions, though the adopted model treated grains with different sizes.[77] The model was of a time-dependent shock and incorporated a self-consistent treatment of the dynamics, the temperatures, the charges on the grains of different sizes and the ionisation structure. The results indicated that near the sputtering threshold shock speed the sputtering depends significantly on the angle θ between the shock propagation direction and the upstream magnetic field. For example, a comparison of results for shocks with speeds of 25 $km\,s^{-1}$ and propagating into upstream media with hydrogen nuclei number densities of 10^5 cm^{-3} showed that about two orders of magnitude more elemental silicon is injected into the gas phase when $\theta = \pi/6$ than when $\theta = \pi/3$.

2.7 Basics of Hydrodynamic Turbulence and of MHD Turbulence

So far, in the consideration of single-fluid media, we have neglected dissipation except when we tacitly assumed that it occurs in J-type shocks, and when we included the standard Ohmic dissipation term in eqn (2.67). In fact, dissipative terms due to compression and expansion as well as shear should be included in the single-fluid momentum eqn (2.10) and (2.57). Corresponding terms and a thermal conduction term should be included in the single-fluid energy eqn (2.16) and (2.65). In many studies of turbulence, the focus is on the momentum equation and simple assumptions about the thermal behaviours of fluids are made. For example, a fluid may be assumed to be isothermal.

If the dissipation terms are sufficiently small, a medium can become turbulent, which can have significant, interesting chemical consequences as addressed much more fully in Chapter 3. Nonlinear coupling between motions with different wavenumbers redistributes the energy on a wide range of scales. The transfer of energy to ever smaller scales down to those at which the dissipation terms become important means that turbulence itself can act as a dissipative mechanism. The thickness of a turbulent boundary layer is governed by the dissipation that the turbulence induces.

The distribution of energy on different scales is not necessarily exactly the same in single-fluid MHD turbulence as in single-fluid hydrodynamic turbulence. In multi-fluid MHD additional dissipation mechanisms can become important, leading single-fluid MHD turbulence and multi-fluid MHD turbulence to differ on small scales.

Turbulence affects the structures and dynamics of many molecular astrophysical sources including star forming regions and protoplanetary discs.

2.7.1 The Reynolds Number and the Kolmogorov Spectrum

The book *Fluid Mechanics* by Laudau and Lifschitz contains treatments of viscous hydrodynamics, incompressible hydrodynamic turbulence and viscous boundary layers.[78]

If the coefficients of viscosity are η and λ, eqn (2.10), in the absence of forces other than those due to thermal pressure and viscosity, becomes

$$\begin{aligned}\partial(\rho v_i)/\partial t + \sum_j \partial(\rho v_i v_j + \delta_{ij} P)/\partial x_j \\ = \partial(\lambda \nabla \cdot \boldsymbol{v})/\partial x_i + \sum_j \partial\{\eta(\partial v_i/\partial x_j + \partial v_j/\partial x_i - 2\delta_{ij}\ \nabla \cdot \boldsymbol{v}/3)\}/\partial x_j. \quad (2.101)\end{aligned}$$

Many studies of turbulence are based on the assumptions that μ and λ are constants and that the fluid is incompressible. Then ρ is constant, $\nabla \cdot \boldsymbol{v} = 0$ and eqn (2.101) becomes

$$\partial \boldsymbol{v}/\partial t + (\boldsymbol{v} \cdot \nabla)\boldsymbol{v} = \nu \nabla^2 \boldsymbol{v} \quad (2.102)$$

where the fluid is assumed to be isothermal and $\nu = \eta/\rho$ is the kinematic viscosity. The Reynolds number is $Re = vL/\nu$, where L is the characteristic length scale. For example, L might be the size of an obstacle around which the fluid flows. Re provides a measure of the relative importance of the second and third terms in eqn (2.102). The expression of the velocity in terms of Fourier components would show that the second term couples components with different wavenumbers, implying that energy can be transferred from one scale to another. In contrast, the third term damps most types of spatial variation in the velocity at a rate that increases as the wavenumber increases.

When *Re* is below a critical value, the flow will be laminar. Depending on the boundary conditions on the flow this critical value may range from being of the order of 10 to being of the order of 10^3. For a somewhat higher range of *Re* a flow may show regular fluctuations. The onset of turbulence usually requires that *Re* is at least of the order of $10^{3.5}$. The full development of turbulence may require even larger values of the Reynolds number.

Take $E(k)\mathrm{d}k$ to be the kinetic energy per unit mass associated with Fourier components having wavenumbers in the range k to $k+\mathrm{d}k$. The smallest relevant wavenumbers are usually those associated with the scales on which the non-uniformities in a flow are driven. For example for a flow past an obstacle of size L, the smallest relevant wavenumbers are of the order of $2\pi/L$. At these wavenumbers $E(k)$ is governed in part by boundary conditions and has no universal form. The general assumption is that most energy is put into the system at small wavenumbers and that it cascades to higher wavenumbers due to the nonlinear coupling associated with the second term of eqn (2.102).

At large wavenumbers dissipation leads to a rapid drop in $E(k)$ with increasing k. In a statistical steady state the rate at which energy is dissipated per unit mass must equal the rate of energy input per unit mass ε occurring at small wavenumbers. If the wavenumber k_D associated with the onset of dissipation depends on only ε and ν, dimensional analysis implies that k_D is proportional to $\varepsilon^{1/4}\nu^{-3/4}$.

The wavenumbers ranging from somewhat larger than the smallest ones to rather smaller than those associated with the dissipation are often said to be in the inertial range. If in the inertial range $E(k)$ is determined solely by ε, dimensional analysis implies that $E(k)$ is proportional to $\varepsilon^{2/3}k^{-5/3}$. This is the Kolmogorov spectrum, which is often observed at intermediate wavenumbers in fully developed laboratory turbulence.

2.7.2 Intermittency

Turbulence is intermittent if at smaller scales the spatio-temporal velocity fluctuations have increasingly larger variations relative to their average value.[79] Intermittency is often studied through the consideration of the probability distribution function of the velocity fluctuations[79–82] defined by

$$\delta\boldsymbol{v}(\boldsymbol{x}, \boldsymbol{r}) = \boldsymbol{v}(\boldsymbol{x}+\boldsymbol{r}) - \boldsymbol{v}(\boldsymbol{x}). \tag{2.103}$$

The pth moment of $\delta\boldsymbol{v}(r)$ is sometimes called the pth structure function and is given by

$$S_p(r) = \langle(\delta\boldsymbol{v}(r))^p\rangle, \tag{2.104}$$

where $\langle\rangle$ indicates the ensemble average. $S_p(r)$ depends only on r, the magnitude of $\boldsymbol{r}$, if the turbulence is statistically homogeneous and isotropic.

The scaling behaviour of the moments is usually assumed to be $S_p(r) \propto r^{s(p)}$. For a Gaussian process $s(2n) = ns(2)$, where n is any positive integer. Experimental and computational studies have demonstrated that the velocity fluctuation moments are not related to one another so simply and fluctuations show large departures from the average far more often than they would for a Gaussian distribution. An analysis of data obtained for the turbulent boundary layer on a smooth flat plate in a wind tunnel[83] yielded 0.7, 1.2, 1.62, 2.00, 2.36 and 2.68 for S_{2n} and n ranging from 1 to 6.

As treated in detail in the following chapter, intermittency affects the spatial and time distributions of dissipation regions in which the temperature can be high enough to drive chemistry very similar to the chemistry occurring in shocks.[79]

2.7.3 Turbulent Viscosity, Boundary Layers and Diffusion

In a turbulent medium the cascade of energy to larger wavenumbers leads to dissipation occurring more rapidly than it would if waves did not couple to one another. Hence, turbulence generates an effective viscosity, and some simulations of astrophysical flows[84] have included sub-grid models that use large scale flow parameters to approximate the effects that turbulence, which cannot be followed computationally, has on the large scale flow.

The kinetic viscosity due to collisions between molecules is of the order of the product of the collision mean free path and the thermal speed in the gas. In analogy, the kinetic viscosity due to turbulence is of the order of the product of the size and the speed associated with the largest turbulent eddies.[78,85] In the modelling of protoplanetary and other discs, the largest eddies are often assumed to have speeds and sizes that are fractions of the disc scale height H and sound speed, and $\nu = \alpha c_s H$ where α is a positive constant less than unity.[85]

To estimate the thicknesses of turbulent boundary layers[86] one could assume that turbulence regulates itself in such a way that the effective kinetic viscosity due to turbulence attains a value that would lead to an effective Reynolds number of the order of 10^3. The thickness of a viscous boundary layer is given by roughly $L/Re^{1/2}$. Consequently, one would expect a turbulent boundary layer to have a thickness of the order of $0.01L$, which is in harmony with results found in a thorough study of the boundary layers of supersonic jets and obstacles.[87] The possibility that CH^+ optical absorption features form in turbulent boundary layers has been proposed.[57] The origin in such boundary layers of H_2 infrared emission features with widths substantially exceeding 45 km s^{-1} has also been proposed.[88]

The effects on chemistry of potential turbulence-induced microscopic mixing of material between different regions have been examined in the context of boundary layers.[89] The chemical consequences of such mixing on more extended scales have also been studied by a number of groups. The effects of such extended mixing on the ionisation structure in protoplanetary discs may be particularly important due to their possible consequences for where turbulence can develop and be a source of the viscosity. The viscosity significantly influences the disc structure and dynamics.[90] For hydrodynamic turbulence the diffusion coefficients used in the chemical rate equations are comparable to the kinematic viscosity. However, in the presence of a magnetic field the microscopic mixing of at least those particles with large Hall parameters with one another might be retarded.

2.7.4 MHD Turbulence

In 1965 Kraichnan published a short, elegant paper[91] in which he investigated the inertial range of incompressible single-fluid MHD turbulence. He began with a form of the single-fluid MHD equations presented in a paper[92] published in 1950 and which is less than a half page long. In it Elsasser showed that for incompressible, isothermal flow

$$(\partial/\partial t + \boldsymbol{q}\cdot\nabla)\boldsymbol{w} = \nu_+\nabla^2\boldsymbol{w} + \nu_-\nabla^2\boldsymbol{q} - \nabla(\boldsymbol{w}-\boldsymbol{q})^2/8, \tag{2.105}$$

and

$$(\partial/\partial t + \boldsymbol{w}\cdot\nabla)\boldsymbol{q} = \nu_+\nabla^2\boldsymbol{q} + \nu_-\nabla^2\boldsymbol{w} - \nabla(\boldsymbol{w}-\boldsymbol{q})^2/8, \tag{2.106}$$

where ν_+ and ν_- are constants and

$$\boldsymbol{q} = \boldsymbol{v} - \boldsymbol{B}/(\mu_0\rho)^{1/2}, \tag{2.107}$$

and

$$\boldsymbol{w} = \boldsymbol{v} + \boldsymbol{B}/(\mu_0\rho)^{1/2}. \tag{2.108}$$

Kraichnan recognised that if the magnetic field consists of a uniform large scale component and small amplitude perturbations and the average velocity is zero but small amplitude velocity perturbations are present, the $\boldsymbol{q}$ waves and the $\boldsymbol{w}$ waves will propagate in opposite directions. This propagation leads to little coupling and little correlation between these two types of waves. No correlation between these waves has the consequence that equipartition exists between the kinetic energy and the magnetic densities associated with the perturbations.

He also considered three waves with values of k nearly equal to one another and much larger than k_0, the wavenumber associated with energy input into the medium. In the absence of a uniform component of the magnetic field or magnetic field components with wavenumbers comparable to k_0, nonlinear interactions would lead to the development of substantial triple correlations between the three waves on a timescale comparable to the local dynamical time, which is of the order of $(E(k)k^3)^{-1/2}$. However, in the presence of a large scale magnetic field of magnitude B_0 the timescale for the relaxation of correlations by propagation is $(\mu_0\rho)^{1/2}(B_0k)^{-1}$. If this timescale is adopted, dimensional analysis implies that $E(k)$ is proportional to $(\varepsilon B_0)^{1/2}(\mu_0\rho)^{1/4}k^{-3/2}$. This is the Iroshnikov–Kraichnan spectrum.

More recent theoretical and numerical studies imply a variety of possible spectral dependences in the inertial range of MHD turbulence.[82] For example, the decomposition of results from simulations of sub-Alfvénic compressible turbulence in a low β medium showed $k^{-5/3}$, $k^{-5/3}$ and $k^{-3/2}$ spectral dependences for Alfvén, slow and fast modes, respectively.[93] One should bear in mind that flows cannot be followed computationally over the many orders of magnitude range in k usually associated with fully developed hydrodynamic turbulence, which is why sub-grid models are sometimes used to study the effects of turbulence, as mentioned in the preceding Section 2.7.3.

In addition to the Reynolds number, two other dimensionless numbers are relevant for single-fluid viscous MHD with Ohmic dissipation. They are the magnetic Reynolds number $Re_M = \nu L \sigma \mu_0$, where σ is the Ohmic conductivity appearing in eqn (2.62) and the magnetic Prandtl number $Pr_M = Re_M/Re$. Dissipation will be important if either Re or Re_M is not sufficiently large.

The results of some simulations for a medium with a large value of Pr_M suggest that overall in the inertial range $E(k)$ scales as $k^{-5/3}$ and that the spectrum $F(k)$ of the magnetic field energy density per unit mass is similar,[94] which is in harmony with Kraichnan's arguments concerning equipartition. The departure of the spectra from the Iroshnikov–Kraichnan form is due at least in part to the turbulence being anisotropic.[79,82] At higher wavenumbers at which dissipation is important, the results show a much steeper drop in $E(k)$ than in $F(k)$ with increasing k. The simulations display intermittency, particularly in the dissipation regime.

Ambipolar diffusion affects the turbulence in many dark molecular astrophysical regions.[95] For the moment consider such a region in which all charged species can be treated as a single fluid and all neutral species can be treated as a single fluid. In such a region an Alfvén wave with a wavevector that is parallel to the large scale magnetic field will damp due to ion–neutral friction at a rate[96] given by $v_{Ac}{}^2 k^2/2\alpha_{ni} n_i$ if $\omega \ll \alpha_{ni} n_i$, ω is very much smaller than the gyrofrequency, the ion's Hall parameter is very large, and ν and $1/\sigma\mu_0$ are zero. For $\alpha_{ni} n_i \rho_n/\rho_i > \omega > \alpha_{ni} n_i$ such a wave does not propagate.[96] One can define a dimensionless ambipolar diffusion Reynolds number Re_{AD}, which in star forming regions and protoplanetary discs is usually much smaller[95] than Re and Re_M. In the limit that the Ohmic conductivity and the Hall conductivity, which is described below, are very large, Re_{AD} is equal to Re_P, a Reynolds number associated with the Pederson conductivity, which is also described below. In many cases studied to date $Re_{AD} = Re_P = \alpha_{ni} n_i\, \nu L/\, v_{Ac}{}^2$. Re_{AD} is of the order of 10–100 in a dense molecular pre-stellar core with $n_H = 10^5\ cm^{-3}$. Given that hydrodynamic turbulence does not usually develop for such small values of Re, one may wonder whether in some circumstances the fluctuations to which astrophysicists refer as turbulence really have the characteristics of true turbulence. However, reality may be complicated. More general investigations of wave dispersion relations than the one cited above, but based on some of the same assumptions, have revealed that in regions with $v_A > c_S$, fast magnetosonic waves, such as Alfvén waves, with $\omega > \alpha_{ni} n_i$ are damped rapidly.[98,99] However, slow magnetosonic waves are not. Consequently, a spectrum of fluctuations with velocity perturbations preferentially nearly aligned with the large-scale magnetic field may persist in the frequency regime in which other types of waves are absent. In regions with $v_A < c_S$, Alfvén and slow magnetosonic waves are damped rapidly at these angular frequencies but fast magnetosonic waves are not.

As mentioned in Section 2.6, in regions in which the grain–neutral drag is important, the Hall parameter for a grain species making a significant contribution to the current may not be large. When the Hall parameter is not large, the components of the current that are perpendicular to the magnetic field are important. As can be shown with the use of the multi-fluid MHD equations, the Hall current is parallel to $\boldsymbol{E} \times \boldsymbol{B}$, and the Pederson current is parallel to the component of the electric field that is perpendicular to the magnetic field. The Hall and Pederson conductivities are the proportionality constants relating the two currents to the corresponding components of the electric field. Using the expression for the Hall conductivity[97] one finds a corresponding dimensionless number

$$Re_H = \mu_0\, v\, L \sum_j \{n_j\, (Z_j\, e)^2 \beta_{Hj}\, \tau_j\, / m_j\, (1+\beta_{Hj}^2)\} = -(\mu_0\, v\, L/B) \sum_j \{n_j\, Z_j\, e/(1+\beta_{Hj}^2)\}, \tag{2.109}$$

where eqn (2.99) was used to derive the final expression. n_j, Z_j, τ_j, β_{Hj} and m_j are the number density, number of elementary charges on each particle, slowing down time due to collisions, Hall parameter and mass of each particle, respectively, of the jth species. In general, Re_{AD} does not equal Re_P, but one can without any loss of generality consider Re_P rather than Re_{AD}. Using the expression for the Pederson conductivity one finds that

$$Re_P = \mu_0\, v\, L \sum_j \{n_j\, (Z_j\, e)^2 \tau_j / m_j\, (1+\beta_{Hj}^2)\} = (\mu_0\, v\, L/B) \sum_j \{n_j\, Z_j\, e\, \beta_{Hj}/(1+\beta_{Hj}^2)\}. \tag{2.110}$$

For a case in which only two charged species are present, only one is important for momentum transfer to the neutrals and both have large values of β_H, the expression for Re_P becomes equivalent to the simpler one given above.

2.7.5 The Effects of Turbulence on Structure

Turbulence has important consequences for the structures and evolution of a variety of astrophysical sources. Its roles in interstellar molecular clouds,[100,101] particularly in the context of star formation, and protoplanetary discs[85] are subjects of considerable investigation.

2.7.5.1 *The Effects of Turbulence on Structure in Interstellar Molecular Clouds*

In the early 1980s bulk sub-cloud scale internal motions within interstellar molecular clouds were considered a means for maintaining a cloud in a self-gravitating equilibrium state.[102] However, starting in the late 1990s a large number of simulations have been more focused on the role that turbulence plays in the creation of substructure, *e.g.* dense prestellar cores, within clouds than on the support of the clouds.[100,101] One particularly clear set of simulations and the analysis of them show that in a region in which $c_s \ll v_A$, the nonlinear evolution of a single fast magnetosonic wave leads to the excitation of a slow magnetosonic density perturbation.[103] When dissipation is unimportant, the slow magnetosonic density perturbation is approximately ${\rho_1}^2 v_A^2/\rho_0 c_s^2$, where ρ_0 and ρ_1 are the density of the background state and the amplitude of the initial density perturbation, respectively. Many simulations with power-law wave spectra have been performed to determine the nature of the structure that emerges when the turbulence is either forced or allowed to decay.[100,101]

2.7.5.2 *Turbulent Viscosity in Protoplanetary Discs*

Discs play vital roles in the accretion of mass onto stars and planet formation. As discussed in Chapter 5, they also are key for launching jets and other outflows associated with young stars.

The equation for mass conservation and the equation for the effect of torque on angular momentum govern the dynamics of a thin axisymmetric accretion disc in vertical hydrostatic equilibrium and in which self-gravity is unimportant.[85] If cylindrical coordinates are used, eqn (2.4) yields

$$\int \partial\rho/\partial t \ \mathrm{d}z + \int r^{-1} \partial(r \ \rho \ v_r)/\partial r \ \mathrm{d}z = 0, \tag{2.111}$$

where v_r is the component of velocity away from the symmetry axis. r, θ and z are the distance from the symmetry axis, the azimuthal angle and the distance above the mid-plane of the disc, which is assumed to possess reflection symmetry about that plane. The column density is defined as

$$\Sigma = \int \rho \ \mathrm{d}z, \tag{2.112}$$

and the mean radial component of the velocity at r is defined as

$$u_r = \int \rho \ v_r \ \mathrm{d}z/\Sigma. \tag{2.113}$$

Eqn (2.111) becomes

$$\partial\Sigma/\partial t + r^{-1}\partial(r\Sigma \ u_r)/\partial t = 0. \tag{2.114}$$

When v_z and all derivatives with respect to θ are zero, eqn (2.101) yields

$$\partial(r^2 \rho\Omega)/\partial t + r^{-1}\partial \ (r^3 \rho \ v_r \ \Omega)/\partial r = r^{-1}\partial(r^3 \nu\rho\partial\Omega/\partial r)/\partial r, \tag{2.115}$$

where Ω is the angular frequency of the orbital motion of the disc at r, z. Integration over z gives

$$\partial(r^2\Sigma\Omega_A)/\partial t + r^{-1}\partial(r^3\Sigma\, u_r)/\partial r == r^{-1}\partial(r^3\nu_A\Sigma\partial\Omega_A/\partial r)/\partial r, \quad (2.116)$$

where ν_A and Ω_A are appropriately weighted averages of ν and Ω, respectively, over z.

If the disc rotates in a Keplerian fashion, $\Omega = (GM/r^3)^{1/2}$ where M is the mass of the central object, and one may use eqn (2.114) and (2.116) to obtain

$$\partial\Sigma/\partial t = 3r^{-1}\partial[r^{1/2}\partial(\nu_A\Sigma\, r^{1/2})/\partial r]/\partial r. \quad (2.117)$$

If viscosity is neglected in the equation for the radial motion, the radial component of the velocity is governed by

$$\partial v_r/\partial t + v_r\partial v_r/\partial r + \Omega^2 r + \rho^{-1}\partial p/\partial r = -GM/r^2 + r^{-2}\partial(r^3\nu\rho\partial v_r/\partial r)/\partial r. \quad (2.118)$$

Often the viscous term on term in eqn (2.118) is neglected. ρ depends on r and z. If the temperature T depends only on r, then

$$\rho(r,z) = \rho_0(r)\exp(-z^2/H^2), \quad (2.119)$$

where $\rho_0(r)$ is the density at the mid-plane and the square of the scale height is given by

$$H^2 = r^3 k_B T(r)/GM\mu m_H. \quad (2.120)$$

The equation of state has been assumed to be

$$P = k_B\rho T(r)/\mu m_H. \quad (2.121)$$

As mentioned in Section 2.7.3, often

$$\nu_A = \alpha c_s H \quad (2.122)$$

is assumed where α is a positive constant less than unity. A model based on this assumption is said to describe an α disc. If the viscous term in eqn (2.118) is neglected and α and the function $T(r)$ are known, then eqn (2.29), (2.112)–(2.114), (2.116) and (2.118)–(2.122) form a closed set. $T(r)$ is sometimes calculated on the assumption that radiative losses balance the heating of the disc by viscous dissipation. Steady α-disc models have been used in some studies of chemistry in proto-planetary accretion discs.

The α prescription given in eqn (2.122) is used because the calculation of the kinematic viscosity from first principles is challenging. The magneto-rotational instability (MRI) is one mechanism that is believed to generate the turbulence causing protoplanetary discs to be sufficiently viscous to evolve on the observationally inferred timescales.[104] This instability has a mechanical analogy. A magnetic field line threading two fluid elements acts somewhat like a spring connecting two bodies. The inner and outer orbiting bodies are body A and body B, respectively. The spring tends to slow body A and speed up body B. However, this causes body A to fall towards, and body

B to move away from, the object that is the dominant source of the gravitational field. The continued stretching of the string increases the rate at which angular momentum is transferred, which causes body A to fall and body B to rise further. In regions called dead zones the ionisation conditions are inappropriate for the growth of MRI and the development of MRI-induced turbulence.[90,104,105] The consequences of the existence of dead zones for planet formation constitute an interesting area of research.

The growth of gravitational instability in some discs in which the self-gravity is important can lead to the production of an effective value of α.[105] Such discs are not well described by thin-disc models for which eqn (2.114) and (2.116) are adopted. Instead full three-dimensional fluid models must be used. Gravitational instability develops when

$$Q \equiv c_s \kappa / \pi G \Sigma < 1. \tag{2.123}$$

Q is called the Toomre stability parameter, and κ is the epicyclic frequency, which is equal to Ω in a Keplerian disc. The growth of the gravitational instability leads to spiral waves and spiral shocks possessing associated vertical hydrodynamic jumps. Heating in such shocks can significantly affect the chemistry, in part by leading to the thermal desorption of material from grains. The reader may recall that this chapter began with a citation to the 1946 paper of Oort and van der Hulst, who addressed dynamics that could result in the return of material adsorbed on grains to the gas phase. This chapter ends, as it began.

References

1. J. H. Oort and H. C. van de Hulst, *Bull. Astron. Inst. Neth.*, 1946, **10**, 187.
2. J. Hartmann, *Astrophys. J.*, 1904, **19**, 268.
3. J. S. Plaskett, *Mon. Not. R. Astron. Soc.*, 1923, **84**, 80.
4. J. S. Plaskett and J. A. Pearce, *Mon. Not. R. Astron. Soc.*, 1930, **90**, 243.
5. A. S. Eddington, *Proc. R. Soc. London, Ser. A*, 1926, **111**, 424.
6. C. S. Beals, *Mon. Not. R. Astron. Soc.*, 1936, **96**, 661.
7. W. S. Adams, *Astrophys. J.*, 1943, **97**, 105.
8. F. P. Dickson, *The Bowl of Night*, N. V. Philips' Gloeilampenfabrieken, 1968.
9. G. Alter, *Vistas Astron.*, 1956, **2**, 1074.
10. A. Li, *J. Phys.: Conf. Ser.*, 2005, **6**, 229.
11. W. Herschel, *Philos. Trans. R. Soc. London*, 1785, **75**, 213.
12. S. Chandrasekhar, *J. Astrophys. Astr.*, 1997, **18**, 3.
13. S. Chandrasekhar and G. Münch, *Astrophys. J.*, 1950, **112**, 380.
14. N. C. Wickramasinghe, *Mon. Not. R. astr. Soc.*, 1965, **131**, 177.
15. T. P. Stecher and D. A. Williams, *Astrophys. J.*, 1966, **146**, 88.
16. T. O. Carroll and E. E. Salpeter, *Astrophys. J., Lett.*, 1966, **143**, 609.
17. P. A. Aannestad, *Astrophys. J. Suppl. Ser.*, 1973, **25**, 223.
18. E. R. Iglesias and J. Silk, *Astrophys. J.*, 1978, **226**, 851.
19. L. Spitzer *et al.*, *Astrophys. J., Lett.*, 1973, **181**, L116.

20. P. A. Aannestad and G. B. Field, *Astrophys. J., Lett.*, 1973, **186**, L29.
21. B. Zuckerman and P. Palmer, *Astrophys. J., Lett.*, 1975, **199**, L35.
22. T. W. Hartquist, A. Dalgarno and M. Oppenheimer, *Astrophys. J.*, 1980, **236**, 182.
23. T. N. Gautier III, U. Finke, H. P. Larson and R. R. Treffers, *Astrophys. J., Lett.*, 1976, **207**, L129.
24. A. Dalgarno, J. H. Black and J. C. Weisheit, *Astrophys. Lett.*, 1973, **14**, 77.
25. W. Roberge and A. Dalgarno, *Astrophys. J.*, 1982, **255**, 176.
26. D. R. Flower, G. Pineau des Forêts and T. W. Hartquist, *Mon. Not. R. Astron. Soc.*, 1985, **216**, 775.
27. G. Pineau des Forêts, D. R. Flower, T. W. Hartquist and A. Dalgarno, *Mon. Not. R. Astron. Soc.*, 1986, **220**, 801.
28. M. Oppenheimer and A. Dalgarno, *Astrophys. J.*, 1974, **192**, 29.
29. B. G. Elmegreen, *Astrophys. J.*, 1979, **232**, 729.
30. L. Spitzer, *Astrophys. J.*, 1941, **93**, 369.
31. W. Pilipp, T. W. Hartquist and O. Havnes, *Mon. Not. R. Astron. Soc.*, 1990, **243**, 685.
32. P. Schilke, C. M. Walmsley, G. Pineau des Forêts and D. R. Flower, *Astron. Astrophys.*, 1997, **321**, 293.
33. B. T. Draine, W. G. Roberge and A. Dalgarno, *Astrophys. J.*, 1983, **264**, 285.
34. M. J. Kaufman and D. A. Neufeld, *Astrophys. J.*, 1996, **456**, 611.
35. D. R. Flower and G. Pineau des Forêts, *Mon. Not. R. Astron. Soc.*, 2010, **406**, 1745.
36. J. Kwan, *Astrophys. J.*, 1977, **216**, 713.
37. W. Roberge and A. Dalgarno, *Astrophys. J.*, 1983, **255**, 176.
38. P. G. Martin, D. H. Schwarz and M. E. Mandy, *Astrophys. J.*, 1996, **461**, 265.
39. D. Hollenbach and C. F. McKee, *Astrophys. J., Letts.*, 1980, **241**, L47.
40. D. A. Neufeld and A. Dalgarno, *Astrophys. J.*, 1989, **340**, 869.
41. T. J. M. Boyd and J. J. Sanderson, *The Physics of Plasmas*, Cambridge University Press, 2003.
42. R. M. Kulsrud, *Plasma Physics for Astrophysics*, Princeton University Press, 2005.
43. N. F. Derby, *Astrophys. J.*, 1978, **224**, 1013.
44. A. Jeffrey and T. Taniuti, *Non-linear Wave Propagation*, Academic Press, 1964.
45. B. V. Somov, *Plasma Physics, Part I*, Springer, 2013.
46. S. Van Loo, S. A. E. G. Falle, T. W. Hartquist and T. J. T. Moore, *Astron. Astrophys.*, 2007, **471**, 213.
47. C. C. Wu and C. F. Kennel, *J. Plasma Phys.*, 1992, **47**, 85.
48. S. Cable, Y. Lin and J. L. Holloway, *J. Geophys. Res.*, 2007, **112**, A09202.
49. S. A. E. G. Falle and S. S. Komissarov, *J. Plasma Phys.*, 2001, **65**, 29.
50. D. E. Osterbrock, *Astrophys. J.*, 1961, **134**, 270.
51. B. T. Draine, *Astrophys. J.*, 1980, **241**, 1021.
52. C.-Y. Chen and E. C. Ostriker, *Astrophys. J.*, 2012, **744**, 124.

53. S. A. E. G. Falle, *Mon. Not. R. Astron. Soc.*, 2003, **344**, 1210.
54. B. T. Draine, *Mon. Not. R. Astron. Soc.*, 1986, **220**, 133.
55. M. Elitzur and W. D. Watson, *Astrophys. J., Letts.*, 1978, **222**, 141.
56. B. T. Draine and N. Katz, *Astrophys. J.*, 1986, **310**, 392.
57. W. W. Duley, T. W. Hartquist, A. Sternberg, R. Wagenblast and D. A. Williams, *Mon. Not. R. Astron. Soc.*, 1992, **255**, 463.
58. D. F. Chernoff, D. J. Hollenbach and C. F. McKee, *Astrophys. J., Letts.*, 1982, **259**, 97.
59. E. F. van Dishoeck *et al.*, *Publ. Astron. Soc. Pac.*, 2011, **123**, 138.
60. S. Leurini *et al.*, *Astron. Astrophys.*, 2014, **564**, L11.
61. S. Leurini *et al.*, *Astron. Astrophys.*, 2013, **554**, 35.
62. A. G. G. M. Tielens, C. F. McKee, C. G. Seab and D. J. Hollenbach, *Astrophys. J.*, 1994, **431**, 321.
63. P. W. May, G. Pineau des Forêts, D. R. Flower, D. Field, N. L. Allan and J. A. Purton, *Mon. Not. R. Astron. Soc.*, 2000, **318**, 809.
64. V. Guillet, A. P. Jones and G. Pineau des Forêts, *Astron. Astrophys.*, 2009, **497**, 145.
65. V. Guillet, G. Pineau des Forêts and A. P. Jones, *Astron. Astrophys.*, 2011, **527**, 123.
66. S. Anderl, V. Guillet, G. Pineau des Forêts and D. R. Flower, *Astron. Astrophys.*, 2013, **556**, 69.
67. P. Caselli, T. W. Hartquist and O. Havnes, *Astron. Astrophys.*, 1997, **322**, 296.
68. M. Wardle, *Mon. Not. R. Astron. Soc.*, 1990, **246**, 98.
69. O. Havnes, T. W. Hartquist and W. Pilipp, in *Physical Processes in Interstellar Clouds*, ed. G. E. Morfill, M. Scholer and D. Reidel, 1987.
70. W. Pilipp and T. W. Hartquist, *Mon. Not. R. Astron. Soc.*, 1994, **267**, 801.
71. M. Wardle, *Mon. Not. R. Astron. Soc.*, 1998, **298**, 507.
72. J. F. Chapman and M. Wardle, *Mon. Not. R. Astron. Soc.*, 2006, **371**, 513.
73. S. Van Loo, I. Ashmore, P. Caselli, S. A. E. G. Falle and T. W. Hartquist, *Mon. Not. R. Astron. Soc.*, 2009, **395**, 319.
74. V. Guillet, G. Pineau des Forêts and A. P. Jones, *Astron. Astrophys.*, 2007, **476**, 263.
75. J. D. Slavin, A. P. Jones and A. G. G. M. Tielens, *Astrophys. J.*, 2004, **614**, 796.
76. I. Ashmore, S. Van Loo, P. Caselli, S. A. E. G. Falle and T. W. Hartquist, *Astron. Astrophys.*, 2010, **511**, 41.
77. S. Van Loo, I. Ashmore, P. Caselli, S. A. E. G. Falle and T. W. Hartquist, *Mon. Not. R. Astron. Soc.*, 2013, **428**, 381.
78. L. D. Landau and E. M. Lifschitz, *Fluid Mechanics*, Pergamon Press, 1959.
79. E. Falgarone, G. Momferratos and P. Lesaffre, *Astrophys. Space Sci. Libr.*, 2015, **407**, 227.
80. Z.-S. She and E. Leveque, *Phys. Rev. Lett.*, 1994, **72**, 336.
81. D. Jou, *Sci. Mar.*, 1997, **61**, 57.
82. A. Beresnyak and A. Lazarian, *Astrophys. Space Sci. Libr.*, 2015, **407**, 163.

83. G. Stolovitzky, K. R. Sreenivasan and A. Juneja, *Phys. Rev. E*, 1993, **48**, 3217.
84. J. M. Pittard, S. A. E. G. Falle and T. W. Hartquist, *Mon. Not. R. Astron. Soc.*, 2010, **405**, 821.
85. J. E. Pringle, *Ann. Rev. Astron. Astrophys.*, 1981, **19**, 137.
86. T. W. Hartquist and J. E. Dyson, *Astrophys. Space. Sci.*, 1988, **144**, 615.
87. J. Cantó and A. C. Raga, *Astrophys. J.*, 1991, **372**, 646.
88. J. E. Dyson, T. W. Hartquist, M. T. Malone and S. D. Taylor, *Rev. Mex. Astron. Astrofis., Ser. Conf.*, 1995, **1**, 119.
89. J. M. C. Rawlings and T. W. Hartquist, *Astrophys. J.*, 1997, **487**, 672.
90. M. Ilgner and R. P. Nelson, *Astron. Astrophys.*, 2008, **483**, 815.
91. R. H. Kraichnan, *Phys. Fluids*, 1965, **8**, 1385.
92. W. M. Elsasser, *Phys. Rev.*, 1950, **79**, 183.
93. J. Cho and A. Lazarian, *Phys. Rev. Lett.*, 2002, **88**, 245001.
94. J. Cho, A. Lazarian and E. T. Vishniac, *Astrophys. J.*, 2003, **595**, 812.
95. E. G. Zweibel, *Astrophys. Space Sci. Libr.*, 2015, **407**, 285.
96. R. M. Kulsrud and W. P. Pearce, *Astrophys. J.*, 1969, **156**, 445.
97. W. G. Baker and D. F. Martyn, *Philos. Trans. R. Soc., A*, 1953, **246**, 281.
98. D. S. Balsara, *Astrophys. J.*, 1996, **465**, 775.
99. T. Ch. Mouschovias, G. E. Ciolek and S. A. Morton, *Mon. Not. R. Astron. Soc.*, 2011, **415**, 1751.
100. M.-M. Mac Low and R. S. Klessen, *Rev. Mod. Phys.*, 2004, **76**, 125.
101. E. Vázquez-Semadeni, *Astrophys. Space Sci. Libr.*, 2015, **407**, 401.
102. R. B. Larson, *Mon. Not. R. Astron. Soc.*, 1981, **194**, 809.
103. A. J. Lim, S. A. E. G. Falle and T. W. Hartquist, *Mon. Not. R. Astron. Soc.*, 2005, **357**, 461.
104. S. A. Balbus, in *Physical Processes in Circumstellar Disks Around Young Stars*, ed. P. J. V. Garcia, University of Chicago Press, 2011.
105. R. H. Durisen, in *Physical Processes in Circumstellar Disks Around Young Stars*, ed. P. J. V. Garcia, University of Chicago Press, 2011.

CHAPTER 3

Non-thermal Chemistry in the Interstellar Medium

3.1 Introduction

The interstellar medium (ISM) is an open system, out of equilibrium, both globally and locally. The gas is organized in cold, warm, and hot regions in atomic and molecular clouds, filaments, and bubbles. The entire system is pervaded by a strong and erratic magnetic field and by cosmic rays, the pressures of which are confined in the disc of the Galaxy by the weight of the interstellar gas. The ISM is regulated by self-gravity, cloud collisions, supernovae and stellar winds, galactic rotation, spiral shocks and magnetic fields. Such processes cycle material and energy from the stars to the ISM. High energy radiation from hot, young stars and cosmic rays heat and ionize the gas, coupling tightly the ISM to the stars in a galaxy. The net result is the establishment of supersonic velocity dispersions, $\sim$5–10 km s^{-1} in both atomic and molecular gases, in fact in all but the very hot gases. Such observed large dispersions, resulting in Reynolds numbers of the order of 10^6–10^8, provide strong evidence that the ISM is turbulent.

In the diffuse ISM, in which gravity is negligible, the thermal pressure corresponding to the balance between heating and cooling is a non-monotonic function of the density.[1] A thermal instability occurs when a compression induces such enhanced cooling that the pressure in the compressed region decreases. Thus, the gas tends to segregate into stable phases at the same pressure, but with very different densities and temperatures. In the standard model, there are four observationally well established phases, that are either almost fully ionized or almost neutral. A possible fifth phase, which is warm and partially ionized (warm partially ionized medium, WPIM), is not generally recognized as such by most astronomers.[2] Cold dense neutral gas (cold neutral

Dynamical Astrochemistry
By David A. Williams, Thomas W. Hartquist, Jonathan M. C. Rawlings, Cesare Cecchi-Pestellini and Serena Viti

Published by the Royal Society of Chemistry, www.rsc.org

medium, CNM) is embedded in a warmer medium that comes in two flavours, either neutral or ionized low density gas (warm neutral medium and warm ionized medium, WNM/WIM), which in turn may be embedded in a much hotter and rarefied ionized component (hot ionized medium, HIM). Such components, in rough thermal pressure equilibrium, are considered in terms of quasi-static morphologies. While the filling factors of the gas phases are poorly known, the warm ionized gas phase is the more pervasive. Indeed, the WIM is a major constituent of galaxies (in particular of star-forming galaxies), consisting of mostly ionized hydrogen. Its origin is due to the feedback of massive stars. Spectroscopic observations indicate a diversity between the diffuse WIM and HII region complexes both in temperatures and the ionization rate. Multi-phase structure and turbulence are two key ingredients in ISM physics.

Under the action of powerful dynamical energy sources, such as HII regions, stellar winds and supernovae, departures from equilibrium frequently occur, triggering severely overpressured gas pockets, which give rise to strong dynamical perturbations. Locally the concept of thermodynamic cold and warm phases still applies at each side of a phase transition front, but because of the turbulent fluctuations in the pressure and the heating rate, the densities and temperatures of these locally stable phases differ in general from one location to another.[3,4] In the extreme case, in which turbulence and the concomitant mixing and dynamical processes dominate thermal instability, the discrete phases become indefinite.

Molecular clouds are condensations in the interstellar gas dominated by molecular rather than atomic hydrogen, typical of the rest of the ISM, mainly because they achieve a sufficiently high column density to be opaque to the ultraviolet radiation that elsewhere dissociates the molecules. When H_2 has formed, other molecules will start to form. These clouds presumably emerge out of the diffuse gas, and evolve by interacting with it. There is quite a large consensus on the major role played by large-scale velocity perturbations in the low density gas, through the interaction between turbulence and gravity. The non-linear development of thermal instability triggered by converging streams of diffuse warm atomic gas leads quickly to the break-up of the WNM into filaments and dense clumps embedded in a more diffuse phase (*e.g.*, Hennebelle *et al.* 2008[5]). When cooling and turbulent compressions provide suitable conditions, gravitational collapse occurs. At this stage the study of molecular gas falls into a different regime (see Chapter 4). Table 3.1[2,6–10] lists some characteristics of thermal phases and some denser regions of the ISM.

The origin of interstellar turbulence is not fully clear. Ubiquitous in the Galaxy, turbulence presents similar amplitudes at large scales that have no correlation to the local properties of the ISM, suggesting that the injection of energy at these scales must be global, possibly galactic,[11] and even beyond when infall is taken into account. There is general agreement that turbulence is driven on scales larger than 100 pc (possibly on kpc scales) by supernovae and spiral shocks, although the relative contributions of the two

Table 3.1 Parameters of idealized ISM phases and some denser regions.

Property	CNM	WNM	WIM	HIM	HIM[6] Orion/ Eridanus superbubble	WPIM[7]	Molecular clouds	Dark clouds
Pressure (cm^{-3} K)	4000[2]	4000[2]	4000[2]	10 000[2]	50 000	2000		
Temperature (K)	50,[2] 30[8]	6000[2,8]	8000[2,8]	1.5×10^6 [2]	1.5×10^6	7000	25[8]	10^5
Hydrogen number density (cm^{-3})	80,[2] 100[8]	0.7[2], 0.3[8]	0.25,[2] 0.1[8]	0.0034[2]	0.015	0.2	300[8]	10 000[8]
Hydrogen column density (10^{20} cm^{-2})	0.5[2]	1[2]	0.08[2]	0.01[2]	0.06	0.06		
Hydrogen column density (10^{20} cm^{-2})[a]	1.5[2]	1.5[2]	1[2]	0.1[2]				
Mass fraction (%)	24,[9] 30[10]	36,[9] 38[10]	23,[9] 14[10]				6[9,10]	11,[9] 12[10]

[a]Line-of-sight at Galactic latitude 90°.

mechanisms are controversial. The turbulent nature of clouds in the Milky Way has been demonstrated by a variety of observations, such as, *e.g.*, the hierarchical, and possibly fractal structures of diffuse and molecular clouds,[12] and the intensity fluctuations and broadened linewidths in channel maps.[13]

Since the disc of the Galaxy is in approximate hydrostatic equilibrium, the total pressure in the midplane of the disc must balance the weight of the material above it. Thus, the scaleheights of the gas components are important parameters for understanding these dynamical processes throughout the Milky Way. In fact, the presence of high-latitude molecular clouds such as Orion requires that molecular and cold HI gas resides at distances above the mid-plane, much higher than expected simply from thermal pressure. The distribution of thermal pressures in the Solar neighbourhood peaks at $\sim$3000 K cm^{-3},[14] with large fluctuations, up to a few 10^4 K cm^{-3}. At large scales, the non-thermal contributions to the total pressure, *i.e.* magnetic pressure, cosmic ray pressure, and dynamical (or ram) pressure are observed in rough equipartition, $\sim$1 eV per cm^3 ($\sim 10^4$ K cm^{-3}). Consequently, the total non-thermal pressure in the galactic plane is of the same order as the largest values of the thermal pressure observed. Evidently, turbulent motions are important contributors to the total pressure, together with the other non-thermal components.[15]

There are four main types of non-thermal energy sources in the diffuse molecular gas: turbulent and magnetic energy, the propagation of shocks, fast particles (*e.g.*, cosmic rays) and high energy radiation (*e.g.*, X-rays), and H_2 formation energy. Large scale flows in the gas are created at a galactic scale, and those flows feed a turbulent cascade of motions down to the smallest scale, where the energy leaks out of the system by cosmic ray diffusion, radiative cooling, reconnections, and dissipation of turbulence. The observed amplitudes of the turbulent motions indicate that the ISM flows from a supersonic turbulent regime at scales of tens to hundreds of parsecs (at which the turbulence is driven) to a subsonic one at smaller scales. The scales where turbulence is subsonic depend on the local physical characteristics. Dense and cold clouds result in subsonic turbulence only at very small scales ($\sim$1 pc), while warmer and more diffuse gas with larger sound speeds show subsonic flows at scales of a few parsecs. Strong evidence for decay of turbulence has been found in the diffuse gas using chemical indicators and H_2 emission lines. Turbulence decay has been observed to occur also in denser regions.[16] The dissipation of kinetic and magnetic energy proceeds through viscous, ohmic, and ambipolar diffusion. Dissipation is irreversible, degrades the energy, and is characterized by an increase in entropy. When the change in entropy is discontinuous, the process is called a shock. This localized heating gives rise to complicated chemistry and dynamics, as well as radiation fields, which perturb the behaviour and interrelation between the different phases of the ISM providing observables. At a microscopic level warm H_2 may constitute a key ingredient for chemistry. The excitation of rotational and vibrational levels of H_2 pumped by either radiative or chemical agents might potentially contribute in the same measure as dynamical effects.

In this chapter we shall describe the chemical effects following the injection of non-thermal energy of dynamical origin in the gas. In the next section we present a concise description of the molecular gas in the ISM, its morphology and energetics. These two aspects are deeply related, as turbulent flows continually move cold and warm gas, and cause their mixing within the fluid. In Section 3.3 we discuss the observational evidences of the turbulent nature of the interstellar regions. In the remainder of this chapter, we will look at the coupling of the chemistry to the dynamics of interstellar gas.

3.2 Molecular Gas

Stars form in a cold, dense, molecular gas organized in coherent, localized clouds. The association of such molecular regions with star formation is observed in all environments, galactic and extragalactic, and back in time as far as possible by current observational sensitivity. Molecular clouds are revealed in molecular line emission, thermal dust emission, or in absorption. Their masses are dominated by molecular hydrogen, with a secondary contribution from He (26%), traces of carbon monoxide, and dust that accounts for some of the material. Locally and generally at near-solar metallicities the mass locked into dust grains is approximately 1% of the gas content.[17,18] Turbulent motions are common to all molecular clouds, regardless of their current star formation rate or surrounding galactic environment.[19] Interactions with shocks are frequent both at the boundaries and in the interiors of the clouds. The physics of these regions is complex, and the description of their structure and evolution is still incomplete.[20,21]

3.2.1 Molecular Clouds

Molecular clouds were discovered in the 1970s through CO rotational spectroscopy, a favourable circumstance that opened the way to surveys. The mapping at millimetre wavelengths of line emissions of CO and other molecules in the galactic disc led quickly to the discovery of giant molecular clouds (GMCs), responsible for the majority of all star formation. GMCs, the most massive entities in a galaxy, have about 10^5–10^6 $M_\odot$, average bulk densities of the order of 10^2–10^3 cm^{-3}, and sizes of tens of parsecs. Possible exceptions are galaxies with gas surface densities substantially higher than that of the Milky Way, where a substantial diffuse H_2 component exists, with little distinction between dense clouds and more diffuse gas.[22] Individual GMCs are now also resolved in nearby external galaxies (see Figure 3.1[23]).

HI envelopes around molecular clouds are quite common. While the CO peaks are always associated with a peak in HI, there are many HI peaks in which CO is not detected.[24] The mass within these HI halos has been estimated to be comparable to or even greater than the mass within the

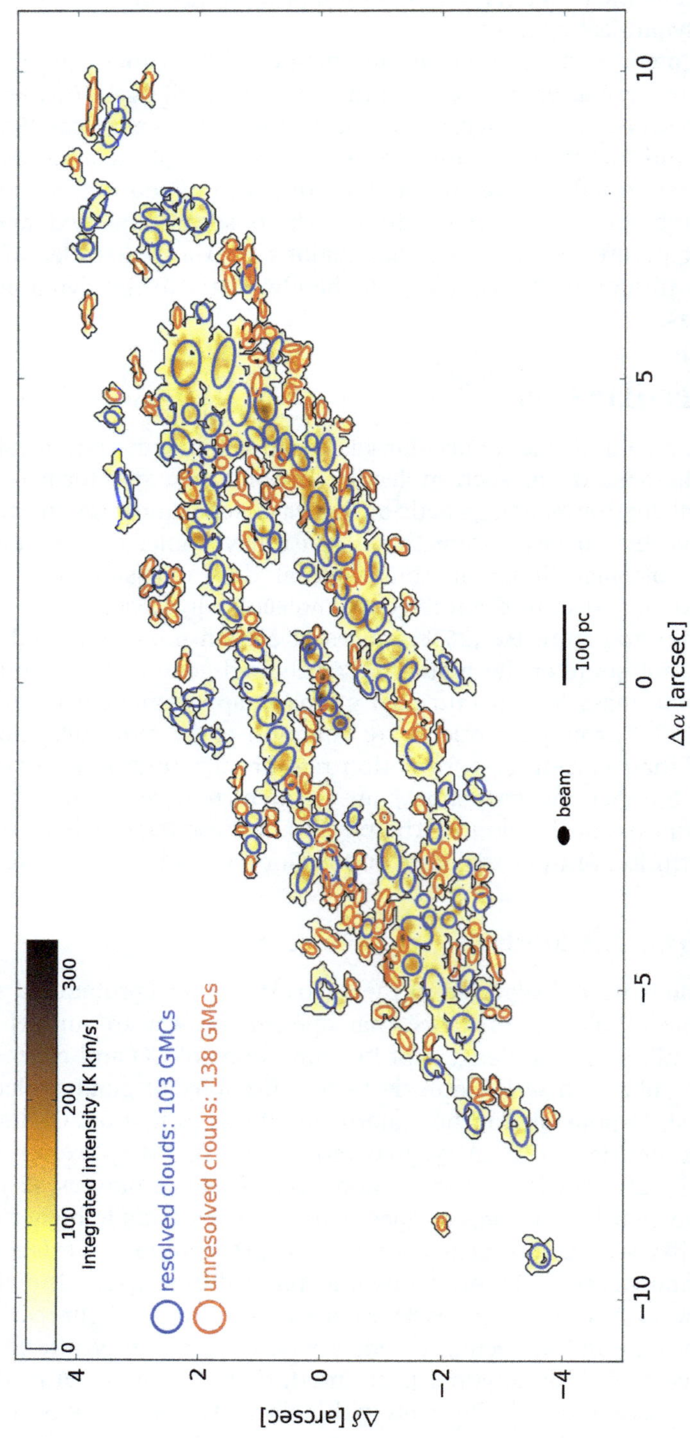

Figure 3.1 GMCs identified in NGC 4526. The beam size and the projected physical size are indicated. The blue and red ellipses mark the location of the resolved and unresolved clouds, respectively. Reproduced with permission from Utomo *et al.*, 2015.[23]

molecular clouds themselves. These envelopes might be photodissociated gas produced by star formation within the clouds, or remnants of atomic clouds that condensed to form GMCs. There is growing observational evidence that molecular clouds are continuously gaining mass during their evolution, accreting gas from HI envelopes. Mass accretion can contribute a significant fraction of the total energy available for driving turbulence.

Although molecular clouds are, by definition, regions in which the gas is primarily molecular, much of their volume is not molecular. The mean number density of the GMC (about 100 cm^{-3} at the 100 pc scale) inferred from their CO column density and size is much lower than the gas density, of the order of $10^3\,cm^{-3}$ or more, required to excite the observed bright CO lines. That is, the filling factor of dense gas is low and the cloud is highly structured with large density variations from one location to another. From the point of view of observers it is convenient to introduce the (meta)concept of clump as a coherent region in position–velocity diagrams, generally identified in the spectral line maps as the CO-emitting brightest structures. Using high dipole moment species (such as CS, CN, *etc.*) that are specific high density tracers, clumps can be decomposed into cores, regions out of which single stars (or multiple systems) form.[25] Clumps have the tendency to blend together at low intensities in low density tracers, emphasizing the vagueness of their definition. In fact, Herschel satellite observations and near-infrared extinction studies have revealed an intricate network of filamentary structures in every interstellar cloud (Figure 3.2[26]). The observed filaments share common properties, such as their central widths, but only the densest filaments (*i.e.* clumps) contain prestellar cores, the seeds of future stars. The pile up of observational evidence has led to a new scenario for star formation in which large-scale supersonic flows compress the gas, giving rise to a universal web-like filamentary structure in the ISM. Next, gravity takes over the control of denser structures whose fragmentation gives rise to prestellar cores and ultimately protostars. Recently, Kainulainen *et al.* (2014)[27] presented an analysis of the stability of clumpy structures in nearby molecular clouds, finding that they are gravitationally unbound entities confined effectively by the turbulent external pressure from the parental molecular cloud. Such pressure-supported clumps appear to form prior to, or at higher rates than, the formation of gravitationally dominated dense cores in the clouds. This suggests that the formation of pressure-confined clumps is a prerequisite for star formation. The confining force exerted upon a clump, however, depends on the isotropy of the turbulent flow surrounding the clump. Supersonic turbulence is locally very anisotropic and may lead to both formation and destruction of filaments and clumps. Small and less massive clumps are then probably transient,[28] merging back in the more diluted filamentary backbone.

While the cores themselves are clearly undergoing gravitational collapse, the gravitational binding of a cloud as a whole is more debatable. In the 1980s and 1990s, GMCs were generally thought to be thermally supported against gravitational collapse, and to be in virial equilibrium. Magnetic

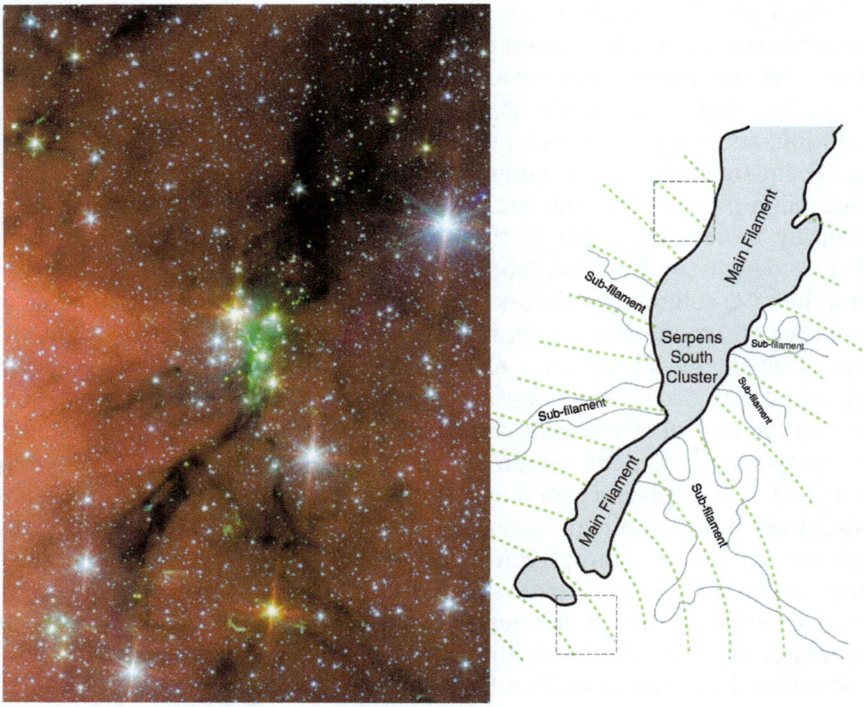

Figure 3.2 Left panel: main and sub-filaments of the Serpens South cloud. This cluster is a relatively dense group of 50 young stars, 35 of which are protostars just beginning to form. Image credit: NASA/JPL-Caltech/ L. Allen (Harvard-Smithsonian CfA) & Gould's Belt Legacy Team. Right panel: schematic drawing including the magnetic field lines represented as green dotted lines. Reproduced with permission from Sugitani *et al.*, 2011.[26]

fields were considered to be a means of stability, while turbulence appeared to be insufficiently powered. More recently, observations challenged such conclusions, revising downward the estimates for magnetic fields, while simulations suggested that clouds may in fact be sustained by turbulence. Another possibility is that GMCs are entirely transient objects that are not supported against collapse at all.[20] Observations of molecular gas in the Large Magellanic Cloud[29] suggest that molecular clouds are undergoing an evolutionary process, in which both their mass and their star formation activity increase in time, going in $\sim$30 Myr from an initial stage in which massive stars are not observed, through a phase with massive stars, compact HII regions but no cluster, to end up with a population of clusters and HII regions. Thus, GMCs may be assembled rapidly from the atomic component, with a prompt onset of star formation.

Molecular clouds are thought to form in dense, post-shock regions when converging flows of lower density gas collide and interact. The process naturally collects large amounts of gas in small volumes, first by compression-triggered thermal instability that promotes fast cooling and then, as a second step by gravitational instability. Simulations of cloud formation in a supernova-driven turbulent ISM without self-gravity show that cold, dense clouds are formed in the expanding shells bounding the supernova remnants, but the induced star formation rate is only 10% of the value required to produce the occurrence frequency of supernovae needed to drive the whole process.[30] The accumulation of cold, dense atomic gas promotes molecule formation. Moreover, clumping of gas preferentially along the magnetic field lines may convey enough mass per unit area to be magnetically supercritical.[31] Thus, the accumulation of mass from converging large scale streams, the transition to the molecular phase, and from magnetically subcritical to supercritical gas, might occur simultaneously, allowing coeval molecular cloud and star formation.

The total mass of the cloud that can be formed by the colliding confluence of WNM regions depends on the mean gas density in the flows, ρ, the cross-sectional area, A, and the velocity correlation length, L, of the converging streams. Assuming the density for WNM gas reported in Table 3.1, a typical galactic GMC, $A \approx 1000$ pc^2, and a size $L \approx 150$ pc, comparable to the molecular gas scale height of the galactic disc, we find a maximum mass ρAL, of approximately a few thousand solar masses, much smaller than the mass of most GMCs. Some additional mechanism is thus required to induce smaller clouds to agglomerate into larger ones. While small molecular clouds can be built by coagulation quite quickly, GMCs with masses of 10^5–10^6 $M_\odot$ require times of the order of 100 Myr.[32] However, for clouds confined in spiral arms, collisions may be much more frequent, greatly reducing the timescale of cloud agglomeration. In the molecule-dominated disc of the galaxy M51 the most massive molecular clouds appear exclusively along spiral arms, while smaller clouds and unresolved molecular emission still dominate over HI in the interarm regions.[33] Such results could be naturally explained by fragmentation of massive clouds upon leaving the spiral arms, suggesting a long lifetime for molecular clouds and molecular gas. This scenario is challenged by observations in the solar neighbourhood, and in some local galaxies, such as M33, and, as mentioned above, the Large Magellanic Cloud. Molecular clouds appear to exist primarily within HI spiral arms with little or no H_2 gas in the interarm regions. The molecular gas should therefore be dissociated during its passage between the spiral arms, setting the cloud lifetimes to be as short as an arm crossing time ~30 Myr (consistent with the estimate of Kawamura *et al.* in the Large Magellanic Cloud[29]). Using CO ($J=1\rightarrow 0$) and HI 21 cm archival data, Koda *et al.* (2016)[34] re-analysed spatial variations of the balance between the atomic and molecular gas in the Milky Way. They found that in the inner part of our Galaxy the gas stays molecular in both spiral arm and interarm regions with small arm/interarm variations, while in the atom-dominated outskirts large variations can occur. The molecular cloud lifetime thus appears determined by the environment, with short lifetimes common in mostly atomic regions, such as the Large

Magellanic Cloud, and extended lifetimes in molecule rich ones such as in M51 and in the inner part of the Milky Way.

From the above considerations, molecular clouds seem to be merely the highest density regions within a globally extended turbulent flow of gas, rather than being well defined entities. This is supported by the existence of CO molecules in extended environments weakly shielded from the ambient ultraviolet field ($A_V <$ a few mag). In addition, tiny CO emitting regions have been detected located in an environment with very low hydrogen column densities. These cloudlets serendipitously found by Heithausen in 2002,[35] cover a very small area on the sky of marginal visual extinction ($A_V < 0.2$ mag). Because these structures have been discovered in a rather limited sample of observations, they might be frequent in the ISM. Kainulainen *et al.* (2009)[36] identified a structural transition in molecular clouds between the diffuse, interclump medium and a population of clumps and filaments that are gravitationally unbound, but significantly supported against dispersal by the external pressure imposed on them by the surrounding medium. Such transition takes place at visual extinction $A_V \approx 2$–4 mag, and occurs when going from a regime dominated by turbulent motions to active star formation driven by gravity. In non-star forming regions the gas resides in the former regime, with star formation preceded by the departure from such structural properties. The diffuse medium comprises a significant mass fraction of molecular clouds, which may be as high as half the mass of the complex. The Heithausen cloudlets might be a leftover of regions stirred by turbulence unable to perform the "transgravity" transition.

Infrared dark clouds (IRDCs) are the darkest, coldest, and densest clouds in our Galaxy. IRDCs are large-scale filamentary structures (several tens of pc) of large hydrogen column density ($\sim 10^{23}$ cm^{-2}), embedded in GMCs of lower extinction containing ten times more mass. These clouds were discovered as shadows against the bright mid-infrared background of the Galaxy. IRDCs are objects in which conditions suitable for high-mass star and star cluster formation are believed to be present or developing, because they harbour clumps of dense gas of the order of 1 pc in which are immersed high-mass starless, possibly prestellar, cores. Such cores usually have high densities ($>10^3$ cm^{-3}), and small sizes ($\sim$0.1 pc), and in some cases show internal structures known as condensations. IRDCs are not quiescent, and show moderately more disturbed kinematics than GMCs. An example of a star forming region in an IRDC is shown in Figure 3.2. Such a region embeds the very young Serpens South cluster, located in the constricted part of the long, filamentary concentration of millimetre emission.

3.2.2 The Origin of Turbulence in Molecular Clouds

Turbulent supersonic motions in molecular clouds are expected to dissipate rapidly in shocks within a few crossing times of the outer scale of the turbulent flow. For clouds of a few pc and internal velocity dispersion of a few km s^{-1} the dissipation time is $\sim$1–5 Myr.[37] Since these motions do not decay, turbulence must be continuously replenished. Many studies, both observational and computational, reach the conclusion that the injection must occur at a scale larger than or equal to the size of the clouds they deal with. This suggests that, at least for a large fraction of clouds, the turbulence driving is external, *i.e.* most of the energy is injected from outside.

One obvious source is the accretion flow from which the cloud itself forms. The accretion-driven turbulence scenario implies that energy is injected into the ISM at large scales in the Galaxy, presumably by a combination of supernovae/bubble expansion and galactic differential rotation, and cascades to smaller scales as the gas accumulates at the molecular cloud level. While accretion flows are one possible source of energy, there are also others. GMCs are embedded in large-scale galactic flows that force them to undergo continuous collisional interactions with the environment, such as cloud–cloud collisions, in which the cloud mass is not necessarily growing. This mechanism is expected to operate in high-surface density galaxies where there is no real distinction between GMCs and other gases.[38]

When enough stars have formed inside molecular clouds, stellar feedback presumably becomes important. There are a number of plausible origins of turbulence, including HII regions, radiation pressure, protostellar outflows, and the winds of main sequence stars, but there is no general agreement on the effective driving rate. Based on numerical simulations some authors have concluded that HII regions are the most efficient in sustaining turbulence,[39] while others have found that the photo-evaporation induced by the HII regions eventually destroys the cloud within less than 10 Myr,[40] consistent with observations showing that star clusters older than 10 Myr are usually gas-free. The origin of these discrepancies is not clear, as the simulations differ in several ways, including the geometry they assume, the size scales they consider, and the way they set up the initial conditions.

Most likely the total contribution of external and internal sources is roughly comparable over the cloud lifetime and possibly coexistent. In fact turbulent motions are clearly detected even in molecular clouds that are not forming stars.

3.2.3 Chemical Transitions in the Diffuse Molecular Gas

Once accumulated, high density gas undergoes the chemical transition from atomic gas, where hydrogen is mostly neutral and carbon is mostly (singly) ionized, to molecular gas characterized by H_2 and CO. The transition region where the gas becomes molecular is called the photon dominated region (PDR). H_2 formed on dust grains with a rate proportional to the gas density is rapidly destroyed in an indirect process involving line excitations of upper electronic states, but the involved ultraviolet photons are quickly exhausted and the molecule is self-shielded. The CO molecule also undergoes line photodissociation, through the absorption into bound electronic states followed by radiation-less coupling with the continuum of a final dissociating state (predissociation). In molecular line observations of low column density molecular clouds, the H_2/CO ratio is found to be variable and much larger than the hydrogen/carbon abundance ratio, suggesting that only a small fraction of carbon is locked in CO. Thus, CO molecules require much more shielding than H_2 before they become the dominant carbon repository. In the solar vicinity the CO column densities corresponding to the molecular transition are $N(\mathrm{CO}) \approx 10^{14}\,\mathrm{cm}^{-2}$.

The location of the transition—or the hydrogen column density at which the transition occurs—between mostly-atomic and mostly-molecular gas within a cloud is linearly controlled by the ratio of the far-ultraviolet radiation field to the gas density, so that H_2 becomes dominant in regions where the gas is dense and the radiation field is attenuated. However, the molecular hydrogen fraction depends exponentially on dust ultraviolet extinction. Such transitions are shifted to higher surface densities in environments with low metallicity and dust abundance, and thus low extinction per unit mass. Because of its less efficient self-shielding the shift in column density is even greater for CO. As a consequence, low metallicity galaxies show significant CO emission only from high-extinction peaks, rather than from the bulk of the molecular material.

There are thus two main chemical transitions, occurring at different points in a molecular cloud. The first of these is the transition between atomic and molecular hydrogen that occurs at small visual extinction ($A_V < 1$ mag). Once the visual extinction of the gas becomes large, CO photodissociation becomes unimportant and so almost all of the carbon in these high extinction regions is found in the form of CO. In clouds subjected to a very high cosmic ray flux, such as those in the Central Molecular Zone of the Milky Way, the destruction of CO proceeds quickly through dissociative charge transfer with He^+ ions produced by cosmic-ray ionization of neutral helium, $He^+ + CO \rightarrow O + C^+ + He$, and the CO fraction can be significantly suppressed, even in well-shielded gas. Cosmic-rays have been shown to be particularly effective in star-forming galaxies, where they can destroy CO (but not H_2) at such a high rate as to render potentially large H_2 gas reservoirs CO-poor and difficult to trace using conventional CO $J = 1 \rightarrow 0$, $2 \rightarrow 1$ line imaging.[41] This effect depends on density, with denser regions showing higher CO/H_2 ratios than low-density ones, the cosmic-ray ionization rate being equal. Turbulent motions, expected to affect strongly molecular gas in star-forming galaxies, may alleviate CO destruction by placing most of this species at high average densities.

CO emission is usually associated to cold, dense, strongly shielded, molecular gas. However, as we noted in the previous section, it becomes more and more evident that a large fraction of the CO emission in our Galaxy comes from warmer, low density, and weakly shielded, diffuse gas. This raises the issue of the CO survival in these environments in which column densities are as low as a few $10^{12}\,\mathrm{cm}^{-2}$, smaller than those required for CO self-shielding, and of the nature of the gas component that CO emission is actually tracing. Different measurements (*e.g.*, Fermi/LAT in the γ-ray domain and Planck in the submillimetre spectral range) identify gas that is dark in CO and other gas in which CO is overluminous per H_2 molecule, with wide variations between sightlines, and at different locations in individual clouds. At small scales such large fluctuations cannot be attributed to variations in the continuum and line shielding, but they must be primarily due to a chemistry whose nature differs significantly from the ultraviolet driven one characteristic of PDRs. In these environments chemistry and dynamics must be linked.

A further unexpected chemical transition to molecular complexity has been identified in diffuse clouds. At column densities lower than $N(H_2) \approx 5 \times 10^{20}\,cm^{-2}$, CO, OH, HCO^+, C_2H, and C_3H are detected, while at larger column densities CS, HCN, NH_3 and H_2CO appear with relative abundances that are strikingly similar to those inferred toward dark clouds, such as TMC-1. Cold chemistry models generally reproduce column densities of OH, C_2, and CH, but underestimate CO, CN, CH^+, SH^+, and HCO^+ by a large factor. Moreover, the agreement between models and observations might be illusory, as equilibrium excitation analyses provide evidence that, *e.g.*, a large amount of CH originates from CH^+ chemistry. Further uncertainty on the nature of diffuse clouds is generated by the discovery along diffuse lines of sight of concentrations of the molecular ion $H_3{}^+$ as large as those observed in dark clouds. Such abundances appear inconsistent with model predictions based on a high removal rate *via* electronic dissociative recombination in a cold gas.

The basic problem for the chemistry is how to incorporate the ambient oxygen and carbon into molecules. Oxygen is ionized by slow charge transfers with H^+, while the abundant ionized carbon does not react rapidly with H_2. Thus, most theoretical models rely on the endothermic reactions $C^+ + H_2 \rightarrow CH^+ + H$ (with a barrier at ~5000 K) and $O + H_2 \rightarrow OH + H$ (~3000 K). The need for energy sources in excess of the average energy density of diffuse and translucent clouds has given rise to models involving localized volumes of warm gas created and sustained by dissipation of gas kinetic energy, as within magnetohydrodynamical (MHD) shocks, or coherent small-scale vortices in MHD turbulence. Thus, the classical picture of the diffuse ISM has been modified incorporating small amounts of warm gas intimately associated with the cold neutral phase. These tiny warm regions might promote the first steps in the chemistry of diffuse clouds, not only by producing the required concentrations of CH^+, but also by inducing the formation of HCO^+, an important chemical precursor of CO. Actually, if the large observed abundance of HCO^+ is inserted into standard chemical models of diffuse clouds, the observed variation in carbon monoxide column density can be explained without resorting to any other assumption. These warm tiny regions may also provide a viable single way to account for the observed large concentrations of both $H_3{}^+$ and CH^+. Such a mechanism would alleviate the constraints posed by the column density of OH, whose concentration increases with the increase in both gas temperature (needed for CH^+ formation) and fractional ionization (needed for $H_3{}^+$ formation), as well as by the column densities of other molecular diagnostics such as HD.

The problem posed by the large abundances of H_3^+ in conventional cold gas models is generally attributed to a slow formation rate. However, the main H_3^+ destruction channel in diffuse clouds, dissociative electronic recombination, decreases significantly with increasing gas temperature, suggesting that transient non-equilibrium chemistry at high temperature might indeed solve the problem.[42]

The molecular richness of the diffuse medium has been further increased by the discovery of small hydrides, the building blocks of interstellar chemistry, with unexpected large abundances. Several species have been observed in abundance ratios providing new scenarios for the chemistry of the ISM. The large excess of OH^+ over H_2O^+,[43] and the ubiquitous HF, suggest formation routes in predominantly atomic gas. Finally, the large abundances of CH^+ and SH^+, both with highly endothermic formation routes, are associated with local molecular fractions fluctuating by large factors,[44] from less than 0.1 to almost unity, further blurring the distinction between atomic and molecular clouds.

What is chemistry telling us? First, there is the clear requirement of superthermal energy in the cold diffuse ISM, and secondly the gas must be very heterogeneous. In Sections 3.4–3.6 we shall examine some possible answers to the chemical questions posed by the observed molecular distributions.

3.3 Observational Evidence of Turbulent Motions

The identification of turbulent motions in the ISM dates back to the 1950s from measurements of velocity dispersions.[45] These observations followed the seminal work of von Weizsäcker in 1951,[46] who recognized a hierarchy in the spatial distribution of dense structures, suggesting their turbulent origin. The discovery went almost unnoticed for 30 years until. Larson[47] put forward a comprehensive analysis of the spectrum of interstellar motions, showing that the internal velocity dispersions of molecular clouds were correlated with their size and mass, and similar correlations held even for the different regions of an individual cloud or complex. The relation found did not differ too much from the 1/3 power spectrum of Kolmogorov subsonic turbulence (see Section 2.7). Since then, the observational evidence for a turbulence-dominated ISM has grown steadily.

While supersonic linewidths and morphological complexity indicate the presence of turbulence in the ISM, little has been learned about the nature of the turbulent flows. One of the main signatures of turbulence in the ISM is the density distribution of its components, as traced by spectral lines and continuum emissions from the dynamical multi-phased system. Key diagnostics are the 21 cm line of neutral hydrogen, line emission from CO isotopologues in molecular clouds, and dust emission and absorption. Emission spectra provide detailed information about the masses, velocities, and sizes of a variety of structures, from the densest star-forming cores, GMCs, and the surrounding, volume filling diffuse gas. Spectroscopic measurements have the advantage of containing information about the underlying turbulent velocity field, whose 3D structure would characterize totally interstellar turbulence. However, observations are always a complex convolution of density and velocity field, affected by many other astrophysical processes, and the separation of different contributions requires the implementation of sophisticated inversion methods. Moreover, one crucial characteristic of turbulence is the presence of fluctuations in the local velocity, energy dissipation, and velocity gradients. Such fluctuations deviate from the homogeneous and isotropic model of turbulence, and have received the name of intermittency (see Section 2.7.2). The transfer of energy from scale to scale is highly non-uniform, with the fluctuations eventually amplified by the cascade. At small scales, the gas is thus subjected to strong variations in the energy injection rate. Analysis of turbulent motions are further complicated by the fact that the ISM is a partially ionized gas in varying physical conditions, contains magnetic fields, and exhibits supersonic velocity dispersions. In other words, interstellar turbulence is compressible, magnetized and multi-phase. In practice, the reconstruction of a true velocity field from observations is impossible given its chaotic, non-linear, and non-deterministic nature. Thus, we have to rely on a statistical approach of the temporal and spatial variability of these radiative fluxes.

An introduction to the needed statistical machinery is given in Section 2.7 (see also the excellent review by Elmegreen and Scalo 2004[48]). In Section 2.7.1 we define the kinetic energy per unit mass associated with Fourier components having wavenumbers in the range k to $k+\mathrm{d}k$, the so-called energy spectrum. In slightly more formal terms, the energy spectrum is the average over all directions in the wavenumber space of the power spectrum $P(k)$, the Fourier transform of the autocorrelation function, this latter being a measure of the relation between neighbouring fluctuations. The energy spectrum in D dimensions gives[48] $E(k)\ \mathrm{d}k = P(k)\ \mathrm{d}\boldsymbol{k}^D = P(k)\ \mathrm{d}\Omega\ k^{D-1}\mathrm{d}k$, which, in the isotropic case, reads as $E(k) = P(k)\ k^{D-1}$. The energy spectra are generally represented as power-law distributions, $E(k) \approx k^{-\beta}$. In the Kolmogorov case described in Section 2.7.1, $\beta = 5/3$. The mean square turbulent velocity, $\langle u_l^2 \rangle$ at size l is determined from the energy spectrum of the velocity fluctuations $E(k)$ over scales smaller than $l = 2\pi/k$. The root mean square turbulent velocity then results as $\langle u_l^2 \rangle^{\frac{1}{2}} \approx l^{(\beta-1)/2}$, which incidentally justifies size–linewidth relations, such as the ones identified by Larson.

The root mean square turbulent velocity is not of course equivalent to an observed width of a spectral line, which depends upon the velocity, density, temperature, and abundance, all varying along the line of sight. However, given a velocity field described by an energy spectrum, some correlation of the measured velocities with projected spatial scale, ΔL (a lag), of the form $\delta u_{\mathrm{obs}} \approx \Delta L^{\gamma}$ is expected. Such measurements are frequently performed on regions in the plane of sky identified in the position–velocity space of CO, $J = 1 \rightarrow 0$ maps (clumps, see Section 3.2.1). Together with size and internal velocity dispersion, the column density (and therefore the mass) can be derived for each of these structures. Figure 3.3[49] is an illustration of the mass–size and linewidth–size scalings of such CO structures. The scaling laws reported in Figure 3.3 depend critically on the environment and the line tracer used. In ^{12}CO the size–linewidth is found to hold over almost five orders of magnitude in lengthscale with a slope around 0.5 (close to Larson's original estimate[47]), although with quite a large dispersion of about a factor of 10, that can be even larger below 0.1 pc (see left panel of Figure 3.3). Moreover, in the size–mass relation two regimes are evident, depending on the gas mass surface density, $\Sigma = M/\pi\Delta L^2$. The transition occurring at a mass surface density threshold of ~100 $M_\odot$ pc^{-2} is ascribed to virial balance between turbulent energy and gravity,[49] with $\delta u^2 \approx GM/\Delta L$ and $\delta u/\Delta L^{\frac{1}{2}} \approx \Sigma^{\frac{1}{2}}$. Below this threshold, the internal velocity dispersion is independent of the mass surface density and scales as $\sim\Delta L^{0.5}$, according to the turbulent nature of the cloud internal motions. Thus, GMCs as traced by CO are virialized, but no solution of continuity is shown in the size–linewidth relation, which is the same in any mass regime. ^{13}CO observations[50] are virtually blind to this threshold because the $J = 1 \rightarrow 0$ line is much thinner than the corresponding one in ^{12}CO.

Figure 3.4[49] shows the spectral index γ of power spectra derived from the emission maps of different regions, plotted against the range of size scales. To construct the diagram several methods and tracers have been exploited (see Hennebelle and Falgarone 2012[49] for details), such as dust emission at 250 μm (Herschel, thick green lines) and 100 μm (IRAS, thin lines), HI emission from WNM (black lines) and absorption from CNM (turquoise), ^{12}CO ($J = 2 \rightarrow 1$) for the whole line emission (thick blue) and for the line-wing emission only (purple), ^{12}CO ($J = 1 \rightarrow 0$) and ^{13}CO ($J = 1 \rightarrow 0$) (blue thin lines) in the filamentary gas of the cloud Polaris Flare, and ^{13}CO line emissions in star-forming regions (red lines). HI emission power spectra are also displayed for the Large Magellanic Cloud (dotted lines) showing a variation in slope from $\gamma \approx 5/3$ on large scales to $\gamma \approx 8/3$ on small scales, consistent with a

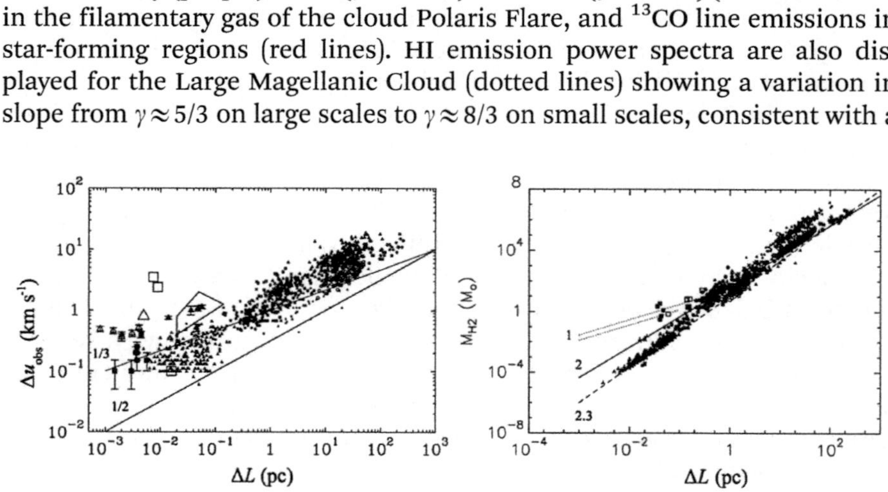

Figure 3.3 Velocity (left panel) and mass (right panel) dispersions as functions of the sizes of a large sample of molecular clouds. Labels indicate the slopes of the correlation lines. Dotted lines represent the scaling laws of isothermal self-gravitating polytropes with size for two gas temperatures 10 K (upper) and 20 K (lower).
The Astronomy and Astrophysics Review, Turbulent molecular clouds, 20, 2012, Hennebelle and Falgarone © Springer-Verlag Berlin Heidelberg 2012. With permission of Springer.[49]

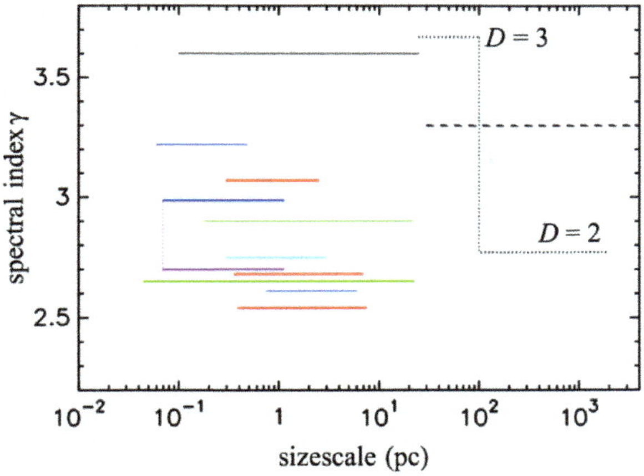

Figure 3.4 Spectral indices γ of emission maps *versus* the range of size scales where the power-law $P(k) \approx k^{-\gamma}$ is observed: dust emission (green), ^{12}CO (blue and purple), ^{13}CO (red) line emission, and HI emission (black). HI absorption is shown in turquoise. The error bars (not shown) on all the values of γ vary between 0.01 and 0.3.
The Astronomy and Astrophysics Review, Turbulent molecular clouds, 20, 2012, Hennebelle and Falgarone © Springer-Verlag Berlin Heidelberg 2012. With permission of Springer.[49]

transition from bidimensional ($D = 2$) to tridimensional ($D = 3$) geometry at about 100 pc, due to a change in the thickness of the HI layer along different lines of sight. Different tracers originate in different phases of the ISM, with varying amounts of small-scale structure that may affect the power spectrum of the density distributions. However, as seen in Figure 3.4, many observations, dust emission included, give power-law indices close to $\gamma = 2.7$, compatible with the near-Kolmogorov turbulence in the diffuse ISM ($D = 2$). At small scale (<1 pc) cold dust emission is flatter, $\gamma = 2.65$, than that of ^{12}CO ($J = 2 \rightarrow 1$), $\gamma = 3$, and very similar to that of CO line-wing emission, $\gamma = 2.7$. This means that at small scale the integrated CO emission morphology is basically featureless, with the presence of structure detectable only in the velocity field, supporting a close link between gas dynamics and the formation of structures. In the WNM, which is not very supersonic on large scales, the HI emission presents an index close to 11/3, as expected in the $D = 3$ Kolmogorov case. CO emission and HI absorption, which both trace colder supersonic media, have power spectra that are much more shallow. Moreover, the trend of the index being higher at smaller spatial scales may also indicate the transition from the supersonic regime to the subsonic one. However, it is highly unlikely that a Kolmogorov-like cascade operates in the ISM, and thus magnetic effects may be important in providing scalings,[51] as those reported in Figure 3.4.

More challenging is the identification of regions of intermittency in interstellar turbulence, since these regions correspond to rare events in time and space, whose localization requires the analysis of large homogeneous statistical samples of the velocity field. Moreover, line emission is integrated along the line of sight, and depends on the velocity field just through its component along the line of sight, *i.e.* the Doppler shifts. This in turn means that spatial variations of the emission in the plane of sky are by essence cross-variations, *i.e.* velocity shears. Then, the velocity information at a given position is not only given by the line profile, but also by its moments, the first moment being the line centroid velocity, *i.e.* the average velocity along the line of sight weighted by the intensity. In principle, line profiles are sensitive to dynamical processes. In the optically thin case, the integrated line profile emerging from a specific line of sight is the histogram of the radial velocities, which in statistical terms may be interpreted as the relative likelihood for the velocity to differ by a given value. In other words, the integrated line profile is equivalent to the probability density function (PDF) of the velocity differences,[52] although obtained by measurements along a single line of sight. In practice, the information contained in the PDF is frequently degraded by systematic dynamical contaminants along the line of sight, such as, *e.g.*, differential rotation, or by radiative transfer effects. An alternative procedure is to estimate the PDF of centroid line velocities. This quantity is assembled by adding measurements between two positions separated by a lag ΔL in the plane of the sky. The method, called centroid velocity increment (CVI) is frequently exploited using ^{12}CO lines because their large optical depth makes them sensitive tracers of rare events, in particular of the gas that emits at velocities in the far line wings (*i.e.* extreme dynamic events).

At large scales the PDF of centroid line velocities is approximately Gaussian, and, as the scale decreases, the wings of the distribution become increasingly stretched, so that large deviations from the average value are present. This PDF, as observed in the Earth's atmosphere and in the solar wind, is known to display extremely non-Gaussian tails, which are often associated with the existence of coherent structures of intense vorticity in the flow. Using a sample of about one million independent CO spectra in a diffuse region, Hily-Blant and Falgarone (2009)[53] identified on statistical grounds the ensemble of positions at which vorticity departs from a Gaussian distribution. The non-Gaussian wings of the PDFs increase as the lag ΔL decreases. The locations at which the PDF departures from Gaussian shapes are more evident form elongated narrow structure ($\sim$0.03 pc wide) that extend over more than one parsec, surrounded by an ensemble of weaker, shorter signatures (Figure 3.5[53]). These coherent regions are proposed to be the manifestations of the intermittency of turbulent dissipation in diffuse molecular clouds. Turbulent dissipation is a possible source of heating of diffuse ISM, opening reaction channels otherwise closed at low temperatures. Such energy injection may be reflected by the H_2 pure rotational emission in diffuse molecular clouds. At the same time such warm H_2 is indispensable for the chemistry to be initiated, through the efficient formation of CH^+, followed by $CH_2{}^+$ and $CH_3{}^+$. The chain stops at this stage with $CH_3{}^+$ preferentially giving rise to HCO^+ *via* reactions with O.

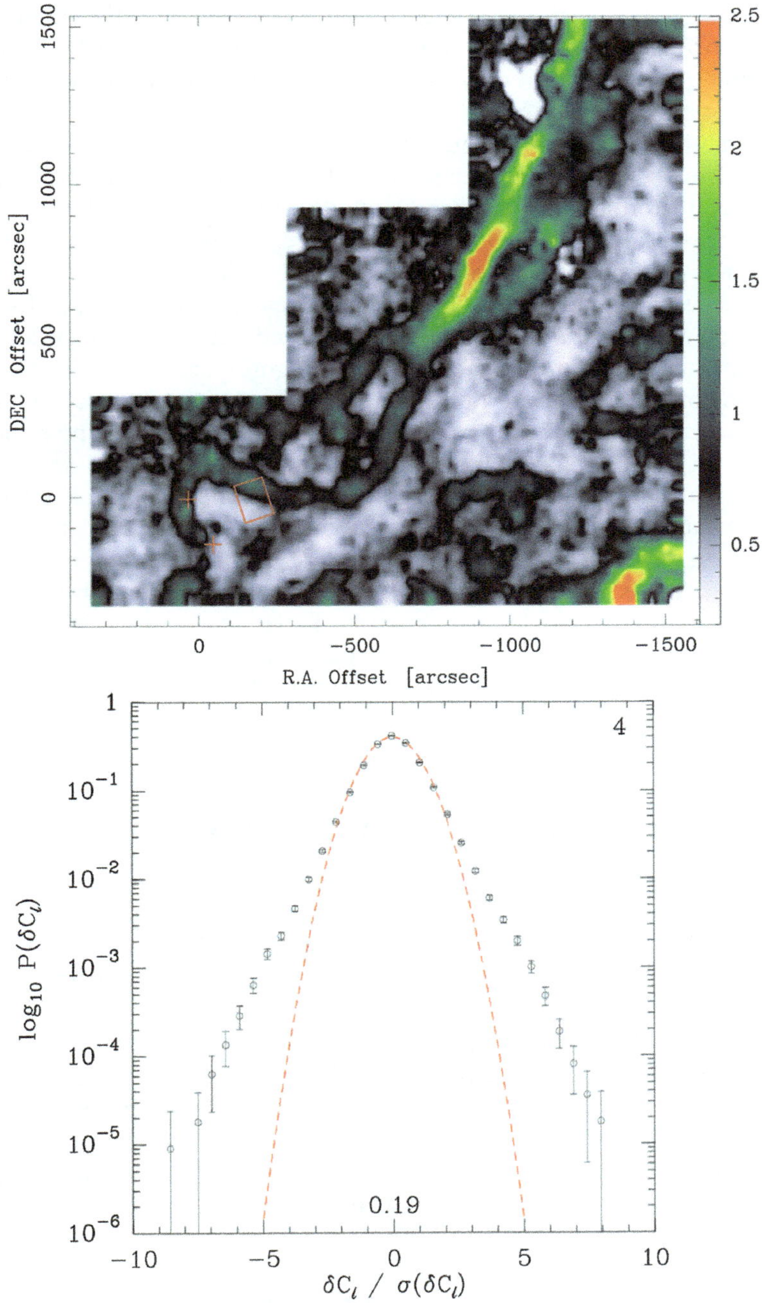

Figure 3.5 Top panel: map of the line CVI (colour scale in km s^{-1}) computed for a lag of 4 pixels (60″). Bottom panel: normalized PDF (filled black circles) compared with a Gaussian distribution (red dashed line) with $\sigma = 0.19$ km s^{-1}. The non-Gaussian wings depend on the lag.
Hily-Blant and Falgarone, A&A, 500, L59, (2009), reproduced with permission © ESO.[53]

Once present in significant concentration HCO^+ becomes the main source of CO through dissociative recombination, solving the CO formation problem.[54] In contrast, in dark clouds CO is the precursor to HCO^+ *via* the reaction with H_3^+.

Information on intermittency and energy flow within the cascade are also contained in higher order statistical functions, such as the structure functions introduced in Section 2.7.2, through the deviation from the Kolmogorov scaling $s(p) = p/3$. The symbol ς_p is frequently used in place of $s(p)$. For instance, measurements of vibrationally excited H_2 emission in OMC-1 containing radial velocity and brightness on a wide range of scales[55] (from 1 pc down to few tens of au) show marked departures of the structure functions at various orders from the Kolmogorov scaling. Since structure functions of high order are dominated by regions of strong velocity difference, such anomalous scaling of the exponent characterizes the regions of largest vorticity or velocity shear. The most successful model so far is the log-Poisson model,[56] which is based on a multiplicative cascade with defects, *i.e.* lack of dissipative events, occurring with a Poisson distribution. It predicts that $s(p) = p/g(1-f) + C_D[1-(1-f/C_D)]$, without any remaining free parameter. C_D is the so-called codimension of the dissipative structure, $g = 2/(\beta - 1)$, and $f = -(\beta - 3)/2$. Vortices in HD turbulence imply $C_D = 2$, $\beta = 5/3$, $g = 3$, and $f = 2/3$, giving $s(2) = 0.696$ (as opposed to 2/3). If we assume that HD turbulence forms dissipative sheets then $C_D = 1$, and we would get $s(2) = 0.741$. The applicability of this model in the compressible regime is still unclear.[57]

3.4 Chemistry in Turbulent Regions

As we have discussed in the previous section, the locations on the CO maps populating the non-Gaussian wings of the velocity gradients might trace locations of enhanced dissipation in interstellar turbulence. These regions are not randomly distributed, and form small-scale elongated structures. In this picture, the cold diffuse ISM contains pockets of hot gas, originating from localized dissipation of non-thermal energy of supersonic turbulence.

The intermittent nature of turbulence in space and time may induce dynamical processes that are locally large enough to drive the heating of the gas and trigger the endothermic reactions at the base of carbon and oxygen chemistries in the diffuse ISM. Such environments would also naturally explain why the diffuse ISM is, occasionally, a strong emitter in the pure rotational lines of H_2. FUSE,[58] ISO-SWS,[59] and Spitzer[60] spectroscopic observations have revealed large populations in excited $J > 2$ levels of H_2, inconsistent with fluorescence generated by ambient ultraviolet pumping. Support for this interpretation is given by the increase in the Doppler parameter of the H_2 lines with the rotational number J, which suggests the existence of a warm component that cannot be heated by ultraviolet photons or cosmic-rays. We noted in Section 2.5.3.1 that the rotational level populations of H_2 provide important constraints on any model for the production of CH^+. The dissipation regions, distributed along the line of sight, may consist of either hundreds of low velocity MHD shocks (<20 km s^{-1}), or thousands of tiny regions of intense velocity shears, heating the gas by ion–neutral friction and/or viscous dissipation. The different timescales of such processes impact on global gas energetics and chemistry. We shall discuss them in the following sections.

3.4.1 Chemistry in a Magnetized Burgers Vortex

We highlight here a series of papers having as their subject the investigation of turbulent dissipation regions (TDRs) based on the property of space-time intermittency of turbulence.[61–64] The purpose of TDR models is to predict molecular abundances in the diffuse ISM transiently formed through the warm chemistry driven by intermittent bursts of turbulent dissipation modelled as small scale (~100 au) magnetized vortices.

The model rests on an analytical solution of the Helmholtz equation for the vorticity, in which the pressure term is suppressed, derived applying the rotor operator to eqn (2.102)

$$\frac{\partial}{\partial t}\boldsymbol{\omega}(\boldsymbol{x},t) + \boldsymbol{u}(\boldsymbol{x},t)\cdot\boldsymbol{\nabla}\boldsymbol{\omega}(\boldsymbol{x},t) = \boldsymbol{\omega}(\boldsymbol{x},t)\cdot\boldsymbol{\nabla}\boldsymbol{u}(\boldsymbol{x},t) + \nu\,\boldsymbol{\nabla}^2\boldsymbol{\omega}(\boldsymbol{x},t) \tag{3.1}$$

where $\boldsymbol{\omega} = \text{curl}\,\boldsymbol{u}$ is the vorticity, and ν the kinematic viscosity. It connects the vorticity structure to the properties of the straining large scale flow. The velocity is decomposed into two-dimensional rotational and three-dimensional potential fields, $\boldsymbol{u}(\boldsymbol{x},t) = \boldsymbol{u}_2(\boldsymbol{x},t) + \boldsymbol{u}_3(\boldsymbol{x},t)$, where $\boldsymbol{u}_2(\boldsymbol{x},t) = (u_x,u_y,0)$, and $\boldsymbol{u}_3(\boldsymbol{x},t) = (-a/2\ x,\ -a/2\ y,\ az)$, a being the strain coefficient. As the rotational part is purely two-dimensional, the vorticity $\boldsymbol{\omega}$ points in the direction of the normal to the plane containing $\boldsymbol{u}_2$. The solution, called the Burgers vortex, corresponds to an equilibrium between the stretching of the vortex by the turbulent strain a, which tends to increase the vorticity and then the orthoradial velocity

$$u_\perp = \int_0^r r'\omega(r')\mathrm{d}r' \tag{3.2}$$

and the diffusion of vorticity. The vorticity displays a stationary Gaussian profile, with the radius $r_0^{\frac{1}{2}} = 4\nu/a$ determined by the ratio of the kinematic viscosity ν and the strain parameter a of the potential velocity field. The amplitude of the vortex is then found by normalizing to the total circulation Λ associated with the vortex, and related to the peak of the vorticity by the relation $\omega_0 = a\Lambda/4\pi\nu$, to give $\omega(r) = \omega_0\ \exp\ (-r/r_0)^2$. To avoid the radial inflow velocity diverging at infinity, an exponential damping term $\exp(-\beta r^2)$ is introduced,[65] which sets the distance Kr_0 at which the vortex stops affecting the dynamics of the gas. The vortex solution is then entirely defined by three parameters, r_0, ω_0, and β (or K).

Because the vorticity is radially non-uniform there is a differential rotation of the fluid within the structure, which induces a viscous dissipation rate.

Ions are initially at rest, and the magnetic field is supposed to be directed parallel to the vorticity. Then, ions are put into motion by the neutral friction in the vortex. This motion generates a magnetic field along the orthoradial direction and a gradient in the original field. At steady state the magnetic field is slightly helical and the ions are almost back at rest, while neutrals participate in the orthoradial motion in the vortex.[64] Such velocity drift dissipates turbulence motions from both viscous and ion–neutral frictions providing an additional heating term. The thermal history of the gas trapped in a vortex may be then described through an effective temperature

$$T_{\rm eff} = T_k + \frac{\mu}{3k_{\rm B}} u_{\rm D}^2 \tag{3.3}$$

where T_k is the gas kinetic temperature powered by the viscous dissipation, μ the reduced mass of the system, and $u_{\rm D} = |u_i - u_{\rm n}|$, the relative ion–neutral drift speed.

The non-equilibrium coupled thermal and chemical evolutions of the gas trapped in the structure are followed as the gas passes through the short-lived active dissipative phase and the long-lasting relaxation period subsequent to any dissipative burst. The duration of the dissipation phase (*i.e.* the lifetime of the vortex), $\tau_{\rm V}$, is an additional free parameter related to the total energy dissipated per unit length of the vortex over its lifetime and volume. It is generally as short as a few hundred years. Once the vortex has vanished, the gas cools down and the chemical signatures imprinted by the active stage persist for a while, and then asymptotically merge back into the cold background. There are thus three regions possibly intercepted along the line of sight (see Figure 3.6[64]): (i) the ambient cold diffuse gas (green in the figure) where the chemistry is driven by the ultraviolet radiation field (PDR-like), (ii) a large number of active vortices (red) maintained by the kinetic energy transferred from the large scales, and (iii) several relaxation stages (blue) whose number is inversely proportional to the dissipation timescale $\tau_{\rm V}$. This random assembling of structures is a rather strong approximation, because while a single Burgers vortex is a solution of the Navier–Stokes equation, a random ensemble generally is not, so that no dynamical information can be deduced from this configuration.

The set of four free parameters (r_0, ω_0, K, $\tau_{\rm V}$) can be transformed in an equivalent group more directly related to chemistry. Defining $\eta_{\rm V}$, the number of the vortices along a line of sight, the filling factor of active regions results $f_L = \eta_{\rm V}\,(2Kr_0/L)$, where L is the length of the path intercepting the dissipation region. Being defined along any random line of sight, it corresponds to a volume filling factor. It thus relates the dissipation rate averaged at large scale to the same quantity averaged over the volume of a vortex $\varepsilon_{\rm V} = f_L\,\Gamma_{\rm turb}$, $\Gamma_{\rm turb}$ being the total turbulent heating rate (viscous dissipation + ion–neutral friction). The remaining three parameters are the fluid strain coefficient a, the maximum orthoradial velocity $u_\perp$, which sets the ion–neutral velocity drift $u_{\rm D}$, and the total energy dissipated per unit length of the vortex over its

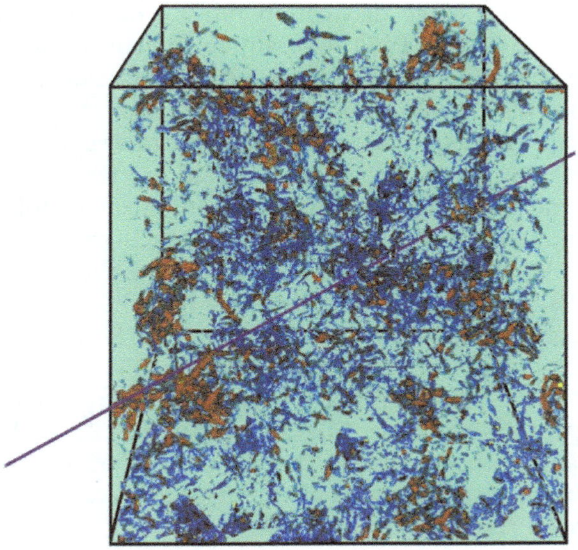

Figure 3.6 Rendering of a region of turbulence dissipation. The line of sight (in purple) passes through unperturbed diffuse gas (in green), active vortices (in red), and scattered relaxation stages (in blue), relics of past active regions.
Godard *et al.*, A&A, 570, A27, (2014), reproduced with permission © ESO.[64]

lifetime and volume $E_V = \pi\ (Kr_0)^2\ \Gamma_{turb}\ \tau_V$, the so-called vortex dissipation integral. These quantities (f_L, $u_\perp$, a, E_V), together with the gas density n_H, the ambient ultraviolet radiation field of strength G shielded by an extinction A_V, the cosmic ray ionization rate ζ, and the elemental abundances, are the boundary conditions on which a particular model is constructed.

Among the final products, the TDR model predicts the column densities of several hundreds of species, and the excitation of the rovibrational levels of H_2. Column densities are constructed integrating chemical abundances arising in the three different regimes (green, red, and blue) shown in Figure 3.6, $N(X) = N_A(X) + \eta_V\ N_V(X) + \eta_V(\tau_R/\tau_V)\ \langle N_R(X)\rangle$, where X is a given species, τ_R the duration of the relaxation phase, $\langle N_R(X)\rangle$ is the time-averaged column density during this phase, and N_A and N_V the column densities in the cold ambient gas and in the active regions, respectively.

CH^+ is the prototypical species formed in superthermal conditions in the ISM, and its chemistry provides a good representation of what is going on in a turbulent dissipative region. The formation of CH^+ is driven essentially by hydrogenation of carbon cations through a reaction endothermic by 0.4 eV (~4600 K), and it is thus very sensitive to the injection of energy in an otherwise very cold gas. CH^+ destruction proceeds mainly through reaction with H, He, H_2, and electrons. The resulting column density is thus inversely proportional to the square of the density n_H and directly proportional to the number of active vortices $\eta_V = (\varepsilon_V/\Gamma_{turb}) \times (L/2Kr_0)$, a quantity which is proportional to the average dissipation rate ε_V, inversely proportional to Γ_{turb}, and independent of E_V. The strain coefficient a and the maximum orthoradial velocity $u_\perp$ control the heating by viscous friction and ion–neutral friction, respectively, and hence the temperature of the gas. The increase in temperature boosts exponentially the rate velocity of the reaction producing CH^+. However, beyond the activation energy this effect drops dramatically, and the resulting column density declines following the decrease in the number of vortices along the line of sight. Subsequent to the formation of CH^+ the production of CH_3^+ occurs *via* the hydrogenation of CH^+ and CH_2^+ by H_2, $CH^+ + H_2 \rightarrow CH_2^+ + H$ and $CH_2^+ + H_2 \rightarrow CH_3^+ + H$. CH_3^+ in turn contributes to the production of *e.g.*, CH and C *via* dissociative recombinations with electrons, both directly $CH_3^+ + e^- \rightarrow CH + H_2$, or through $CH_3^+ + e^- \rightarrow CH_2 + H$ followed by the neutral–neutral reactions of CH_2 with atomic hydrogen. The abundance of CH is then the result of both cold and hot chemistries. The comparison of model outputs with observed values sets an upper limit for the average dissipation rate $\varepsilon_V \leq 10^{-23}$ ergs cm^{-3} s^{-1}.

A similar chemistry gives rise to SH^+, which is formed through hydrogenation of S^+, a process that is almost twice as endothermic as the CH^+ production. Sulfur is unusual among the abundant elements because its hydrides and hydride cations have relatively small bonding energies. Consequently, none of the species S, SH, S^+, SH^+, or $SH_2{}^+$ can undergo an exothermic H atom abstraction reaction with H_2. For suitable values of T_{eff}, this is irrelevant and hydrogenation reactions also dominate the production of $SH_2{}^+$ and $SH_3{}^+$. All these molecular ions are removed *via* dissociative recombination. At chemical equilibrium the $N(SH^+)/N(CH^+)$ column density ratio scaling as $\exp(-\Delta T/T_{eff})$, with $\Delta T = 5220$ K being the difference in the endothermicity of the two leading reactions, depends strongly on the velocity drift between C^+ and S^+ and H_2. The observed column densities require the ion–neutral velocity drift to be $u_D \leq 3.5$ km s^{-1}.

We note that in dissipative regions the chemistry is not in equilibrium and the electron fraction is substantially smaller than that in the ambient cold gas. This gives support to the suggestion that the observations of the ion $H_3{}^+$ sample, at least partially, active dissipative regions.[42] In fact, in addition to the increase in temperature that slows down the rate coefficient of dissociative recombination, the electron deficiency further contributes to a decrease in the destruction rate of $H_3{}^+$.

Hydrogenation of carbon has implications for part of the oxygen chemistry. In ultraviolet driven chemistry the formation of HCO^+ proceeds *via* the oxygen hydrogenation chain followed by the two reactions: $OH + C^+ \rightarrow CO^+ + H$ and $H_2O + C^+ \rightarrow HCO^+ + H$. In TDRs the abundant CH^+ and $CH_3{}^+$ molecular ions both lead to the formation of HCO^+ either directly or through CO^+ formation. Then, in a dissipative region it is not the hydrogenation chain of oxygen that matters, but that of carbon. The opening of the carbon-based additional pathway boosts the abundances of HCO^+ by several orders of magnitude and that of CO by a factor of up to 100. Direct endothermic hydrogenation of atomic oxygen gives rise to the formation of water and H_3O^+. The carbon and oxygen chemistries driven by the ultraviolet radiation field and turbulent dissipation are compared in Figure 3.7.[64]

However, the contributions of active regions along the line of sight fail to reproduce the CO column densities by more than one order of magnitude, although providing enough HCO^+. On the other hand, observations show that the measured CO column densities arise naturally from the observed HCO^+ abundances in a standard cold chemical scenario acting throughout the whole cloud. Thus, such CO deficiency may be only apparent. The chemistry in a vortex reaches the equilibrium in times that are instantaneous with respect to the timescales of cold interstellar chemistry. In contrast, CO molecules produced efficiently in the turbulent regime are not instantaneously removed by the gas phase as a turbulent dissipation phase switches off, but decay to their equilibrium values in the cold gas with a characteristic timescale that is generally much longer than the vortex life time. If the continuous switch on and off of the hot regions in an otherwise cold gas is fast enough, the cloud will be rapidly and densely populated by their chemical relics. Thus, CO traces the relaxation phase, which provides abundances lower than in active regions, but with a much larger filling factor. The relaxation phase has a typical lifetime of roughly 100 000 years, widely independent on the vortex lifetime. However, a long vortex lifetime, which is linearly dependent on the vortex dissipation integral E_V, would make negligible the impact of the relaxation phase on the chemical abundances. Thus, contrary to the case of CH^+, CO and (to a lesser extent) HCO^+ depend on the vortex dissipation integral.

As the velocity scales with the rate of strain a so does the viscous dissipation rate (which depends linearly on the derivatives $\partial u_i/\partial x_j$), while the equilibrium radius r_0, which is inversely proportional to a, decreases inducing a decrease in the vortex filling factor. In addition, as r_0 decreases the vortex lifetime τ_V increases. Thus, both the column densities of HCO^+ and CO decrease with increasing strain parameter, but for different reasons. HCO^+ and CO depend differently on the standard gas parameters, the first one sensitive to the gas number density and almost unaffected by variations in the visual extinction, whereas the reverse is true for the latter (see Figure 3.7).

The predictions of the TDR model are compared with the observed column densities of CO and HCO^+ in Figure 3.8,[64] using the best value for the vortex dissipation integral $E_V = 1.4\times10^{-13}$ ergs cm^{-1}, which corresponds to a vortex lifetime of $\sim$100 year. Along each track reported in the figure, the rate of strain a varies between 10^{-11} and 10^{-10} s^{-1} in going from the right to the left hand side.

Finally, the TDR model also reproduces the populations of the first five rotational levels of H_2 observed in the local diffuse matter through UV absorption spectroscopy (*e.g.*, Gry *et al.* 2003[58]).

The species mainly produced in the active phase can be used as signatures of the dynamics of the dissipation regions; those formed in the relaxation

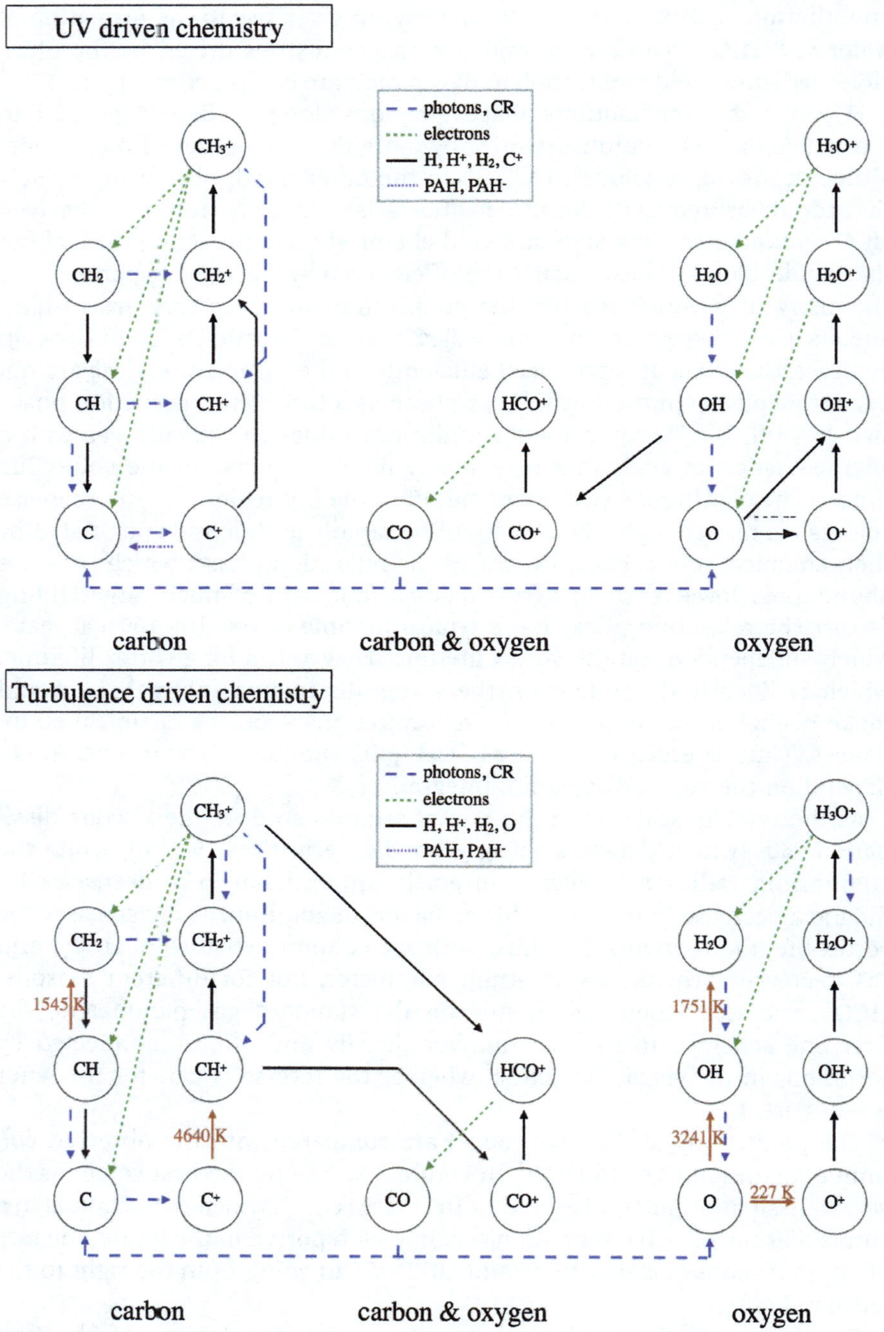

Figure 3.7 The main chemical routes of carbon- and oxygen-based species driven by ultraviolet radiation (top panel) and turbulence dissipation (bottom panel). The endothermicities and energy barriers of endothermic reactions (reported only for the turbulence-induced chemistry) are indicated in red. Godard *et al.*, A&A, 570, A27, (2014), reproduced with permission © ESO.[64]

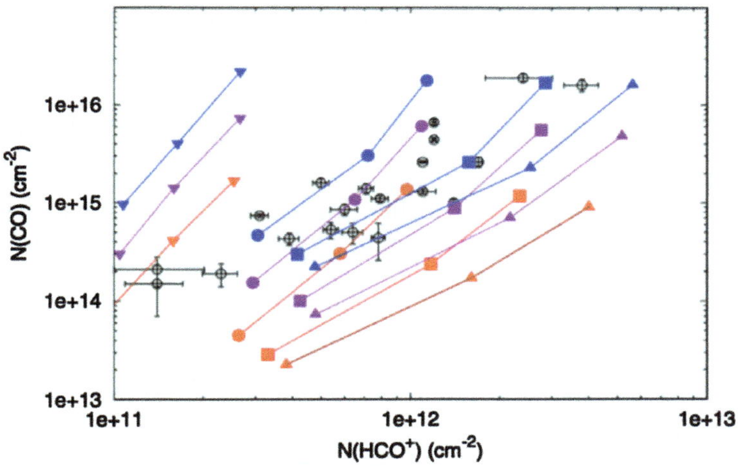

Figure 3.8 Comparison of observations with the prediction of a TDR model, with $E_V = 1.4 \times 10^{-13}$ ergs cm^{-1}. The results (filled symbols) are shown for: $A_V = 0.2$ (red), 0.4 (magenta), and 0.8 (blue) mag; $n_H = 30$ (triangles), 50 (squares), 100 (circles), and 300 cm^{-3} (upside-down triangles). The total column density is $N_H = 1.8 \times 10^{21}$ cm^{-2}. Along each curve, the rate of strain a varies between 10^{-11} (right) and 10^{-10} s^{-1} (left).
Godard *et al.*, A&A, 570, A27, (2014), reproduced with permission © ESO.[64]

Table 3.2 Parameters of the TDR standard model.

Property			Property		
Hydrogen density	n_H (cm^{-3})	50	Rate of strain	a (s^{-1})	3×10^{-11}
Radiation field scaling factor	G	1	Average dissipation rates	ε_V (erg cm^{-3} s^{-1})	1×10^{-24}
Visual magnitude	A_V (mag)	0.4	Maximum orthoradial velocity	$u_\perp$ (km s^{-1})	3
H_2 cosmic-ray ionization	ς_{H2} (s^{-1})	1×10^{-16}	Vortex dissipation integral	E_V (erg cm^{-1})	3×10^{-13}

phase are sensitive to the relative timescales between the active dissipation and the relaxation; those produced in the ambient medium are unaffected by turbulent dissipation. The variety of chemical routes then allows the independent determination of free parameters characterizing interstellar turbulence: the rate of strain a, the large scale dissipation rate ε_V, the maximum orthoradial velocity $u_\perp$, and the dissipation integral E_V (or the dissipation timescale τ_V). The inferred values are generally consistent with the current knowledge of turbulence in the diffuse medium, and show internal coherence (see Table 3.2). The TDR model requires that a substantial part of the turbulent cascade energy to dissipate in very intermittent small-scale structures. Thus, a small and transient volume has tremendous impact on the chemical distribution of the ISM.

3.4.2 Non-equilibrium Chemistry in Magnetized Shocks

In the previous section the dissipation of a significant fraction of the turbulent energy in incompressible flows occurs in coherent dynamical structures, such as vortices, temporarily heating a small fraction of the gas. Such space-time bursts of local heating may be also described in terms of shocks, which then constitute another way of dissipating supersonic turbulence.[16] Simulations of compressible MHD turbulence shock waves suggest that shock waves may dissipate up to 50% of the turbulent energy.[66]

Turbulent motions in compressible media will generally include both solenoidal components, for which the compression is zero but vorticity is non-zero, and compressive components. Even when compressive motions are initially ignored (as usually done in subsonic turbulence), solenoidal turbulence produces pressure fluctuations of amplitude $\sim \rho u^2$, which, if the sound speed c_S is finite, will rapidly generate a compressive component.[67] The presence of compressive modes is an important issue also for H_2 formation and excitation, given the strong density dependence of these processes (see Section 3.4.4).

Shock formation and evolution were detailed in Chapter 2, Section 2.5.1. A shock may be regarded as a thin region where the relative kinetic energy between the fluid parcels is converted to magnetic, radiative, thermal and internal energy. This region divides the gas flow into pre-shock and post-shock sides. The velocity of the incoming fluid in the reference system of the interface region defining the shock, the shock speed, must be compared with the magnetosonic speed in the ionized fluid and the slower Alfvén speed, v_A in the neutral gas. When the shock velocity is greater than both these speeds, we have a *J*-type shock in which the kinetic energy is dissipated viscously in a very sharp velocity jump. This occurs because the shock front is travelling faster than any signal-bearing speed in the medium, and thus, the pre-shock medium cannot receive any hint of the approaching disturbance. A smoother transition occurs when the shock velocity is intermediate between these two speeds, with the kinetic energy continuously degraded into heat and photons *via* ion–neutral friction and cooling, giving rise to the continuous structure of a *C*-type shock. In other words, a *C*-type shock occurs as long as the shock velocity remains below the propagation speed of the magnetic precursor. In a *J*-type shock, ions and neutrals remain coupled at all times. The separation between these two classes of shock processes is not sharp. While shock velocities larger than the magnetosonic speed always characterize a *J*-shock, lower flow velocities may originate *C*-shocks, *J*-shocks, or even a combination of the two.

In both shocks and velocity shear vortices, the dynamical timescales can be short compared to the time needed to reach the equilibrium between the rates of molecular formation and destruction. Their different timescales, however, make these processes impact differently on the global gas energetics and chemistry. The heating of compressed gas in a thin post-shock region provides distinguishing features to chemistry and radiative cooling. Thus, the radiative characteristics of turbulent dissipation will be strongly shaped by cooling in shocks. Weak shocks in a large range of velocities are responsible for the majority of dissipation in simulations of decaying MHD turbulence. Furthermore, low-velocity MHD shocks are required to explain the observed excitation of the pure rotational lines of H_2. The observational signatures of the low-velocity shocks have been studied by Pon *et al.* (2012),[68] who considered *C*-type shocks travelling at speeds of 2–3 km s^{-1} perpendicularly to the magnetic field, and in a more comprehensive way by Lesaffre *et al.* (2013)[69] who took a statistical approach by computing diagnostics due to a distribution of *C*- and *J*-type, perpendicular, MHD shocks at velocities ranging from 3 to 40 km s^{-1}.

The shock model exploited in the work of Lesaffre and collaborators[69] is based on a one-dimensional, steady-state, two-fluid plane-parallel description of *J*- and *C*-type shocks. The MHD equations are solved in parallel with a moderately large chemical network of about 100 species and approximately 1000 reactions, with pre-shock conditions representative of the cold diffuse interstellar gas in the Milky Way, therefore including the treatment of ultraviolet heating, photo-ionization and photo-dissociation. At each fixed pre-shock density, the model determines the stationary chemical state of the pre-shock medium before the actual shock calculation is performed. The main input parameters of the model are the number of fluids, the pre-shock magnetic field strength, the pre-shock hydrogen density, and the shock velocity, in addition to the initial H_2 *ortho*/*para* ratio, the initial gas and dust temperatures, visual extinction, and the cosmic ray ionization rate. *C*-type shocks are employed when the fluid velocity is below the critical velocity for their existence, a fact that depends on the boundary conditions; otherwise *J*-type shocks are computed. The populations of the H_2 rovibrational levels and the cooling lines of the most abundant atomic metals and their ions are outcomes of the simulations.

The formation of species relying on endothermic reactions is enhanced by several orders of magnitude as the shock velocities increases to over 7 km s^{-1}. Most chemical properties of *J*-type shocks vary within an order of magnitude for velocities in the range 7–30 km s^{-1}, where H_2 dissociation sets in, and the molecular content drops. *C*-type shocks display a more gradual molecular enhancement with increasing shock velocity. This is easily understood when looking at the temperature profile in a post-shocked region, and in particular at its peak value. In weakly magnetized shocks the peak temperature is proportional to the square of the shock velocity. Since the viscous front between the peak temperature and the pre-shock is adiabatic, the maximum temperature reached in a *J* shock is given by $T_M = 2(\gamma - 1)\mu/k_B(\gamma + 1)^2 \times u^2 = 53\text{K} \times (u/1\ \text{km s}^{-1})^2$ in which the numerical value is obtained by setting $\gamma = 5/3$, and $\mu = 2.33$ a.m.u., as appropriate for a

Table 3.3 Barrier or endothermicity and the corresponding velocity of *J*-type shocks for some key reactions.

Reaction	Barrier (K)	Shock velocity (km s^{-1})
$O+H_2 \rightarrow OH+H$	2980	7.5
$C^+ + H_2 \rightarrow CH^+ + H$	4640	9.4
$S+H_2 \rightarrow SH+H$	9620	13.5
$S^+ + H_2 \rightarrow SH^+ + H$	9860	13.6
$C+H_2 \rightarrow CH+H$	14 100	16.3
$Si^+ + H_2 \rightarrow SiH^+ + H$	14 310	16.4
$N+H_2 \rightarrow NH+H$	14 600	16.6
$N^+ + H_2 \rightarrow NH^+ + H$	168[a]	<0.5
H_2 dissociation energy	52 000	31.3

[a]This barrier is uncertain and may not exist.

dilute fully molecular gas, in which only the lowest energy levels of H_2 are populated (at the highest velocities $\gamma = 7/5$ for shocks propagating into sufficiently dense media, see Section 2.3.1). This relation allows us to link the activation energies of critical endothermic reactions to velocities in *J* shocks. In Table 3.3 the main reactions involving H_2 and a single atom or monoatomic ion are listed along with their characteristic barriers and the corresponding shock velocities needed to overcome them. The width of the viscous front between the peak and the pre-shock temperature is close to the viscous length $\lambda_V \approx \sigma_V n_H$ where $\sigma_V = 3\times10^{-15}$ cm^2 is the viscous cross-section,[70] whose value is based on H_2–H_2 collision data computed by Monchick and Schaefer (1980)[71] for a velocity dispersion of 1 km s^{-1}. For $n_H = 100$ cm^{-3} the viscous length is approximately $\lambda_V \approx \text{few}\times10^{13}$ cm. Beyond the peak, gas temperatures return to their pre-shock values within 10^{15} cm, with higher velocity shocks not surprisingly cooling down at a smaller distance than low-velocity shocks (see Figure 3.9,[69] left upper panel). The range of flow time through these shock structures spans about one order of magnitude, from 400 year to 2000 year, depending on the velocity. While in the weakly magnetized case all shocks are virtually *J* shocks, in a strongly magnetized gas both types of perturbations may be present. The transition from *C* shocks to *J* shocks depends on the strength of the magnetic field, which acts to limit the compression in the shock and the collisional processes take longer to occur. In the case shown in Figure 3.9 such transition occurs slightly above 20 km s^{-1}. In fact, here the viscous length for *J* shocks is approximately one order of magnitude wider than in the previous case. Consequently, the flowing time across the shock structures varies only from 6000 to 8000 year for *C* shocks and from 1000 to 1500 year for *J* shocks. The lifetime of a dissipation event is therefore generally larger than in regions of velocity shears. For higher densities, the picture is much the same, and in particular timescales vary little within one category of either *J*- or *C*-type shocks. However, shocks that strongly dissociate H_2 (such as dense shocks at moderate velocity) have a wide H_2 reformation plateau that provides the necessary heat to keep the gas warm.

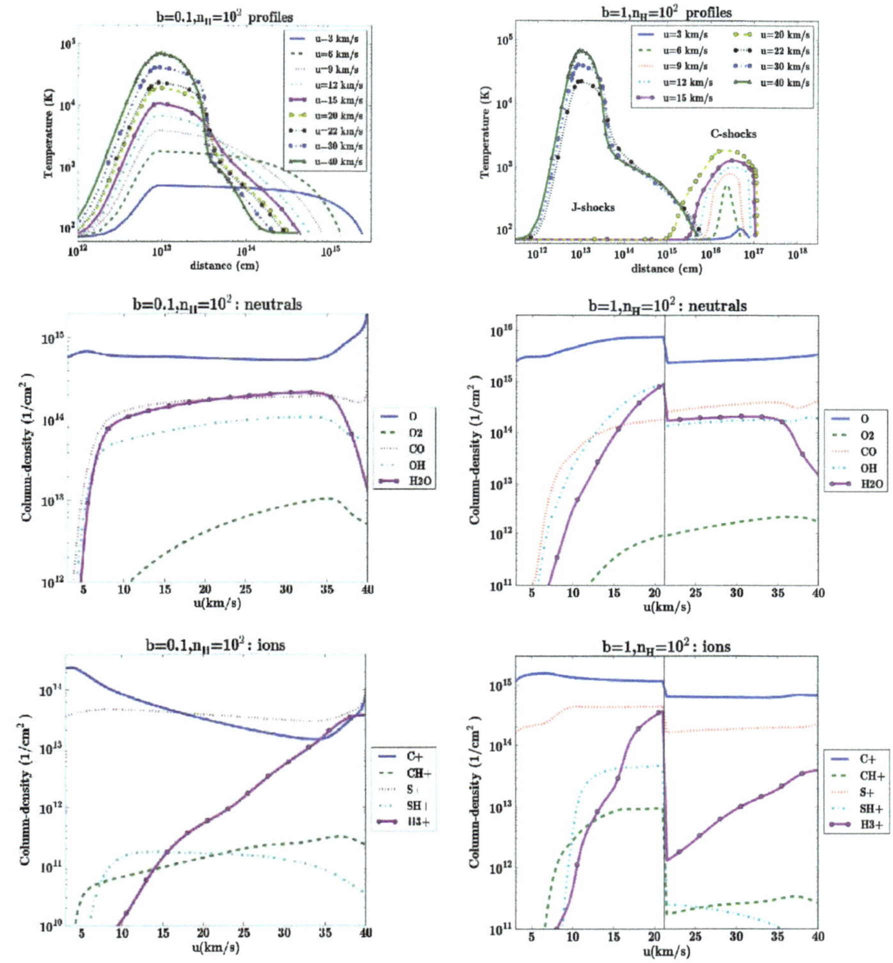

Figure 3.9 Top panels: maximum temperature in weakly (left) and highly (right) magnetized shocks. The fluid flows from left to right with the pre-shock on the left and the post-shock on the right; the pre-shock density is $n_H = 100\ \text{cm}^{-3}$. In the left panel only *J*-type shocks exist for this low value of the magnetic field. In the lower panels molecular column densities of some neutral (middle panels) and ionic (bottom panels) species of interest are shown as functions of the shock velocity.
Lesaffre *et al.* A&A, 550, 106, (2013), reproduced with permission © ESO.[69]

In strongly magnetized shocks, the kinetic energy is continuously transformed into thermal energy *via* ion–neutral friction. As a result, in *C* shocks the heating is generated in a much more extended region than for *J* shocks, and the peak temperature is hence much lower (Figure 3.9 right upper panel). However, reactions between neutral and ion species benefit from the ion–neutral drift, and the effective temperature driving the reaction rate is higher (see eqn (3.3)). Figure 3.9—middle (neutrals) and lower (ions)

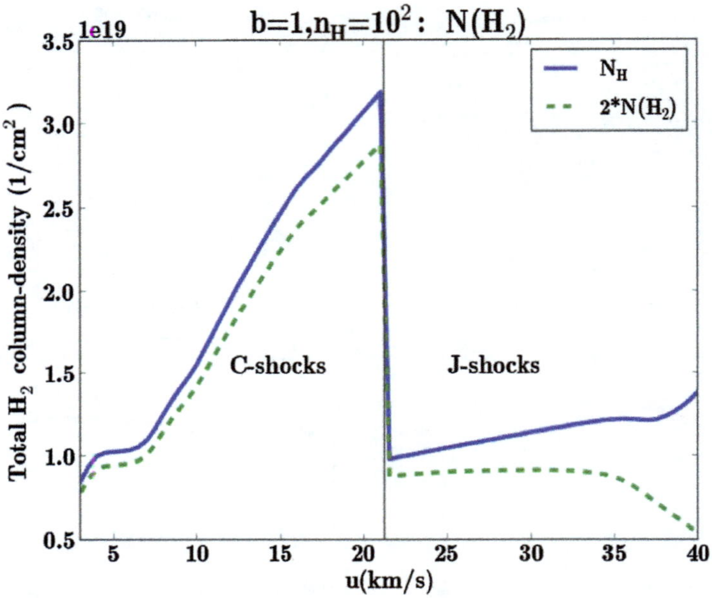

Figure 3.10 The total hydrogen nuclei and H_2 molecule column densities (cm^{-2}) as functions of the shock velocity ($km\,s^{-1}$) in highly magnetized shocks. The pre-shock density is $n_H = 100$ cm^{-3}.
Lesaffre *et al.* A&A, 550, 106, (2013), reproduced with permission © ESO.[69]

panels—shows the molecular production over a wide range of shock velocities at $n_H = 100$ cm^{-3}. The column densities of most species fall sharply at the transition between *C* and *J* shocks, which indicates higher molecular formation efficiency in *C* shocks. The major difference controlling the chemistry between these two types of shocks is the ion–neutral drift, which helps to overcome reaction barriers of ion–neutral reactions, such as the prototypical carbon hydrogenation $C^+ + H_2 \rightarrow CH^+ + H$. Moreover, the resulting frictional dissipation provides enough heating to oppose the cooling with the result that gas remains warm throughout the shock region. As shown in Figure 3.9 (upper right panel) a *J* shock is very hot at the peak temperature immediately behind the viscous front, but quickly cools down in the following relaxation layer. This produces a net increase in the total gas column density in *C* shocks as compared to *J* shocks (Figure 3.10[69]). Such enhanced column densities affect significantly the evolution of molecular species produced through neutral–neutral channels (*e.g.*, O-bearing species). Indeed, as shown in Figure 3.9, species such as H_2O increase their abundances much more gently in *C* shocks (middle right panel) than in *J* shocks (middle left panel) reflecting the slower rise of the maximum neutral temperature. The chemical enhancement in the shocks could be even higher if the medium is shielded. For standard irradiation C^+ is the dominant carbon gas species, which is also a result that depends on shielding.

In weakly magnetized shocks H_2 emission dominates the cooling by more than one order of magnitude for shock velocity in the range 5–40 km s^{-1}. At lower velocities atomic O or C^+ cooling takes over, depending on the density. At larger velocities the dissociation of molecular hydrogen slows down the cooling, giving rise to a temperature plateau where atomic oxygen and partly H (Lyman-α) emission are the dominant coolants. Increasing the density, and for shock velocities below 40 km s^{-1} molecular cooling (*e.g.*, H_2O) may be important (see Section 2.3.3). With moderate to high strengths of the magnetic field, C^+ is the most efficient coolant when the shock velocity is lower than 5 km s^{-1}; otherwise H_2 emission controls the cooling. However, below 10 km s^{-1} the kinetic energy flux stops being radiated away, and it is converted into a magnetic energy flux because of the compression of the field. Due to the presence of radiation fields, CO is photodissociated and the resulting atomic carbon ionized, so that their contributions to gas cooling are not particularly important. Thus, in shocks propagating through an ultraviolet irradiated region, C^+ produces the most significant line cooling. At higher densities, important low-velocity coolants become O and water. Molecular cooling is responsible for the release of less than one per cent of the total kinetic power, except at very high densities where H_2O in *C*-type shocks can contribute as much as 10%.

Even in the case of dissipative vortices examined in the previous paragraph, the emission in the pure rotational lines of H_2 is by far the dominant coolant in the layers, where the temperature of neutrals is larger than 200 K, while the cooling rate due to the ionized carbon C^+ decreases in the warmest layers as the abundance of C^+ reduces in response to the chemical evolution. Once the active stage stops, the abundance of C^+ rapidly increases during the relaxation phase, which generally lasts much longer. In this period of time that lasts for more than ~10 000 year the abundance of C^+ is high and the gas still enough warm to induce C^+ emission. Thus, the energy dissipation that takes place in vortices may be responsible for an enhancement of C^+ emission above the background level.

Simulations and experiments of driven turbulence suggest a variety of velocity PDF shapes for the statistics of shocks, such as *e.g.*, decaying exponentials, power-laws, Gaussians, and piece-wise exponential fits. Fitting specific lines of sight, a statistical distribution of shock velocities and shapes is recovered. The resulting combination should then be considered as a phenomenological description of the complex statistical properties of turbulence dissipation. For the shock models discussed above, the comparisons with observational data indicates the need for a (at least) bimodal distribution of shocks, statistically biased towards low velocities. This may be an intrinsic property of supersonic turbulence dissipation, or may simply reflect the inadequacy of data, since *e.g.*, H_2 emission lines are scarcely affected by shock velocities above 20 km s^{-1}. An interesting point is the diversity of regimes of energy dissipation in which low- and moderate-velocity shocks operate: the low-velocity shocks are of *C*-type and dissipate energy *via* ion–neutral drift, whereas the higher velocity shocks are of *J*-type and are subjected to viscous dissipation. Low-velocity shocks are less efficient than moderate-velocity shocks, but they are more numerous. Thus, they could potentially account for a higher fraction of the excitation and formation of molecules.

One of the major shortcomings of shock models is their simple geometry. However, since only a reasonably small chemistry is required to study the dynamics, reliable 3D shock models may be within the reach of numerical experiments in the near future. The rest of the chemistry can be added *via* post-processing. Another frequent approximation in such studies is the assumption that shocks are steady, while some flow velocities are subjected to instabilities, and in that case even quasi-steady models are barely suitable when not invalid. Another important issue is the resolution of the simulations, which, when too low, acts to smear out the temperature, creating a nearly isothermal shock in which the resulting temperature is not enough to trigger molecule formation.

3.4.3 Non-equilibrium Chemistry in Ideal MHD Simulations: the Sulfur Problem

The intermittent dissipation of turbulence described above, either through shocks and vortices, deals with individual dissipation events, paying little attention to relating these events to the properties of the large-scale turbulence. The simulations frequently assume very simple (if not simplistic) velocity, density, or magnetic field distributions with no link to the large scale dynamics of the gas. Recently, a complementary approach to such models has been put forward by post-processing ideal 3D MHD turbulence simulations.[72,73] While some details are missed, this point of view has the advantage of drastically reducing global assumptions, such as the nature of the intermittent structures or the number of dissipation events along a line of sight. This is an important point when looking for data coherence in correlated events, because *e.g.*, as we noted in Section 3.4.1, a random ensemble of vortices is not in general a solution of the Navier–Stokes equation. A significant problem in these types of analyses is raised by the numerical resolution at which the simulations were performed, which generally do not exceed the spatial scale of a few hundredths of parsecs. Such a figure is definitely much larger than the dissipation scale provided by the kinematic viscosity of interstellar gas $\lambda_V \approx 10$ au, with the net result that these "large-scale" simulations underestimate the intermittency of the energy dissipation, and thus fail to describe the distribution of the heating rate. However, ion–neutral friction removes energy from the turbulent cascade at a rate that is comparable or even larger than viscous dissipation (up to 70%[74]), and this occurs at the length scale set by the ambipolar diffusion that, for a cloud of size ΔL, $\lambda_{AD} = \Delta L/R_{AD}$, $R_{AD} = [4\pi\rho_n\rho_i\delta u/(m_n + m_i)]\ \langle\sigma u\rangle_{in}\ \Delta L/B^2$ being the ambipolar diffusion Reynolds number,[75] and δu the velocity dispersion (see Section 3.3). Assuming the gas in the cloud to be characterized by the linewidth-size relation $\delta u \approx (\Delta L/\mathrm{pc})^{\frac{1}{2}}$ km s^{-1}, and considering

C^+ and H_2 as typical colliders we obtain $R_{AD} \approx 5000 \times (n_H/30\ cm^{-3})^{\frac{1}{2}}$, when reasonable values for the magnetic field in a diffuse gas with number density $n_H < 100\ cm^{-3}$, are used, *i.e.* $B \approx few \times \mu G$ ($1\ \mu G = 1 \times 10^{-10}$ T). For a 10 pc cloud with gas volume density $n_H = 30\ cm^{-3}$, $\lambda_{AD} \approx 0.01$ pc, and thus ion–neutral friction becomes significant at a much larger scale than viscous dissipation, a scale large enough to be comparable to the cell resolution. Although this procedure under estimates the intermittency for 30% of the energy dissipation, it is more accurate for the remaining 70% than the opposite assumption that all the energy in a cascade makes it to 10 au scales. The implications for the distribution of the chemical products arising through endothermic reaction channels are obvious.

To estimate the effects of the ambipolar diffusion on the production of species driven by ion–neutral reactions with barriers *e.g.*, CH^+, a prescription is needed for the ion–neutral drift velocity u_D. If the system is weakly ionized, then the Lorentz force and the ion–neutral drag force dominate all the other terms in the ion momentum equation (*e.g.*, the ion inertia). The resulting equation yields

$$(\nabla \times \boldsymbol{B}) \times \boldsymbol{B} = 4\pi\ \rho_n \rho_i \frac{\langle \sigma u \rangle_{in}}{m_n + m_i} u_D \tag{3.4}$$

The equation is then solved assuming that all resistivities associated with the electrons can be neglected. Starting from the hypothesis that detailed dynamical and chemical models of intense dissipation events provide a reliable description of the chemical properties of diffuse lines of sight, Myers *et al.* (2015)[72] re-assessed self-consistently the origin of CH^+ determining the properties of the intermittent turbulence that produces this species, such as the rate of strain a in the dissipation events, and their frequency in the ISM. Such quantities were assumed to be not self-consistently calculated in the models of turbulence dissipation reported above (see Section 3.4.1). For post-processing these authors developed a chemical solver able to calculate the chemical equilibrium for a given set of physical conditions. They also included the energy balance equation and solved it cell-by-cell for the temperature, finding a widespread extension of cold gas with interspersed warm islands heated to temperatures in excess of 1000 K filling less than 1% of the cloud. In addition, using eqn (3.4) the drift velocity was mapped throughout the cloud. While on average the drift is negligible, in isolated regions it can reach values equal to or exceeding $u_D \approx 4\ km\,s^{-1}$. Since both these factors combine to form CH^+, the observed abundances of this species were readily obtained for a turbulent molecular cloud with typical physical conditions for the ISM. CH^+ production is controlled by drift velocities higher than $3\ km\,s^{-1}$ which, although rare, are enough to produce CH^+ column densities at the level observed. Turbulence favours the production of CH^+ both by heating a small percentage of the gas directly and by inducing large ion–neutral drift velocities within localized regions of intense dissipation. Myers and coworkers concluded that the drift is more important, providing ~2/3 of the CH^+ abundance, but the contribution from the heated gas is never negligible.

In a subsequent work Valdivia *et al.* (2016)[73] exploited a similar technique in addressing the related problem of assessing the roles of two physical phenomena that have previously been invoked to lead to the production of CH^+ under realistic physical conditions: the presence of warm H_2 and the increased formation rate due to the ion–neutral drift. These authors pointed out that several caveats were present in the work of Myers and coworkers related to the simplifying assumptions that might indeed be influencing their results. The main points regarded firstly the ionization fraction assumed to be constant and equal to the ionization fraction in diffuse clouds (*i.e.* dominated by C^+), which might overestimate the drift velocities, and secondly the strong assumption of freezing the hydrogen molecular fraction to 10% throughout the entire region. These constraints likely combine to boost artificially high CH^+ abundances.

In fact, the results of Valdivia and coworkers are scarcely consistent with the ones of Myers *et al.* The major conflicting point is the distribution of drift velocities that peaks at 0.4 km s^{-1} without a high-velocity tail. In other words drift velocities higher than 1 km s^{-1} are extremely rare events. The impact on the effective temperature, which depends quadratically on the drift velocity (see eqn 3.3), is rather limited and thus the observed abundances of CH^+ are underpredicted by a factor of the order of six. While this may reflect the existence of some hidden or ancillary mechanism other than the dissipation of turbulence, at the same time it highlights the importance of a reliable description of small-scale physical processes and quantities, such as the electron density, as well as the description of the main ion–neutral interactions, in order to avoid unrealistic outcomes. However, this not the last word, nor it is the final epitaph for ion–neutral drift, as high resolution (numerical) effects seem to have a strong impact on the high-velocity tail.

The simulations also show that molecular hydrogen plays a fundamental role in the chemistry, and that this role changes with the evolution of molecular hydrogen. In particular, the formation of CH^+ seems to be more efficient in regions where H_2 is not expected at equilibrium. We recall that the turbulence simulations of Valdivia and coworkers were performed in ideal MHD, including the heating and cooling of the multi-phase ISM, and following the formation and destruction of H_2 and its thermal feedback. No other chemical species is computed on the fly. The initial conditions include colliding streams of atomic gas that form a highly structured molecular cloud with a wide range of physical conditions in density, temperature, and shielding parameters. H_2 molecules may end up in the warm phase of the gas as a consequence of turbulent mixing and shielding against photodissociation provided by the multi-phase structure. The presence of H_2 in a warm environment increases its excitation and leads to substantial column densities of high-lying rotational levels comparable to those observed in space. In the post treatment of such simulations, the chemical solver returns the abundances of the species included in the chemical network at the chemical equilibrium, allowing for the possibility that the abundance of H_2 can be fixed beforehand.

Figure 3.11[73] shows the abundances of CH^+ as a function of the gas total column density under two different numerical configurations, *i.e.* using either the out of equilibrium H_2 abundances obtained in the MHD simulation or calculating the full chemical equilibrium for all the species, including H_2. In the first case, to compute the composition of the gas at chemical equilibrium with a fixed abundance of H_2 it is necessary to verify that the timescales required to reach equilibrium are smaller than that of H_2, which is generally longer than that of any other species. The typical timescale for H_2 formation is of the order $t_F \approx 10^9/(n_H/\mathrm{cm}^{-3})$ year, which would lead to ages older than 10 Myr for molecular clouds of mean densities of the order of 10–100 cm^{-3}. In fact, any attempt to speed up the evolution of the H_2 fraction would require densities well beyond the range usually considered typical for diffuse gas.[76] However, the density compressions created by supersonic turbulence allow H_2 to form more rapidly than in a static medium. The validity of overall chemical equilibrium is much more questionable, since the chemical timescales of each species involved in the reaction network should also be smaller than the typical timescales of the variation of dynamical quantities, *e.g.*, density and temperature. In the WNM and CNM the perturbations of physical conditions propagate at the speed of sound

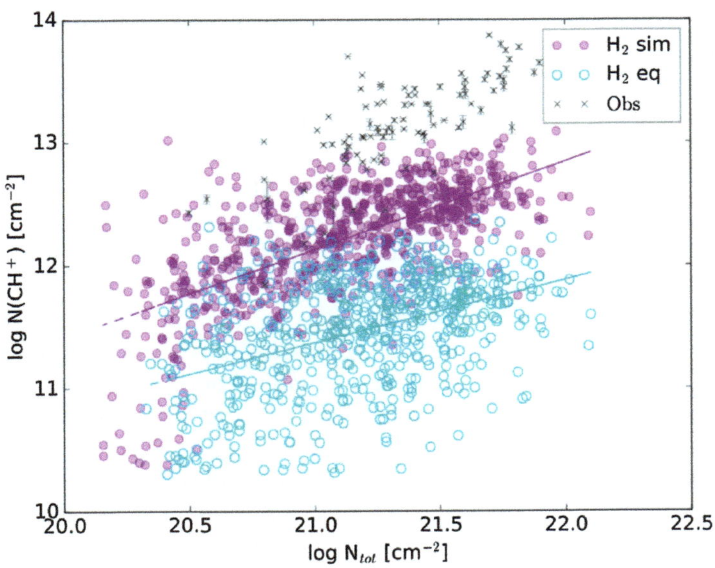

Figure 3.11 Observed (crosses) and modelled CH^+ column densities as functions of the total gas column density derived when H_2 abundances are fixed from the simulation (solid circles) and calculated at equilibrium (open circles). Valdivia *et al.* A&A, 600, A114, (2017), reproduced with permission © ESO.[73]

over times of 10 000 year or so, and thus an accurate treatment would require following the time dependent evolution of the entire chemistry, which is still currently hopeless. With this limitation clearly in mind, the distribution of CH^+ abundances show that this species is produced in regions that present specific characteristics favourable to its formation, with fractions of H_2 higher than those predicted at equilibrium, and temperatures of the order of several hundreds of K, most likely of the order of 1000 K. However, as noted above, the dissipation of turbulence fails to reproduce the observed CH^+ abundances, suggesting that small-scale processes, and other non-thermal processes not included in the present MHD simulation, may be of crucial importance, while higher resolution studies with better controlled dissipation processes are needed to understand the role of ion–neutral friction.

While these studies are suggestive and provide the astrophysical framework in which to accommodate specialized points of view, they all face the same methodological problem of computing the chemistry in post-processing, taking the magnetic field, velocity, temperature, and density data from an ideal MHD turbulent box. It is impossible to assess whether such results would have been different had the chemistry been computed self-consistently and allowed to affect the subsequent flow. Unfortunately, non-ideal MHD simulations are out of the reach of current computational capabilities. Moreover, these studies are not able to discriminate among different mechanisms, nor can the chemistry. Kinematics may leave an imprint on the spectral line profiles. In TDRs, line shapes of species formed in the active phase are expected to be wider than those of species mainly produced by the relaxation phase. In *C*-type shocks the chemical production may induce a velocity shift between the line profiles of *e.g.*, CH and CH^+. However, taking a random distribution of shocks with different orientations will drastically reduce the line shifts. Whatever the case may be, as we have already mentioned, the dissipation of turbulent motions is not likely to occur through a single specific mechanism, and the real interest in future works would be to assess the fraction of the energy dissipation that takes place in vortices, shocks, or other processes such as current sheets in plasmas.

Ultimately, the dissipation of turbulence in one (or in a combination) of its manifestations, has proven to be successful in reproducing not only the abundances of CH^+, but also those of many other species, including *e.g.* SH^+, CO, HCO^+, and their correlations observed in the diffuse medium under the constraint of the mechanical energy injected at large scale. However, they seem to fail for some components of the sulfur family. Rather ironically the chemistry of interstellar molecules containing sulfur is such that none of the species S, SH, S^+, SH^+, or H_2S^+ can react exothermically with H_2 in a hydrogen atom abstraction reaction, $X + H_2 \rightarrow XH + H$. This feature greatly inhibits the formation of sulfur-bearing molecules in gas at low temperature. In a recent work Neufeld and coworkers[77] presented the observations of five sulfur-bearing molecules, SH, H_2S, CS, SO, and H_3S^+, in foreground diffuse molecular clouds lying along the sight-lines to bright continuum sources. The resulting column densities were then tested against current astrochemical models.

In Figure 3.12[64,77] the main production and destruction channels of a suite of 19 sulfur- and carbon-bearing molecules relevant in PDRs, and in warm regions heated by turbulent dissipation in magnetized shocks or in vortices, are illustrated. Using this reaction network within a TDR model, Neufeld and colleagues were able to account for the observed CS/SH and SH/NH column density ratios when the ion–neutral drift $u_D \approx 6\ \mathrm{km\ s^{-1}}$, somewhat larger than that invoked to explain the CH^+/SH^+ column density ratio.[64] On the other hand TDR models systematically under-predict the column densities of both H_2S and SO by approximately one order of magnitude unless the cloud radiative properties overflow in the translucent regime. Moreover, they overestimated additional observational constraints, such as the CS/HCS^+ ratios reported by Lucas and Liszt (2002).[78] The comparison between predictions and observations are shown in Figure 3.13[77] (left panel). Similar discrepancies are found when the predictions of shock models discussed in Section 3.4.2 are compared with the data. As in the TDR case, the predicted H_2S/SH and SO/SH column density ratios are underestimated by a factor of about 10 for clouds with $n_H = 100\ \mathrm{cm^{-3}}$ and small shielding $A_V \leq 1$. The CS/HCS^+ ratio is also underestimated, except for high density values. Unlike TDR models, however, shock models underestimate the CS/SH ratio by a factor of about 5 for $A_V \leq 1$ (see Figure 3.13, right panel).

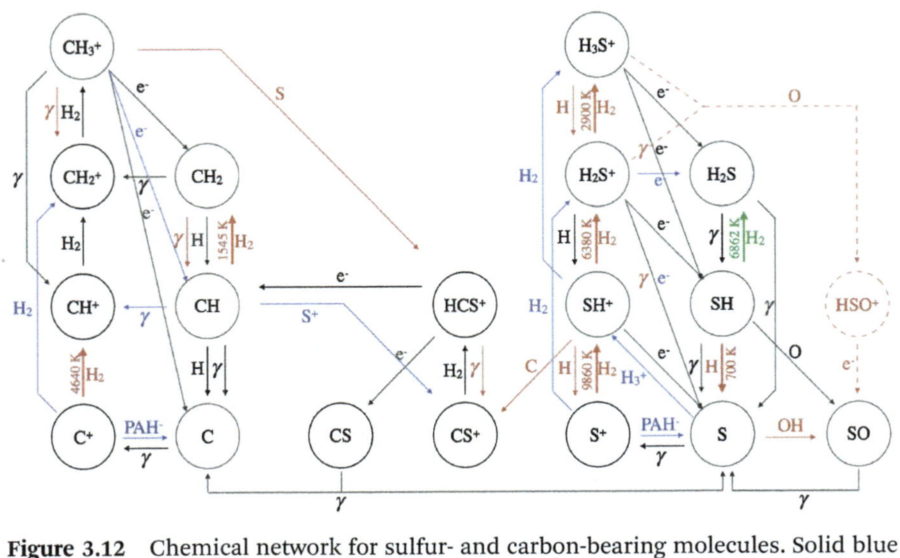

Figure 3.12 Chemical network for sulfur- and carbon-bearing molecules. Solid blue arrows indicate reactions that are only important in PDRs while solid red arrows indicate reactions that are only important in TDRs or shocks. Green and dashed red arrows denote the additional processes discussed at the end of Section 3.4.3. Activation energy barriers are indicated by temperatures marked next to the relevant reaction arrow. Neufeld *et al.*, A&A, 577, 106, (2015), reproduced with permission © ESO.[77]

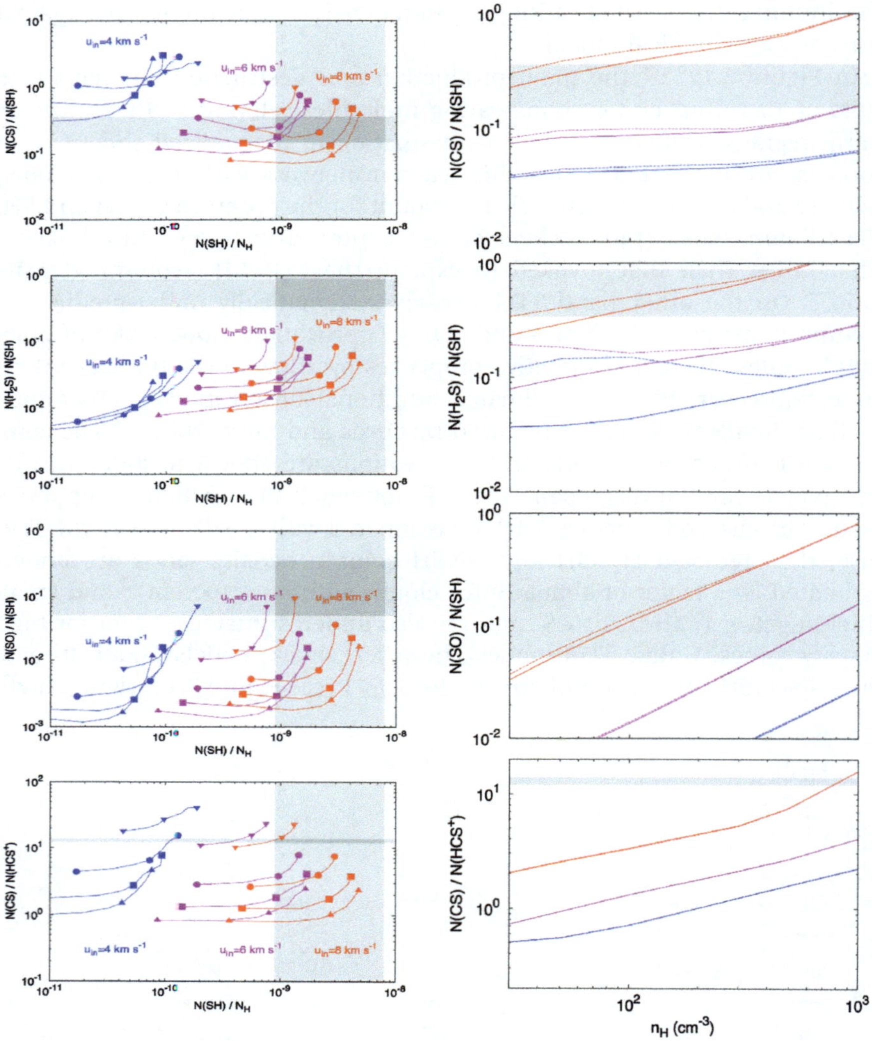

Figure 3.13 Column density ratios predicted by TDR (left panels) and shock (right panels) models. Grey shaded regions indicate the observed range of values. In the case of TDRs the column densities are shown as functions of the fractional abundance of the species SH, while shock models are shown as functions of pre-shock density. TDR models: rate of strain $a = 3\times10^{-11}$ s^{-1}; $u_D = 4$ (blue), 6 (magenta), and 8 (red) km s^{-1}; $n_H = 30$ (triangles), 50 (squares), 100 (circles), and 300 (upside down triangles) cm^{-3}. The total hydrogen column density is taken as $N_H = 1.8\times10^{21}$ cm^{-2}; A_V varies between 0.1 and 1.0 in going from left to right. Shock models: $A_V = 0.1$ (blue), 1 (magenta), and 2 mag (red). Velocity memory effects are included for SH assuming a momentum transfer rate coefficient equal to 1×10^{-10} (dashed lines) and 5×10^{-10} cm^3 s^{-1} (solid lines).
Neufeld *et al.*, A&A, 577, 106, (2015), reproduced with permission © ESO.[77]

Searching for physical effects not previously included in shock standard models, Neufeld and coworkers "corrected" the chemistry with a process resulting from a combination of events occurring at astronomical and microscopic levels, which involved the radical SH: the newly born SH molecule produced by the dissociative recombination of H_2S^+ or H_3S^+ is injected in the gas with a fraction of the velocity of the ionized parents; such translational excitation may significantly enhance the production of H_2S *via* the neutral–neutral reaction $SH + H_2 \rightarrow H_2S + H$, which is endothermic by 6862 K. Such a mechanism, called the *velocity memory effect*, depends on the momentum transfer between SH and H_2, which unfortunately for this system has never been accurately estimated. Taking into account this process the discrepancy between the observed and computed abundance of H_2S may reduce to a factor of ~ 3, for reasonable values of the momentum transfer.

3.4.4 Formation and Excitation of Molecular Hydrogen in Turbulent Dissipation Regions

While most of the molecular hydrogen is in the form of low temperature gas, it is now well known that there exist large amounts of fairly warm H_2; so warm that neither ultraviolet pumping nor formation excitation can account for the observed population of the high-lying ($J > 2$) rotational levels. The H_2 column density has been estimated along several sightlines mainly with the Copernicus[79] and FUSE[58,80–83] satellites. As we saw in the previous sections, turbulence dissipation may, in principle, deliver enough heat locally to the diffuse gas to drive H_2 excitation at the observed level. However, the whole process of formation and excitation depends on the nature of turbulent velocity fields, because compressible forcing leads to denser clumps than solenoidal forcing, for which the motions are less compressible. This is another clear manifestation of the role of dynamics in shaping the chemistry, at least at the scale of a molecular cloud.

The problem may be described by coupling the fluid equations governing the behaviour of the gas to the ones that control the formation, destruction, and the excitation of the H_2 molecules. However, a complete description of the chemistry is still out of reach of any numerical simulation, while the numerical complexity in computing the rovibrational structure increases boundlessly with the number of included levels. Micic and coworkers[84] explored the sensitivity of H_2 formation to the nature of the turbulent velocity field. They addressed this problem by performing purely HD simulations, *i.e.* neglecting any complications introduced by magnetic fields or the effects of self-gravity, and reducing drastically the number of chemical routes determining the mass balance of molecular hydrogen to just 10 reactions among 4 species, H_2, H, H^+, and e^-. Level excitations were not considered, either collisionally or radiatively, while the mean column density was supposed to be sufficiently high to adequately shield H_2 against photodissociation for standard ultraviolet fields. Within these approximations the evolution of the H_2 number density n_2 is described by the following relation:

$$\frac{\partial n_2}{\partial t} + \nabla \cdot (n_2 \boldsymbol{u}) = k_f n_H (n_H - 2n_2) - k_d n_2 \quad (3.5)$$

where k_f is the formation rate on dust grains, and k_d indicates the exit rate occurring through both thermal (H, H_2, e^-) and non-thermal (cosmic-rays) collisional dissociations.

To excite a turbulent flow with a specified rms turbulent velocity $\boldsymbol{u}$, Micic *et al.* (2012)[84] included a stochastic forcing term in the gas momentum eqn (2.10), and described the evolution of such forcing in the space of wave numbers, through a stationary solution of the Langevin equation in which a random process generates Gaussian random vector deviates. The assumed model is the so-called Wiener process that describes a random, but continuous motion, of a particle, subjected to the influence of a large number of chaotically moving colliders. The Wiener process is thus a natural model of Brownian motion. The perturbation may be decomposed in the fully solenoidal and the fully compressive modes, tuned by a coupling parameter that determine the power of the compressive modes with respect to the total forcing power.[85]

To quantify the rate at which H_2 forms, eqn (3.5) is integrated over the simulation volume on the assumption of periodic boundary conditions and that destruction routes are negligible. This last assumption is justified as long as $n_2 \ll n_H$. The final step is to consider the H_2 mass-weighted mean molecular fraction $\langle x_2 \rangle = M_H{}^{-1} \int (2m_H n_2)\,dV$, M_H being the mass enclosed by a volume V, as being uncorrelated with gas density n_H, *i.e.* assuming the turbulence to perfectly mix the gas on a timescale much shorter than the chemical timescale $\langle x_2 \rangle \approx \langle n_2 \rangle / \langle n_H \rangle$. The study is performed starting with a gas of uniform "low" ($n_H = 30$ cm^{-3}) and "high" ($n_H = 300$ cm^{-3}) densities. The time required to evolve a significant mean molecular fraction decreases as the background density or the strength of turbulent driving increase, with compressively driven turbulence leading to more rapid formation of H_2, particularly at early times. In Figure 3.14[84] (left panels) it is shown how the mass-weighted H_2 fraction varies with density. For a given initial density there is substantial scatter in the molecular fraction, but there is also a clear trend with density in the distribution. The differences between the compressive and solenoidal case are mainly due to the much wider spread of densities arising from compressive forcing with respect to solenoidal forcing, as is clear looking at how the density PDF varies as the mean density changes (Figure 3.14, right panels). The density PDF is estimated by constructing a histogram of an ensemble of measurements of n_H at specified locations. Since the transition rate to the molecular phase scales almost linearly with density when the H_2 concentration is small, compressive

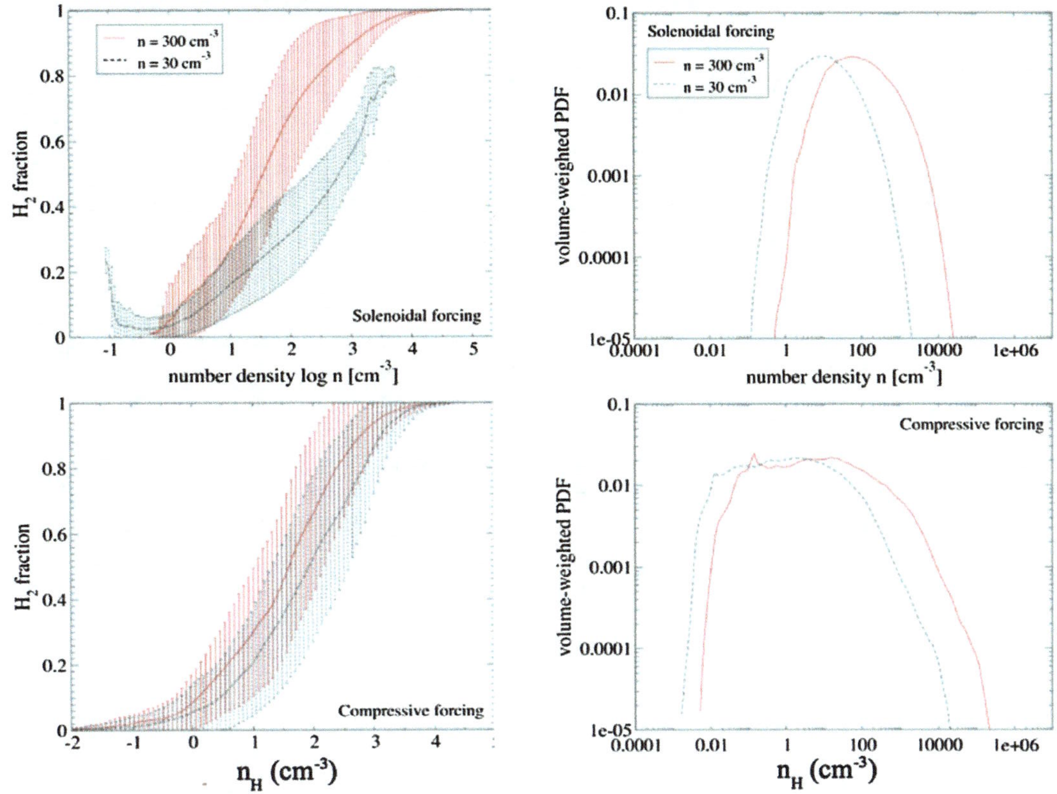

Figure 3.14 Left panels: H_2 mass-weighted mean molecular fraction $\langle x_2 \rangle$ plotted as a function of the gas number density n_H, in the case of pure solenoidal (top panel) and compressive (bottom panel) forcing. The standard deviation in the value of $\langle x_2 \rangle$ in each bin is indicated by the error bars. Right panels: density PDF for solenoidal (top panel) and compressive (bottom panel) forcing. Results are given for the two initial densities $n_H = 30$ (black dashed lines) and 300 cm^{-3} (red solid lines), and at a time corresponding to half turbulence crossing time $t = 0.5 \times \tau_C$ (see text).
Reproduced with permission from Figure 3.4 and 3.5 in Modelling H_2 formation in the turbulent interstellar medium: solenoidal *versus* compressive turbulent forcing, Micic *et al.*, *Mon. Not. R. Astron. Soc.*, 2012, **421**,[84] by permission of Oxford University Press on behalf of the Royal Astronomical Society.

turbulence forms H_2 much more rapidly at early times. The simulations have been shown at a time $t = 0.5\times\tau_C$, with τ_C the turbulence crossing time[86] given by $\tau_C = 10\ \mathrm{Myr}\times(L_d/100\ \mathrm{pc})\ (u/10\ \mathrm{km\,s^{-1}})^{-1}$ in which L_d is the cloud size, and u the magnitude of the rms turbulent velocity. The clear correlation between the hydrogen molecular fraction and density seen in Figure 3.14 suggests that turbulence cannot merge overdense clumps into their lower density surroundings on very short timescales. Consequently the assumption of uncorrelated H_2 mass-weighted mean molecular fraction and number gas density n_H is incorrect as long as the final, equilibrium state of the gas is the goal of the study, while it remains reasonably accurate in the estimate of the locations in which the ISM becomes mainly molecular, and the corresponding timescales.

From the above results it is clear that H_2 is significantly influenced by the dynamics of the flow in which it forms. A more recent work of Valdivia and coworkers (2016)[87] exploited high-resolution MHD colliding flow simulations, and included, in addition to the formation and destruction of H_2, the calculation of the column densities of a few excited rotational levels. In eqn (3.5) the main mechanism that destroys the H_2 molecule is photodissociation by absorption of UV photons in the Lyman and Werner transitions. The destruction efficiency depends on the dust extinction and molecular self-shielding.

Calculations regarding molecular hydrogen are made in post-processing, and they need the total hydrogen number density, the H_2 number density, the He number density, the temperature, and the *ortho*-to-*para* ratio. The final requirement means that intersystem collisions (such as those with protons) are not incorporated. The size of the synthetic cloud is $L = 50$ pc, and the gas is initially atomic with a number density $n_H = 1\ \mathrm{cm^{-3}}$ and temperature $T = 8000$ K, typical of the WNM. At the beginning the gas is uniformly magnetized by a moderate magnetic field of a few μG (few$\times10^{-10}$ T). The cloud lifetime was set at 20 Myr. The colliding streams induce the transition from rarefied to moderately dense gas, then, when enough gas has been accumulated, gravity becomes important and triggers infall, affecting an ever growing fraction of the cloud, and increasing the filling factor of the dense gas. Figure 3.15[87] (left panel) shows the hydrogen molecular fraction along any line of sight included in the simulations as a function of total column density. These model fractions are plotted against observed lines of sight spanning a wide range of column densities, resulting in a fairly good agreement, although a number of data points are not reproduced by the simulations. This is not particularly worrying as molecular hydrogen is affected by ultraviolet radiation, and thus every line of sight should entail a detailed radiative transfer modelling. However, simulations reproduce well the global trend, specifically the vertical transition branch at column densities between 1×10^{20} and $3\times10^{20}\ \mathrm{cm^{-2}}$ as well as the higher column density region. In the right panel of Figure 3.15 the density PDF at different evolutionary times is shown.

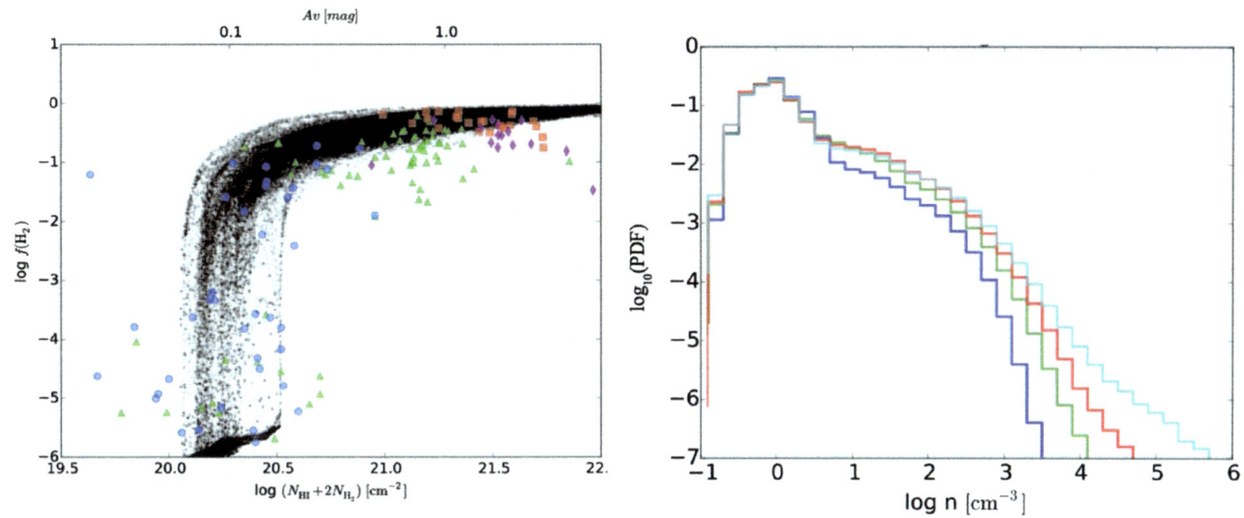

Figure 3.15 Left panel: molecular fraction as a function of total hydrogen column density; colour points indicate observational data. Right panel: density PDF evolution; colour lines show results at $t = 5$ (blue), 10 (green), 15 (red), and 20 Myr (light blue). Valdivia *et al.* A&A, 587, A76, (2016), reproduced with permission © ESO.[87]

Because the two simulations are very different in nature it is not easy to compare their results. The hydrodynamic (HD) density PDFs obtained by Micic *et al.* (2012)[84] are shown at a time t that is half of the turbulent crossing time. For a cloud size $\Delta L \approx 20$ pc and rms velocity $u = 2$ km s^{-1}, $\tau_C \approx 10$ Myr, and thus $t = 5$ Myr. The MHD density PDF at 5 Myr (reported in Figure 3.15, right panel) bears little similarity to the density distribution generated by purely compressive forcing in hydrodynamical models whose PDFs are very broad, while it is intermediate between low and high initial densities for purely solenoidal forcing. Increasing the evolution time, the solenoidal density PDF should tend to broaden as the densest gas becomes fully molecular, until the effect of the density increases the number of formed H_2 drops in response to the shortage of atomic gas. However, in the magnetic case the high-density tail of the distribution is much less steep than the one in pure HD fluids.

In the Valdivia and coworkers simulations, during the first 5 Myr, the H_2 fraction exponentially increases to ~0.1. Such a formation time would correspond to densities of the order of 100–1000 cm^{-3} using the standard formation scaling law $t_F \approx 10^9/(n_H/\text{cm}^{-3})$ year. However, the mean density of the simulated cloud at that time would be about one order of magnitude lower. Thus, such a relatively short amount of time needed to form an appreciable quantity of molecular hydrogen is somewhat surprising. This suggests that the presence of H_2 at low density is dependent on the rapid formation of H_2 in denser parts of the ISM.[87] Possible explanations may be the expansion of dense gas merging with the more diffuse background through sonic expansion or evaporation, as well as the nesting of diffuse gas pockets into denser regions providing strong self-shielding. Interestingly, in addition to dense and diffuse regions, a significant fraction of H_2 forms even in the warm interclump medium. This component contributes to the thermal balance of the gas, and in the range of density around 10 cm^{-3} its cooling rate is similar to the standard cooling rate of the ISM. H_2 emission and absorption thus reveal the interaction between the warm and cold gas—which recall the percolation of water through sand—with the emergence of hot (cold) pockets into the CNM (WNM).

We note that photodissociation effects may be important for H_2 trapped in hot gas pockets. Initially, the hydrogen molecules are efficiently shielded by the embedding cold gas, but, as the increased collision rates begin to populate higher-lying levels, self-shielding rapidly declines following the opening of a large number of destruction channels with large photodissociation rates in excited rovibrational states.[88] Thus, the level populations, and consequently the H_2 volume density decrease in response to an increased photodestruction rate. Such an effect may significantly impact on the chemistry in long-lived perturbations (older than a few 10 000 year). As in the case of chemical products, the imprint of the heating burst on the populations of the first two levels of H_2 persists for a period of time that, depending on the model assumptions, might be comparable to the lifetime of diffuse interstellar clouds.[88] This may have implications, among others, for the ortho/para ratio estimates.

3.5 Dust in Magnetized Turbulence

From just after formation in stellar outflows until incorporation into the dense regions of proto-stellar cores, dust grains play a key role in many aspects of the evolution of the ISM. Since the ISM is a dynamical system in which as time passes matter may move from one phase, *i.e.* one type of region, to another, dust is subjected to powerful processing through gas–grain, grain–grain and photon–grain interactions. The extent and nature of such processing depends on the rate and the energy of these interactions, both of which are related to the density structure and dynamics of the ISM.

Interstellar turbulence, because it is generally supersonic, creates in the ISM a texture of shocks and localized intense vortices, which might affect dust evolution more frequently and more significantly than the faster supernovae shock waves. Being cycled continuously through a variety of physical conditions, dust grains experience growth mechanisms and processes that may reduce or entirely remove the solid dust component (see Section 2.5.3.4). One such process is the shattering of grains in grain–grain collisions at high energy; such collisions redistribute grain mass into units of smaller size, generating a size distribution that favours small dust grains over large ones ($\sim$0.1 μm, Section 2.6). Shattering may also cause vaporization of some of the dust material as atoms or molecules into the gas phase, and may remove smaller grains entirely. Gas–grain collisions at sufficiently high energy may lead to sputtering; this is a process in which the grain surface is eroded and atomic or molecular material is ejected to the gas phase. At lower velocities, coagulation in grain–grain collisions occurs, making larger grains from smaller ones. Since the number density of grains in the interstellar space of the Milky Way is very low on average (typically, one large grain in a million cubic metres of the averaged interstellar medium) this process requires huge enhancements in the number density of grains for the process to be effective on reasonable timescales. Grain motions are very sensitive to coupling to turbulent eddies, which depends on the surface area-to-mass ratio of the particles. At low velocities, grain collisions will lead to the growth of very open and fluffy structures, while at intermediate velocities more compact aggregates form. At very high velocities, destruction will halt the growth. There have been suggestions that the effects of intermittency would boost the relative velocities of small grains to make their mutual coagulation the dominant process in the early evolution of the dust size distribution.[89]

In contrast, in the standard scenario large grains move faster than the smaller ones resulting in a gradual removal of all small grains during coagulation. Although forms and ways are at issue, it is clear that coagulation does happen. Low-velocity grain collisions may have dramatic chemical consequences by triggering grain mantle explosions. Grains may also grow in the ISM by the accretion of atoms and molecules from the gas phase. All these processes are in competition with the exposure of dust to energetic radiation fields and fast particles. Here we will not discuss in detail dust evolution; interested readers may refer to the book by Williams and Cecchi-Pestellini (2016)[90] for an extensive discussion of dust response to the various "stresses" in the ISM. In this section we shall limit ourselves to a description of some aspects of the effects of turbulent motions on dust grains that are relevant for chemistry, the most obvious example of this being the accumulation of ice mantles on the surfaces of dust grains. Since dust grains move fast through the turbulent interstellar gas, grain accretion could be modified, providing significant effects upon interstellar chemistry.

We shall firstly discuss the shattering of carbonaceous dust grains in interstellar turbulence as the formation mechanism of polycyclic aromatic hydrocarbons (PAHs). Apart from a few dissenting voices (*e.g.*, Kwok[91]), the PAH hypothesis has become extremely popular with those in the astronomical community. Such molecules have many appealing characteristics: they absorb ultraviolet radiation effectively, convert it with high efficiency to vibrational excitation, and release it as infrared emission, in bands qualitatively matching some observed features. Purely energetic considerations, based on the overall integrated emission in the mid-infrared, and assuming unit conversion efficiency and ultraviolet absorption cross-sections, led to the estimate that about 15–20% of the available interstellar carbon must be locked in PAHs to produce the observed infrared emission fluxes. This implied that PAHs must be taken into account as a necessary ingredient of interstellar extinction models. PAHs are also very resistant to photodissociation (provided they are large enough), and can possibly form in C-rich outflows of evolved stars (with chemical pathways inspired by combustion chemistry), but their injection rate in the ISM appears to be very low, being approximately (2 Gyr)-1. Here we examine the hypothesis that PAHs (or PAH-like carbonaceous pieces) form through the collisional fragmentation of carbonaceous grains, or carbonaceous mantles in core–mantle grains. There are two major circumstances for shattering in the ISM: one is supernova shocks, and the other is turbulence in the diffuse ISM. However, the grain velocities in supernova shocks are probably too high to allow the survival of the fragments as well defined molecular entities.[91] Dust motions in turbulence are moderate, and PAHs may possibly survive.

3.5.1 Dust Motions in Magnetized Turbulence

MHD turbulence induces both fluid motions and magnetic fluctuations. Relative grain–grain motions arising from MHD turbulence are discussed by *e.g.*, Yan *et al.* (2004).[8] These authors based their analysis on the analogy between the dynamics of charged grains and the dynamics of cosmic-rays, modifying the existing tools used for the description of cosmic rays. The turbulent acceleration is modelled as the acceleration due to a spectrum of MHD waves that can be decomposed into incompressible Alfvénic modes and compressible magnetosonic modes. While the fluid motions accelerate grains through hydrodynamic drag due to the frictional interaction with the gas (*hydrodrag*), the electromagnetic fluctuations accelerate grains through energy exchange involving resonant interactions between the particles and the waves (*gyroresonance* and *transit accelerations*).

A resonance occurs when $\omega - k_{\parallel} u_{gr} \mu = n\Omega_L$ with $n = 0, \pm 1 \cdots 2, \ldots$, where ω is the wave frequency, $k_{\parallel}$ the parallel component of wavevector $\boldsymbol{k}$ along the magnetic field, u_{gr} the velocity of the dust grain, μ the cosine of the pitch angle relative to the magnetic field,[93] and $\Omega_L = q_{gr}B/m_{gr}c$ the Larmor frequency, q_{gr} and m_{gr} being the charge and mass of the dust grain, respectively. In other words, resonance occurs when the Doppler-shifted frequency of the wave in the grain's guiding centre rest frame $\omega_{gr} = \omega - k_{\parallel} u_{gr} \mu$ is a multiple of the grain gyrofrequency Ω_L. The sign of the harmonic number n denotes the polarization of the wave. The + sign represents left-hand polarization, whereas the − sign right-hand polarization. Transit acceleration ($n = 0$) requires longitudinal motions and only operates with compressible modes. For low speed grains (of interest here) gyroresonance is the dominant MHD interaction. Gyroresonance occurs when the resonance condition is fulfilled and, in addition, the rotating direction of the electric wavevector is the same as the direction for the Larmor gyration of the grain. The particle will see an accelerating electric field along its transverse motion over a substantial fraction of a cyclotron period. If the resonance condition is not satisfied then the interaction time will be very short and the wave will have a negligible effect on the particle motion. The gyroresonance scatters and accelerates the particles. Since the electric field, which accelerates the grain, is perpendicular to the magnetic field the velocity of grains is larger along this direction.

Grain velocities are obtained through the use of the Fokker–Planck formalism, which describes the diffusion of grains in momentum space. The treatment needs to be simplified because the solution to the Fokker–Planck equation including, simultaneously, hydrodrag and resonances, is rather difficult. Thus, the velocity of grains is estimated considering the two effects separately, and adding them in quadrature. However, when the accelerations arising from the two processes are comparable, the situation is more complicated, as the gaseous friction to be considered, for *e.g.*, gyroresonance calculations, is, in general, affected by fluid motions. Another important simplifying assumption regards the possibility of separating the MHD perturbations that characterize turbulence as being due to distinct modes (see the discussion in Cho and Lazarian in 2002[94] and references therein). Their results may be summarized as follows: (i) motions perpendicular to the magnetic field lines are essentially hydrodynamic, (ii) Alfvén modes in compressible regime exhibit scalings and anisotropy similar to those in incompressible regime, with eddies elongated along the magnetic field;

(iii) slow modes passively mimic Alfvén modes; (iv) fast modes show isotropy and a scaling similar to turbulence caused by interacting sound waves; (v) the production of slow and fast modes by Alfvénic turbulence is suppressed. The lack of the energy transfer among the various modes allows a significant simplification of the problem since the different components of the MHD turbulence cascade can be assumed as independent.

The resulting grain relative velocities for silicate and graphite grains are shown in Figure 3.16[8] as functions of the grain size for a typical CNM: $T = 100$ K, $n_H = 30$ cm^{-3}, $B = 6$ μG (6×10^{-10} T), and (somewhat high) fractional ionization 1.5×10^{-3}, with corresponding Alfvén velocity $v_A = 2$ km s^{-1} and the ratio between gas and magnetic pressures $\beta = P/P_B \approx 0.4$, with $P_B = 2c_s^2/v_A^2$. Gyroresonance induced by fast modes tends to dominate grain motion, and can drive grains to supersonic velocities. Fast modes also provide the major contribution to hydrodrag, either in the CNM or in other environments such the WNM and dark clouds, although the contribution of Alfvénic modes may increase.[8]

An important point that affects differentially the dust grain size distribution is the damping cut-off in the turbulence spectrum. At small scales the turbulence spectrum is truncated by damping. In a medium in which damping is dominated by the kinetic viscosity (see Section 2.7.1) the damping rate is $t_D = (\nu k^2)^{-1}$. For primarily molecular regions ν is the order of the neutral mean free path times the mean thermal speed of the neutral particles. As noted in Section 2.7.4, the Pederson conductivity plays an important role in the damping of waves in many astronomical molecular sources, and the Hall conductivity does as well in some sources. However, studies are progressing in a step-wise fashion, and it is instructive to

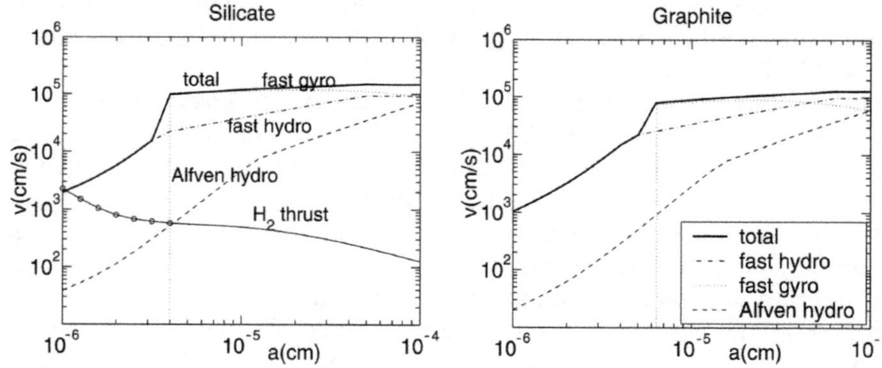

Figure 3.16 Relative velocities as a function of radii (thick solid lines) in CNM for silicate grains (left) and graphite grains (right). The hydrodrag contribution is represented by dashed lines (Alfvén modes), and dash-dotted lines (fast modes). A further velocity component induced by H_2 formation (thin solid line) is also plotted. At small grain sizes (open circles) this contribution is incorrect because two critical dynamical pieces of information are missing (see also ref. 8).[94]
Reproduced by permission of the AAS from Yan *et al.* (2004).[8]

consider the case in which damping is due solely to the kinetic viscosity. The turbulence is assumed damped at the scale l when t_D is of the order of the eddy turnover time $\tau_l \approx l/u_l$. In the case of Alfvénic turbulence, the eddies are elongated along the magnetic field, $l \approx k_\perp^{-1}$, and $\tau_l \approx (k_\perp u_l)^{-1} = (k_\parallel v_A)^{-1}$. The equality arises by the assumption of a *critically balanced cascade*,[95] which is based on the intuitive assumption that any increase in $k_\perp$ must at some point start affecting $k_\parallel$. The turbulent cascade of fast modes is expected to be slow, and, in the absence of collisionless damping fast modes, should persist over a longer time than Alfvén or slow modes. In this case $\tau_l \approx (v_A/u_l) \times l/u_l$. Using the dispersion relation for fast modes $\omega = v_A k$, and setting $l \approx k^{-1}$ (fast modes are isotropic), $\tau_l \approx \omega/(k^2 u_l^2)$. The relation $t_D \approx \tau_l$ defines the cut-off scale k_C, which is different for different modes. The cut-off scales are $k_{\parallel,C} = 4 \times 10^{-16}$ cm^{-1}, and $k_C = 7 \times 10^{-15}$ cm^{-1}, in the Alfvén and fast mode regimes, respectively. Then, by equating the cut-off scale to the resonant scale k_{res} ($\bar{a}_{gr}$, ρ_{gr}), the minimum scale required for gyroresonance, we can get the critical grain size $\bar{a}_{gr}$ over which grains are effectively gyro-accelerated by specific modes, for dust grain of a given density ρ_{gr}. For instance, the critical size for the silicate grains shown in Figure 3.16 (left panel) is $\bar{a}_{gr} = 10^{-5}$ cm for Alfvén modes, and $\bar{a}_{gr} = 4 \times 10^{-6}$ cm for fast modes. Cut-off scales and critical grain sizes may, of course, vary according to the modified physical conditions (see Table 3.4).

Subsequent studies[96] have shown that the stronger waves of the Alfvén modes show a smaller anisotropy than that supposed under the assumption of a critically balanced cascade, suggesting that imbalanced Alfvénic turbulence may be efficient in the interaction with grains. Such studies are, however, not very quantitative, and furthermore there is not yet a conclusive theory for imbalanced fast modes, which are the most important for dust acceleration. Finally, imbalanced turbulence develops under unequal energy fluxes coming from opposite directions, as *e.g.*, in the vicinity of a star, and not in the general ISM.

3.5.2 Formation of PAHs by Turbulence-induced Shattering Events

Seok *et al.* (2014)[97] proposed that shattering of carbonaceous dust grains in interstellar turbulence might be a source of PAHs. The rate increases sharply above a critical metallicity at which shattering becomes efficient. Such a proposal was not new. About 20 years earlier, Jones *et al.* (1996)[98] suggested that in interstellar shocks with velocities larger than 50 km s^{-1} about 10% of the initial graphite grain mass might be converted into fragments whose sizes correspond to the size range of PAHs. This conclusion was revised by Micelotta *et al.* (2010),[92] who pointed out that interstellar PAHs were either completely destroyed or lose a substantial fraction of carbon atoms in high velocity shocks. Lower velocity shocks (approximately a few tens of km s^{-1}) may lead to significant production of small grains, but only in a rather high density gas.[99] Shattering conducive to PAH, or small grain formation, occurs

Table 3.4 Physical conditions associated to grain motion calculations.[a]

Property		CNM	WNM	WIM	Molecular cloud	Dark cloud	
Electron number density	n_e (cm^{-3})	0.03	0.03	0.0991	0.1	0.01	0.001
Magnetic field	B (μG)[b]	6	5.8	3.35	11	80	
Injection scale	L (pc)	0.64	100	100	1	1	
Alfvén speed	v_A (km s^{-1})	2	20	20	1.2	1.5	
Cut-off scale wavenumber	k_C (cm^{-1})	7×10^{-15}	4×10^{-17}		4.5×10^{-14}	5.3×10^{-15}	5.3×10^{-17}

[a]Number density and temperature are reported in Table 3.1.
[b]1 G = 10^{-4} T.

in the turbulent diffuse ISM, because grains obtain moderate velocities, $\sim 10\ \mathrm{km\,s^{-1}}$ (see Section 3.5.1, and Table 3.4).

Possible shattering events following the collisions between two dust grains (k target, j projectile) are shown in Figure 3.17 (left panel).[100] When the two particles collide at sufficiently high velocity, strong shock waves are driven within both particles, compressing them to very high pressures. Two distinct stages develop during the collision. In the first one shocks arising at the interface between the particles hit both projectile and target at approximately constant velocity and pressure. When the shock reaches the back surface of the projectile it is reflected as a rarefaction wave, which when it enters into the target gives rise to the second phase. At this stage there are two separate flow fields in the target, a strong "detached" shock that can lead to the vaporization and melting of the target material, and the excavation flow that produces a crater (Figure 3.17, right panel[98]).

A solution to the problem of shock wave propagation in solids has been developed by Tielens and coworkers[98,101] through a self-similar approach. The important quantity is the mass M shocked to a pressure P_1 sufficiently high to produce cratering in a target of mass m_k. The critical pressure P_1 is the only free parameter contained in the model, and corresponds to a minimum collision velocity of $\sim 1\ \mathrm{km\,s^{-1}}$. Let's then consider the impact of a projectile of mass m_j and relative velocity v_r on the target. The shock velocity v_S and the velocity of shocked material (the target) in the laboratory reference system v_1, are given approximately by $v_\mathrm{S} = c_\mathrm{S} + s \times v_1$, where s is a constant close to unity characteristic of the collider material (either the target or the projectile). For a shock in a solid of negligible initial pressure P_0 and internal energy ε_0 the Rankine–Hugoniot conditions are identical to those for a strong shock in a gas (see Section 2.2.3). The pressure just behind the shock results $P_1 = \rho_0\, v_1 v_\mathrm{S}$. The Mach number corresponding to such a pressure is given by $M_1 = 2\Phi_1/[1 + (1 + 4s\Phi_1)^{1/2}]$, with $\Phi_1 = P_1/\rho_0 {c_\mathrm{S}}^2 = v_1/c_\mathrm{S}$. The mass of the target material shocked to the pressure P_1 is

$$\frac{M}{m_j} = \frac{1 + 2R}{(1 + R)^{9/16}} \frac{M_r^2}{M_1^2 \sigma_1} \sigma_r^{-1/9} \tag{3.6}$$

with $M_\mathrm{r} = v_\mathrm{r}/c_\mathrm{S}$, $R = (s_k \rho_0 / s_j \rho_{0,j})^{1/2}$. The quantity σ is a weak function of the shock parameters $\sigma = 0.30 \times (s + C_M - 0.11)^{1/3}/(s + C_M - 1)$ with $C_M = 1/M_1$ for σ_1, and $(1 + R)/M_\mathrm{r}$ per σ_r, respectively; $R = 1$ for colliding grains of the same material.

Based on this theory, Hirashita and Yan (2009)[100] proposed a recipe to describe the shattering event (see Figure 3.17, left panel): (i) the entire projectile is assumed to fragment in the collision, $m_\mathrm{frag} = m_j$, m_frag being the total mass of the fragments; (ii) if more than the half of the target is shocked, $M > m_k/2$, the entire target is shattered $m_\mathrm{frag} = m_k$, otherwise, only a fraction of the target mass $m_k^p = f_\mathrm{e} V_\mathrm{c}\, \rho_\mathrm{d}$ is ejected. V_c and f_e are the crater volume and the fraction of this volume that is vented out of target, respectively. The detailed analysis performed by Tielens and coworkers for the collision of a finite solid particle (the projectile) on a semi-infinite layer (the target) shows that $f_\mathrm{e} \approx 0.4$.

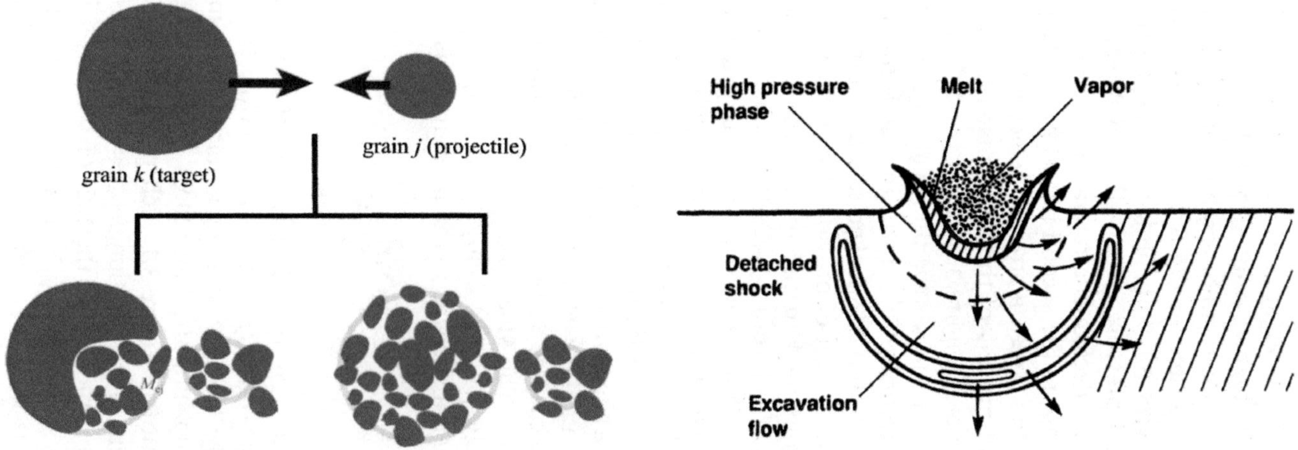

Figure 3.17 Left panel: schematic view of shattering events following the collisions between two dust grains (k, target; j, projectile). Following the impact the entire projectile is assumed to fragment in the collision. If more than half of the target is shocked, the entire target is shattered, otherwise only a fraction of the target mass is ejected. Reproduced with permission from Figure 3.1 in Shattering and coagulation of dust grains in interstellar turbulence, Hirashita and Yan, *Mon. Not. R. Astron. Soc.*, 2009, 394,[100] by permission of Oxford University Press. Right panel: a pictorial representation of the effects on the target due to the impact; the shock can lead to vaporization and melting of the target material, and produce an excavation flow, which produces a crater. Reproduced by permission of the AAS from Jones *et al.* (1996).[98]

Since real grains are of limited sizes we assume that that is given by $V_c \rho_d \approx m_k$, and thus $m^p{}_k = 0.4\ m_k$. The same analysis gives the dimensional spectrum of the fragments that appears to follow a power-law distribution with an exponent very close to the standard MRN index for the size distribution of dust grains in the ISM[90] (see Section 1.2), $n_{\rm frag}(a)\,{\rm d}a = K_{\rm frag}\ a^{-3.3}\ {\rm d}a$, where the constant $K_{\rm frag}$ is found by normalizing to the total mass of fragments $m_{\rm frag}$

$$m_{\rm frag} = \int_{a_{\rm frag}^-}^{a_{\rm frag}^+} m_{\rm frag}(a) n_{\rm frag}(b) {\rm d}a. \tag{3.7}$$

Such integration requires the knowledge of the upper and lower bounds of the fragment radii for both target and projectiles. When the particle is totally destroyed, the break is assumed to be caused by spalling processes rather than by cratering flow. Such cases apply to the projectile and to the target when $M > m_k/2$. The maximum fragment size is determined by the critical spalling collision velocity given by $v_{\rm C} = [m_k/(1+2R)m_j]^{9/16} \sigma_1^{1/2} \sigma_{\rm r}^{1/16}\ (1+R)\ M_1$ leading to $a_{\rm frag}^+ = 0.22(v_{\rm C}/v_{\rm r}) \times a_x$, with $x = j$ or k. Somewhat arbitrarily, the minimum grain size, $a^-{}_{\rm frag}$ is assumed to be few per cent of the maximum size. When $M < m_k/2$, the total ejected volume is $V_k = m_k^p/\rho_{\rm d} = 0.4\ m_k/\rho_{\rm d}$. The maximum grain size is determined from the equivalence $4\pi/3\ (a_{\rm frag}^+)^3 \approx V_k/50$. The minimum size is assumed to be $a_{\rm frag}^- = a_{\rm frag}^+\ (P_1/P_{\rm V})^{1.47}$, where $P_{\rm V}$ is the critical pressure for vaporization.

Coagulation counters the effects of shattering and is assumed to occur when the relative velocity is lower than the threshold velocity for coagulation[8,102]

$$v_r^{\rm th} = \frac{21.4}{\chi^{1/3} \rho_{\rm d}^{1/2}} \left[\frac{a_k^3 + a_j^3}{(a_k + a_j)^3}\right]^{1/2} \times \left[\frac{(a_k + a_j)^{\Gamma}}{a_k + a_j}\right]^{5/6} \tag{3.8}$$

where Γ is the surface energy per unit area, and $1/\chi = (1-\kappa_k)/E_{\rm k} + (1-\kappa_j)/E_j$, with E Young's modulus and κ the Poisson ratio, *i.e.* the amount of transversal expansion divided by the amount of axial compression of the materials.

The corresponding time evolution of the grain size distribution may be evaluated considering spherical grains under the assumption of total mass conservation. In other words shattering and coagulation reshuffle the number density and the mass of grains in the specified size and mass ranges. Practically it is convenient to consider N discrete bins for the grain radius, $\{a_i^{\rm b}\}$. Any bin is represented by an effective grain of size $a_i = (a_{i-1}^{\rm b} + a_i^{\rm b})/2$ and mass $m_i = 4\pi/3 a_{\rm i}^3 \rho_{\rm d}$. The mass density of grains contained in the ith bin then results in $\rho_i^{\rm (M)} = m_i\ n(m_i)\ [m_i^{\rm b} - m_{i-1}^{\rm b}]$ and its time evolution

$$\begin{aligned} \frac{{\rm d}\rho_i^{\rm (M)}}{{\rm d}t} = &- m_i \rho_i^{\rm (M)} \sum_{k=1}^{N} \alpha_{ki} \rho_k^{\rm (M)} + \sum_{j=1}^{N} \sum_{k=1}^{N} \alpha_{kj} \rho_k^{\rm (M)} \rho_j^{\rm (M)} m_{kj}^{\rm sh}(i) \\ &- m_i \rho_i^{\rm (M)} \sum_{k=1}^{N} \beta_{ki} \rho_k^{\rm (M)} + \sum_{j=1}^{N} \sum_{k=1}^{N} \beta_{kj} \rho_k^{\rm (M)} \rho_j^{\rm (M)} m_{kj}^{\rm cg}(i). \end{aligned} \tag{3.9}$$

In eqn (3.9) $\alpha_{ki} = \sigma_{ki} v_{ki}/m_k m_i$ if v_{ki} the relative velocity between k and i effective grains is larger than the critical velocity for shattering $v_{sh} = P_1/\rho_0 \, v_S$, and zero otherwise; $\beta_{ki} = \sigma_{ki} v_{ki}/m_k m_i$ if v_{ki} is lower than the critical velocity for coagulation, eqn (3.8), and zero otherwise. σ_{ki} is the geometrical collision cross-section. The mass distribution of the shattered fragment m_{kj}^{sh} depends on the velocity v_{kj}, and it is given in eqn (3.7), but now the integration limits are set by the bin boundaries, a_{i-1}^{b} and a_i^{b}. As we pointed out in the Section 3.5.1, grain velocity dispersions induced by gas drag and gyroresonance in a turbulent medium are size-dependent. The coagulation term depends on the total mass of colliders $m_{kj} = m_k + m_j$: $m_{i-1}^{b} \leq m_{kj} \leq m_i^{b}$ then $m_{kj}^{cg} = m_i$, and zero otherwise.

Starting from an initial dust size distribution, *e.g.*, the MRN distribution,[90] by means of eqn (3.9) it follows that the evolution of the grain dimensional spectrum is shattering and coagulation take place in various ISM regions. The difference between classical dust and PAHs is only ascribed to different size domains separated by a somewhat artificial threshold (Seok *et al.*, 2014[97] place it at 2 nm), so that the total mass of the two populations are found by just integrating the same distribution over two different, but contiguous, size ranges.

By using the time evolution of the PAH density $\rho^{(M)}$ by shattering, one may evaluate the shattering timescale as $[d\rho^{(M)}/dt]_{sh} = \rho_d^{(C)}/\tau_{sh}^{(i)}$, where $\rho_d^{(C)}$ refers to carbonaceous dust, and τ_{sh} depends on the ith region of the ISM. Shattering effects are relevant only in CNM or WIM. In WNM, because of the ion–neutral collisions, the fast modes are damped on a larger scale than in WIM. Thus, gyroresonance is not efficient for small grains, and only large grains with $a > 0.2$ μm can be accelerated to a velocity large enough for shattering. The shattering timescale is inversely proportional to the time-dependent carbonaceous dust-to-gas ratio, and thus depends on dust and metal formation in AGB stars and supernovae. The total shattering timescale of the ISM is given by $\tau_{sh} = 1/(\sum_i f_i/\tau_{sh}^{(i)})$, where f_i is the mass fraction of each ISM region (see Table 3.1). In the same way, the timescale of coagulation in each ISM region can be derived as $-[d\rho^{(M)}/dt]_{cg} = \rho^{(M)}/\tau_{cg}^{(i)}$, implying that PAH removal by coagulation is a simple exponential decay. For evident reasons, coagulation is not efficient in the WIM and WNM. The coagulation timescale is also inversely proportional to the dust-to-gas ratio. Grain aggregates in dense regions cannot be released into the general ISM before a time that is of the order of the lifetime of the cloud, although the coagulation inside the cloud is completed. As a consequence the coagulation timescale cannot be lower than typical lifetimes of dense clouds (~3–5 Myr). Finally, the total coagulation timescale of the ISM is found in a similar way as for shattering, by weighted summation for the mass fraction of each ISM region. According to the calculations of Seok *et al.* (2014)[97] at solar metallicity total shattering and coagulation times are ~130 and ~3 Myr, respectively, for the MRN size distribution. Using a different size distribution these timescales can either increase or decrease.

PAHs may be destroyed in supernova shocks on a timescale that is given by $\tau_{SN} = M_{ISM}(t)/[\gamma_{SN} \int \varepsilon(v_S)\, dM_S\,(v_S)]$, where γ_{SN} is the supernova rate, $\varepsilon(v_S)$, the PAH destruction efficiency as a function of the shock velocity v_S, and M_S the mass of the ISM swept up by a shock. Typically, in shocks large PAHs can be destroyed *via* both electron and ion collisions, while smaller PAHs are preferentially destroyed by electrons.[92]

It is evident that the problem of PAH mass evolution can only be solved within the context of the chemical evolution of a galaxy. Since dust grains consist of metals, one of the key factors involved is the ratio of the metal mass to the ISM mass; the metallicity. In general, galaxies are believed to evolve from a state with a very low metallicity and a very small amount of dust rather than one with higher amounts of metal and dust. Galactic chemical evolution models provide a useful framework in which one may interpret the large and ever expanding body of observational data concerning elemental and isotopic abundances in stars of various ages, galaxies, and even the intergalactic medium, both in the local and far Universe. Chemical evolutionary models follow the time evolution of the gas and dust content and the chemical elements produced in stars across the galaxy lifetime. The basic ingredients of such models are: the initial mass function (IMF); the star formation rate (SFR); stellar yields and feedback, *i.e.* the amount and type of elements ejected from the dying stars and elements that deplete into dust; gas flows, or any other processes in addition to the birth and death of stars, such as interactions with the ISM, intergalactic medium and other galaxies; the boundary conditions, *i.e.* the galaxy as an open or closed system. The simplest models to solve, the so-called "closed box models", assume that the galaxy does not lose or accrete material from the surrounding medium. Hence, the evolution of the mass of gas and dust is completely determined by stellar processes. In a pristine galaxy, the gas is progressively turned into stars at the SFR, with the star masses having a time-dependent IMF distribution. At the end of its lifetime a star returns a part of its mass to the ISM, either through stellar winds (in the case of low and intermediate mass stars) or through supernova explosions. The ejected material is enriched in elements synthesized by nuclear reactions in the stellar interiors, while some fragile isotopes, like deuterium, are absent from its composition. Thus, the ISM of galaxies is progressively supplied with elements heavier than H, while its deuterium content is reduced. New stellar generations are formed from this ISM, their composition being progressively more enriched in heavy elements, *i.e.* with increasing metallicity. Dust may be explicitly considered as a separate component. On this basis, many models have been put forward, displaying unavoidable tension among their results and the observations. A comprehensive guide to galactic chemical evolutionary models has recently been published by Matteucci (2012).[103]

In the simplest formulation galactic chemical evolution is described by a set of four coupled equations tracing the history of the total stellar mass M_*, ISM mass M_{ISM}, metal mass M_Z, and dust mass M_d, in a way consistent with the stellar evolution and the ISM recycling. For the evolution of dust mass, we have to take into account dust supply from AGB stars and supernovae, and dust mass growth by the accretion of metals onto dust grains in the ISM, as the sources of dust, and supernova shocks as the destruction mechanism of dust. This equation may be split to follow different materials, *e.g.*, silicates and carbonaceous dust if one separately considers the different stellar dust sources. The carbonaceous dust equation may be further refined if one treats the classical dust and the molecular component distinctly. The time evolution of the total PAH mass $M^{(M)}$ in the galaxy is then simply described by the equation $dM^{(M)}/dt = -M^{(M)}(t) \times (\tau_{SF}{}^{-1} + \tau_{SN}{}^{-1} + \tau_{cg}{}^{-1}) + M_d/\tau_{sh} + M_d/\tau_{sh}$, where τ_{SF} is the star formation timescale, which may be considered as a free parameter.

Figure 3.18[97] (top panels) shows some examples of the variation of PAH and dust abundances, $Z_{PAH} = M^{(M)}/M_{ISM}$ and $Z_{dust} = M_d/M_{ISM}$ with respect to

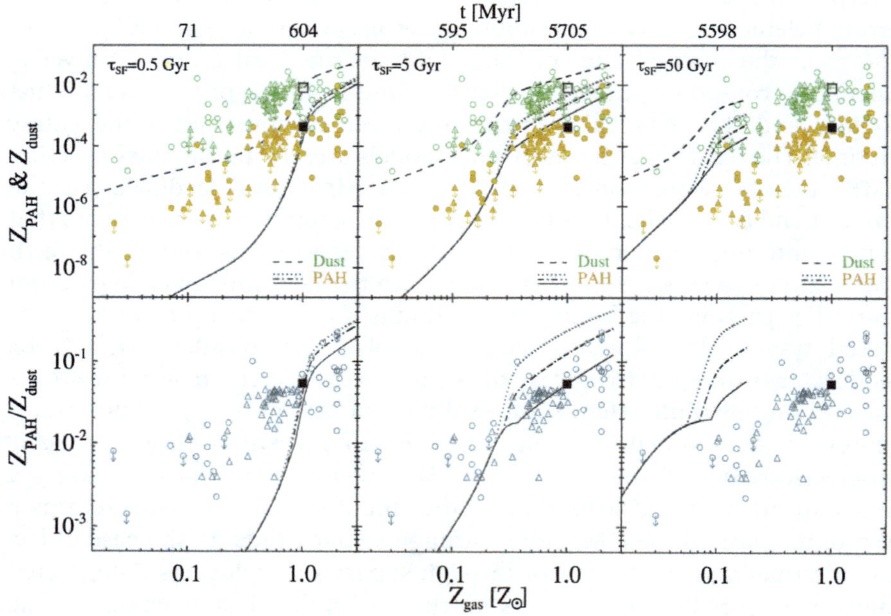

Figure 3.18 Top panels: PAH-to-gas mass ratio $Z_{PAH} = M^{(M)}/M_{ISM}$ as a function of the gas metallicity $Z_{gas} = M_Z/M_{ISM}$. Lower panels: total mass ratio of PAH to dust Z_{PAH}/Z_{dust}. Calculations have been performed for three values of the star formation timescale, $\tau_{SF} = 0.5$ (left panels), 5 (middle panels), and 50 Gyr (right panels), three dense cloud lifetimes, $\tau_{DC} = 1$ (solid lines), 3 (dot-dashed lines), and 10 Myr (dotted lines), and initial MRN size distribution for grains. Dashed lines trace the evolution of the dust-to-gas ratio $Z_{dust} = M_d/M_{ISM}$. Observational data (circles and triangles) are overlaid for comparison. In the top panels filled and open squares mark the Galactic values for Z_{PAH} and Z_{dust}, respectively, while in the lower panels the filled square indicates their ratios.
Reproduced with permission from Figure 3.5 in Formation history of polycyclic aromatic hydrocarbons in galaxies, Seok *et al.*, *Mon. Not. R. Astron. Soc.*, 2014, **439**,[97] by permission of Oxford University Press on behalf of the Royal Astronomical Society.

the gas metallicity $Z_{gas} = M_Z/M_{ISM}$ for three values of the star formation timescale, $\tau_{SF} = 0.5$, 5, and 50 Gyr, three dense cloud lifetimes, $\tau_{DC} = 1$, 3, and 10 Myr, and initial MRN size distribution for grains. In the lower panels of the figure the PAH-to-dust mass ratios are also reported. In metal-poor environments $Z_{gas} < 0.1\ Z_{\odot}$, PAH formation by shattering is ineffective due to the low dust abundance. Above a certain metallicity, called critical metallicity dust growth in the ISM, regulation by metallicity becomes the main source of dust in galaxies. As the ISM is enriched with metals, PAH formation by shattering becomes effective because of the rapid boost of dust abundance by dust growth. The critical metallicity is the switching point in which dust stellar sources decline while grain growth in the ISM takes over. The PAH abundance increase slows down at larger metallicities as most gas-phase metals have already been accreted onto dust, and thus grain growth becomes ineffective. In the standard case, $\tau_{DC} = 3$ Myr and $\tau_{SF} = 5$ Gyr, this occurs at $Z_{gas} = 0.3\ Z_{\odot}$. As the metallicity exceeds its critical value, the dust-to-gas ratio increases faster, and the coagulation lifetime decreases. However, since τ_{cg} is forced to be lower than τ_{DC}, it remains frozen in long-lived dense clouds. In contrast, for short τ_{DC} (1 Myr) coagulation strongly affects the PAH abundance. Interestingly, Z_{PAH} grows rapidly compared to the dust abundance (see Figure 3.18, lower panels). The PAH-to-dust mass ratio is very low at low metallicities and shows a drastic increase at the critical metallicity although the rapid rise in the PAH abundance and the dust abundance occur simultaneously; shattering is a collisional process, it is then sensitive to the dust abundance. These results are consistent with the observed trends of the PAH abundance in galaxies, that show *e.g.*, a deficiency of PAHs in low metallicity galaxies, and a metallicity dependence of PAH abundances in higher metallicity galaxies.

3.5.3 Dust Accretion in Turbulent Flows: Effects on Chemistry

In standard gas–grain chemical models, the accretion of gas onto grain surfaces is usually considered assuming steady grains embedded in a homogeneous gas. In these conditions the accretion is solely determined by the thermal motion of gas-phase species. We have seen in the preceding sections that grain motions in turbulent media are strongly affected by both hydrodrag and gyroresonance acceleration, which can accelerate dust grains to supersonic motion relative to the gas. Since grains move quickly through the turbulent interstellar gas, significant effects upon interstellar chemistry may be induced by the changes in the modes of gas accretion.[104]

The importance of accretion in interstellar chemistry is witnessed by observations of some particular astronomical locations leading to the conclusion that the dust grains in them accumulate mantles of ices containing a limited variety of fairly simple molecules, mainly water, carbon monoxide and carbon dioxide, with smaller amounts of some other simple molecules.

CO dark regions in correspondence with dust emission peaks support this conclusion. While these unprocessed ices are certainly interesting *per se*, their chemical complexity can be dramatically enhanced, apparently in some form of solid-state chemistry, with the more complex products enriching the gas phase. These more complex product species, such as ethanol, acetic acid, and glycolaldehyde, are detected at relatively high abundances in various interstellar locations, especially in regions of star formation, and are considered to be related to the emerging subject of astrobiology.

To follow accurately the accretion mechanism one important point is the description of grain charging processes, which is determined by the photo-electron ejection balanced by the interactions of grains with free-flying electrons and ions. The first process was pioneered by Draine in 1978[105] and later improved by Draine (see ref. 17). The emission rate per grain is the superposition of the rate for the photoionization of valence electrons, with a second term including the photo-detachment rate for "extra" attached electrons on negatively charged grains. The important quantity to be determined is the photo-electron yield, the product of the probability to produce a photo-electron after the absorption of an ultraviolet photon, the probability for the absorbed photo-electron to travel out of the bulk of the grain, and finally the probability for the electron to escape from the grain surface to infinity. The opposite process is electron attachment to both neutral and charged grains. Ions can be accreted onto both neutral and charged grains, but in cold and dark regions the opposite process, the ejection of ions from grains, is assumed to be impossible. Photodesorption of neutral species, *e.g.*, CO has been instead proved to be feasible.[106] Ions and electrons are connected through ionization and recombination.

The accretion process of ions onto grains is mediated by different long range interactions according to the neutral, heterocharge and homocharge nature of dust particles. In the interaction of ions with neutral grains the geometrical cross-section is modified by the polarizability of the grain. Cross-sections will be significantly either enhanced for heterocharge grains or reduced for homocharge grains because of Coulomb interactions. In this latter case, accreting ions may be subjected to "tunnelling discharge" leading to their neutralization with oppositely charged grains through quantum tunnelling before touching the grain surface. During the process these species may directly return to the gas phase. Multi-atom ions whose neutral counterparts are unstable (*e.g.*, H_3^+) may break up into smaller, more stable pieces after their charges are neutralized with hetero-charge grains.

Ge *et al.* (2016)[104] investigated the effects of fast grain motions on the chemistry in the CNM, and in molecular and dark clouds (MCs, DCs). To model the chemistry in these environments, they assumed that dust grains have a constant radius $a = 10^{-5}$ cm, a value typically used in gas–grain models. The average turbulent grain velocities for grains are obtained from the analysis described in Section 3.5.1. For grains with radii $a = 10^{-5}$ cm these velocities give $u_{gr} = 1.20$, 0.49, and 0.60 km s^{-1} for the CNM, MCs, and DCs, respectively.

The accretion rate of a gas-phase neutral species i onto grain surfaces is given by $R_{acc}(i) = K_n(i)\ n(i)$ where K_n is the rate coefficient $K_n(i) = s_n(i)\ \pi a^2\ v_r\ n_d$, $s_n(i)$ the sticking coefficient, v_r the relative velocity, and $n_d = n_H m_p Z_{dust}/(4\pi/3)a^3\rho_d$ the number density of grains, m_p being the proton mass. The relative velocity, usually taken as the average thermal speed of species i, $v_i^{th} = (8k_BT/\pi m_i)^{1/2}$, is modified to $v_r = [8k_BT/\pi m_i + u_{gr}^2]^{1/2}$ when the stochastic grain velocity u_d in turbulent interstellar gas is taken into account. The charge accretion rate coefficient for an ionic species j with charge q_i is $K_c(j,Z) = s_c(j)\ \pi a^2\ (8k_BT_k/\pi m_p m_j)\, J_Q$, where $s_c(j)$ is the sticking coefficient, and J_Q is a factor describing the modification to the charge accretion cross-section due to Coulomb interactions. The values of J_Q have been given for steady grains by Draine and Sutin (1987)[107], in terms of the two parameters $\nu = Ze/q_i$ and $\tau = ak_BT/q_i^2$ where Z is the number of charges on the grain. For charge accretion onto a neutral grain $\nu = 0$, and $J_Q = 1 + (\pi/2\tau)^{1/2}$. In the case of charge neutralization between an ion or electron and a heterocharge grain $\nu = -\ |Z| < 0$, and $J_Q \approx [1 - \nu/\tau] \times [1 + 2/(\tau - 2\nu)^{1/2}]$. Finally, when an ion or electron accrete onto a homocharge grain, $\nu = |Z| > 0$ and $J_Q \approx [1 + (4\tau\ + 3\nu)^{-1/2}]^2 \times \exp[-\nu/\tau\ (1 + \nu^{-1/2}]$. This process is important only when the grain charge is not very large and the gas temperature is high enough. With moving grains the Coulomb factor J_Q is modified by the involvement of kinetic temperature in the parameter τ, which now becomes $\tau = a(k_BT + m_i m_p u_{gr}^2)/q_i^2$. The net effect is an increase in the relative velocity v_r together with either a decrease in the ion/electron accretion cross-section for neutral and heterocharge grains or an increase in homocharge grains.

In Figure 3.19[104] shows the effects of grain motion on the grain accretion rate coefficients of neutral (left panel) and ionic (right panel) species. The models in which grain motions are taken into account are labelled "v". Neutrals greatly enhance their accretion rates at high masses in particular, with the net result that the accretion rate coefficient of most species is almost constant. The grain motion effect for the neutral species is the largest at low gas temperatures, *i.e.* in dark clouds, because the relative impact velocities depend essentially only on the additional motion of the grains, $v_r \approx u_{gr}$. When the accreting species is charged the grain motion effect is strongly sensitive to the number of charges on a grain. Ge *et al.* (2016)[104] set these charges to be $-e$ in DC models, $+e$ in MC ones, and $+20e$ in the CNM, and considered only cations. In the DC case, the charge accretion is controlled by Coulomb attraction between the impinging cations and the negatively charged grains, and thus the inclusion of the grain motion makes the Coulomb attraction more difficult. In MC models the interaction between the cation and the homocharged grain is regulated by a potential barrier to charge accretion. Since the kinetic energy of the impinging cations is comparable to such a barrier, the additional grain motion helps the impinging cations to overcome the Coulomb screen. The exponential dependence of the charge accretion rate on the ion–grain relative motion produces a large increase in the charge accretion rate coefficient. In the CNM kinematic interactions are pretty much the same as in MCs, but the physical conditions in the CNM are such that the high average grain speed of 1.2 km s^{-1} can help the impinging cations to gain a kinetic energy comparable to, or even higher than, the Coulomb barrier, resulting in a huge exponential rise (up to twelve orders of magnitude) in the charge accretion rate coefficient. Thus, the general effect of grain motion is to increase the accretion of neutral species on to grains, to slow down ion–grain charge neutralization, but to also boost exponentially the accretion of ions onto homocharge grains. Accretion of ions onto neutral grains will be instead reduced by the grain motion effect, as in the case of ion–grain charge neutralization.

We shall now discuss the impact of modified accretion rates on both gas-phase and surface chemistries in dark clouds. For the gas-phase species, both neutral (*e.g.*, CO) and ion (*e.g.*, HCO^+) abundances are lowered by 2–3 orders of magnitude due to the enhanced grain accretion. Most of the surface ice species are also sensitive to turbulent grain motions with a significant increase in their abundances up to two orders of magnitude. Some ionic species, *e.g.*, C^+, appear to be very insensitive to kinetic effects because their production and destruction processes mainly involving neutral species, and are reduced to similar factors by the grain motion effect leaving the abundances almost unchanged. By contrast, some other species such as *e.g.*, CH_3 and HeH^+ show enhanced abundances. This is due to the subtle interplay between production and destruction. In the ionic case the enhancement is due to the decrease in their neutral destroyers, while some neutral species have their abundances enhanced because they are mainly produced through surface reactions, whose rates are raised in response to the increased

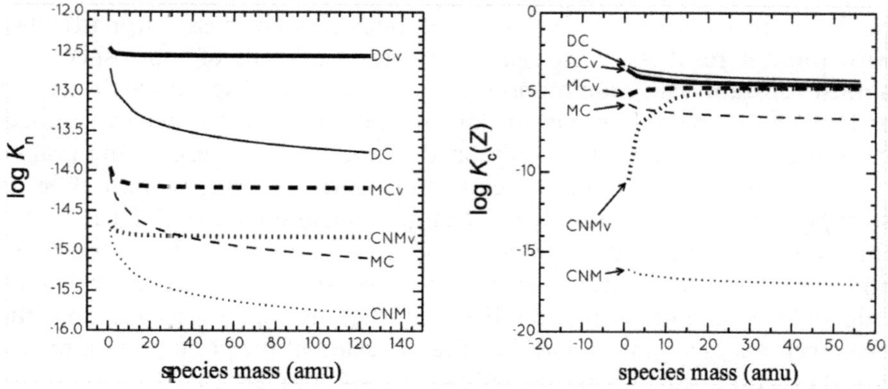

Figure 3.19 Grain accretion rates of neutral (left panel) and cation (right panel) chemical species in DC (solid lines), MC (dashed lines) and CNM models (dotted lines). Rate coefficients with turbulent grain motions are labelled with the suffix "v".
Reproduced with permission from Figure 3.2 in Effects of turbulent dust grain motion to interstellar chemistry, Ge *et al.*, *Mon. Not. R. Astron. Soc.*, 2016, **455**,[104] by permission of Oxford University Press on behalf of the Royal Astronomical Society.

abundances of surface reactants. As for some gas-phase species, a number of surface molecules, *e.g.*, H_2O seem to be insensitive to the inclusion of grain motion following the cancellation between the increase and decrease in abundances of two or more building blocks in the ice of the grain surface reactions. Ammonia ices unexpectedly show a decrease in the abundance by the grain motion effect, caused by the opening of a large number of surface reactions involving atomic hydrogen, which is a crucial ingredient in the NH_3 recipe.

However, distinctions exist in the chemical effects of grain motion among the model results for dark clouds, molecular clouds, and the CNM. In Figure 3.20[104] shows the histograms of chemical effects induced by grain motions in these three environments. Gas-phase neutral and ionic species, and surface species are considered separately to highlight their different responses to the grain motion. Most of the gas-phase species show abundances decreasing in the DC models (with $\log(X_v/X)_{max} < 0$), slightly increasing in the MC models, and only non-negative $\log(X_v/X)_{max}$ changes in the CNM models. There is thus a general agreement with the trends of the accretion rate coefficients that push down the abundances of both the neutral and ionic species in DC models, but push them up in the MC and CNM. Such differences arise from the rapid formation and desorption of neutral species on the grains in MC and CNM, that tend to dominate the molecular distribution in the gas-phase. At low temperature and high gas density, *i.e.* in dark clouds, the dominant kinematic effect is the reduction of the gas-phase abundances, while at higher temperature and low gas density, *i.e.* in the CNM, the gas-phase abundances increase *via* grain surface formation channel which is more efficient in the CNM models because of the highest grain velocities. Surface species are enhanced in all regions, with the largest increase in CNM models (by a factor 10^2–10^6), intermediate in DC models (by factors up to 10^3), and smallest in MC models (by factors only up to 10). This is a direct consequence of the different physical conditions characteristic of these environments, with neutral species decreasing from DC to CNM, through MC and ions increasing along the same sequence with decreasing gas number density and extinction, and increasing average charge of grains and gas temperature. Grain motions increase the accretion rates of neutrals on all grains, and of cations onto positively charged grains. However, neutral accretion is weaker in the MC models than in the DC models, because of the lower amount of neutral gas in the former models and their lower sensitivity to grain motion due to higher gas temperature. Moreover, cation accretion is also weaker in the MC models than in the CNM models, because of the lower amount of ionized gas in the MC models and the smaller change in the accretion rate coefficients due to slower grain motions.

3.6 Chemistry in Transient Small Scale Regions

The gaseous component of diffuse and translucent interstellar clouds has been probed and found to contain microstructures, which take the form of

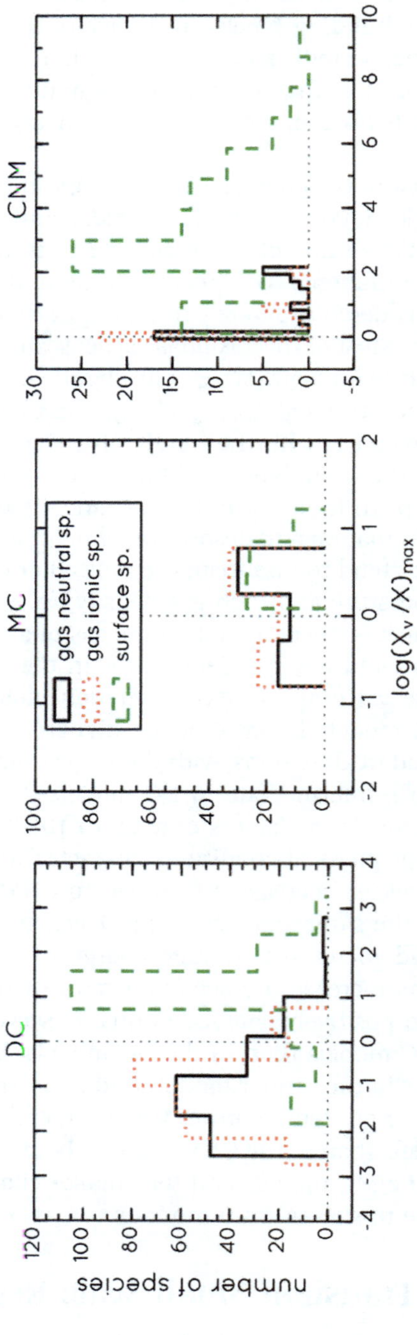

Figure 3.20 Histogram of the abundance change due to grain motion $\log(X_v/X)_{max}$, where X_v and X are the concentrations of the species X in models with and without grain motion, arising in DC (left panel), MC (middle panel) and CNM models (right panel). The black solid, red dotted and green dashed lines are for neutral, ionic and surface species, respectively. Reproduced with permission from Figure 3.10 in Effects of turbulent dust grain motion to interstellar chemistry, Ge *et al.*, *Mon. Not. R. Astron. Soc.*, 2016, **455**,[104] by permission of Oxford University Press on behalf of the Royal Astronomical Society.

dense knots of gas or filamentary and/or sheet-like density fluctuations, with a size-scale comparable with that of the solar system. This small-scale structure has been revealed in optical line intensities on timescales of about 10 years in atomic and ionic species, in angular variability along neighbouring lines of sight to binaries and globular clusters, and also time variability in lines of several species, including CH, CH^+, CN, and diffuse interstellar absorption band carriers. These tiny regions must be over-pressured with respect to the background, and therefore they are not in dynamical equilibrium. The small-scale structures are thus expected to evaporate quickly, so a mechanism must be invoked to replenish their population.

The origin of microstructures has been the subject of many studies. Some authors argue that dynamical condensation and fragmentation processes of the ISM driven by thermal instabilities in a shock may form small-scale clumps of CNM behind a shock front in WNM. In very simple terms, since the cooling rate depends on the density and temperature, it is then possible that in a isobaric medium if a density fluctuation shows a larger cooling rate than the surrounding gas, the density enhancement must necessarily grow. Other studies assert that microstructure cannot originate through non-magnetic hydrodynamics. In (isothermal) plasmas, there are 3 types of MHD wave: Alfvén, slow, and fast waves. Alfvén modes are incompressible while slow and fast modes are compressible (see Section 3.5.1). The passage of an Alfvén wave does not affect the density. Like sound waves in non-magnetized fluids, the slow mode and fast mode waves may instead have associated density fluctuations. In ideal MHD, from the linearized continuity and induction equations it results that, in low β ($= P/P_B$) plasmas, density fluctuations are dominated by slow waves, $(\delta\rho/\rho) \approx \delta u/c_S$,[94] with δu the average rms velocity. This scheme was pursued by Falle and collaborators,[108,109] who proposed that the microstructure may be excited in regions of high magnetic pressure by slow-mode magnetosonic waves generated at the boundaries of a diffuse interstellar region hit by a shock. A viable mechanism for the excitation of the fluctuations involves the generation of waves through the Kelvin–Helmholtz instability. When a rarefied material impacts on a denser region it drives a shock through it, and flows around the interface. The resulting shear gives rise to small-scale perturbations that draw kinetic energy from the mean flow, causing waves to propagate into the region.

The transience of the microstructure implied by the high overpressure must affect the chemistry. The high density can evidently compensate for the short timescales and low extinction, so that a significant chemistry can develop even in this apparently unfavourable region of parameter space. For instance, the nature of the microstructure may be reflected in the apparent dichotomy between excitation and formation of the dicarbon radical C_2. There are both theoretical and observational indications that C_2 formation in the diffuse gas occurs in clumps that are colder and denser than the average gas. On the other hand, these findings may be, in contrast with gas densities, inferred by excitation analysis of the C_2 rotational ladder requiring generally a much less dense thermal bath. Since overpressured tiny regions are expected to merge back into the embedding rarefied gas on a very short timescale, there should be a stage in their chemical evolution in which large molecular fractional abundances coexist with low density.[110]

The chemistry in diffuse clouds with transient microstructure was explored by Cecchi-Pestellini *et al.* (2009),[111] exploiting a model of diffuse interstellar gas in which the standard cold chemistry is supplemented by chemistry in many transient and tiny perturbations. These perturbations are assumed to be a population of transient density enhancements maintained by some external and unspecified source. Ambipolar diffusion is allowed to occur within them, together with the possibility of a temperature rise in the perturbation due to viscous energy dissipation. This is reminiscent of the passage of a MHD wave. The purpose of the authors was to determine the extent to which a model of many tiny scale transient density fluctuations in diffuse clouds, together with chemistry in the quiescent gas, can account for the variety and abundances of observed species.

The parameters adopted are largely conventional, apart from the treatment of the perturbations. CH^+ (100%) and HCO^+ (95%) are made predominantly in the perturbations, which also provide significant contributions for CH, OH, and H_2O (50%), H_3^+ (10%), and at a few per cent level for C_2, C_2H, CN, and CO. A higher temperature favours the activation of endothermic channels, while the rise of gas density alleviates the problem posed by the very short lifetimes of the transient perturbations, as chemical rates scale with reactant abundances. In addition, ion–neutral endothermic reactions benefit from the contribution of the ion–neutral drift, described through the effective temperature reported in eqn (3.3). The chemistry also produces abundant ionic species such as HS^+, OH^+, and H_3O^+, together with hydrocarbons of moderate complexity, up to CH_3^+, produced *via* exchange reactions with H_2 triggered by the formation of CH^+. Photodissociation of CH_3^+ ends the reaction chain leading back to CH_2^+.

An important point regards the formation of OH, which in the perturbations proceeds *via* neutral–neutral reactions of molecular hydrogen with atomic oxygen. The reaction has a barrier of about 3000 K. Therefore, a moderate increase in the gas kinetic temperature of the perturbations to 500 K limits the OH concentration. As a consequence, OH and CH^+ appear to be dynamically linked, but chemically decoupled, since OH is formed through neutral reactions while CH^+ is formed through ion–molecule exchanges. This behaviour reflects one of the major differences between the chemistry of transient warm microstructures and shock chemistry, where the bulk of the gas is non-selectively heated to (usually) much larger temperatures, for a much longer time.

The model relies on the presence of about 100 warm perturbations in the line of sight. For an assumed velocity $u_w \approx v_A$, and the lifetime $\tau_w \approx$ few hundred years, the average size of the perturbation results in $\Delta L_w \approx u_w \times \tau_w \approx$ few hundred au. The passage of one perturbation into or out of the line of sight will then occur once every few years. This is consistent with observations of secular changes in ionized and neutral microstructures in the diffuse ISM.

Interestingly, the prediction of the model that several ions, such as *e.g.*, OH^+ and H_2O^+, may be detectable has been confirmed by their discovery in the later Herschel observations in the diffuse line of sight towards W49N,[43] a galactic star forming region known to intersect a large collection of foreground gas clouds.

References

1. G. B. Field, *Astrophys. J.*, 1965, **142**, 531.
2. C. Heiles and M. Haverkorn, *Space Sci. Rev.*, 2012, **186**, 293.
3. E. Audit and P. Hennebelle, *Astron. Astrophys.*, 2005, **433**, 1.
4. E. Vazquez-Semadeni, *ArXiv e-prints*, 2009, **0902**, 0820.
5. P. Hennebelle, R. Banerjee, E. Vázquez-Semadeni, R. S. Klessen and E. Audit, *Astron. Astrophys.*, 2008, **486**, 43.
6. Z. Guo, D. N. Burrows, W. T. Sanders, S. L. Snowden and B. E. Penprase, *Astrophys. J.*, 1995, **453**, 256.
7. S. R. Redfield and J. L. Linsky, *Astrophys. J.*, 2008, **673**, 283.
8. H. R. Yan, A. Lazarian and B. T. Draine, *Astrophys. J.*, 2004, **616**, 895.
9. B. T. Draine, *Physics of the Interstellar and Intergalactic Medium*, Princeton Series in Astrophysics, 2011.
10. A. G. G. M. Tielens, *The Physics and Chemistry of the Interstellar Medium*, Cambridge University Press, 2005.
11. D. Diego Falceta-Gonçalves, in *The Labyrinth of Star Formation*, ed. D. Stamatellos, S. Goodwin and D. Ward-Thompson, 2014, p. 115.
12. B. Burkhart, A. Lazarian, A. Goodman and E. Rosolowsky, *Astrophys. J.*, 2013, **770**, 141.
13. A. A. Deshpande, K. S. Dwarakanath and W. M. Goss, *Astrophys. J.*, 2000, **543**, 227.
14. E. Jenkins and T. Tripp, *Astrophys. J.*, 2011, **734**, 65.
15. D. P. Cox, *Annu. Rev. Astron. Astrophys.*, 2005, **43**, 337.
16. R. L. Larson, N. J. Evans II, J. D. Green and Y.-L. Yang, *Astrophys. J.*, 2015, **806**, 70.
17. B. T. Draine, *Annu. Rev. Astron. Astrophys.*, 2003, **41**, 241.
18. A. Rémy-Ruyer *et al.*, *Astron. Astrophys.*, 2014, **563**, A31.
19. M. H. Heyer and C. M. Brunt, *Astrophys. J.*, 2004, **615**, L45.
20. C. L. Dobbs *et al.*, *Protostars and Planets VI*, 2014, p. 3.
21. R. S. Klessen and S. C. O. Glover, in *Star Formation in Galaxy Evolution: Connecting Numerical Models to Reality, Saas-Fee Advanced Course*, 2014, vol. 43, p. 85.
22. J. Pety *et al.*, *Astrophys. J.*, 2013, **779**, 43.
23. D. Utomo *et al.*, *Astrophys. J.*, 2015, **803**, 15.
24. T. Wong *et al.*, *Astrophys. J.*, 2009, **696**, 370.
25. L. Blitz and J. P. Williams, *Nato ASI Ser. Ser. C*, 1999, **540**, 3.
26. K. Sugitani *et al.*, *Astrophys. J.*, 2011, **734**, 63.
27. J. Kainulainen, C. Federrath and T. Henning, *Science*, 2014, **344**, 183.

28. O. Morata, J. M. Girart and R. Estalella, *Astron. Astrophys.*, 2005, **435**, 113.
29. A. Kawamura *et al.*, *Astrophys. J., Suppl. Ser.*, 2009, **184**, 1.
30. J. R. Dawson *et al.*, *Astrophys. J.*, 2013, **763**, 56.
31. F. H. Shu, R. J. Allen, S. Lizano and D. Galli, *Astrophys. J.*, 2007, **662**, L75.
32. L. Blitz and F. H. Shu, *Astrophys. J.*, 1980, **238**, 148.
33. J. Koda, N. Scoville and T. Sawada, *Astrophys. J.*, 2009, **700**, L132.
34. J. Koda, N. Scoville and M. Heyer, *Astrophys. J.*, 2016, **823**, 76.
35. A. Heithausen, *Astron. Astrophys.*, 2002, **393**, L41.
36. J. Kainulainen, H. Beuther, Th. Henning and R. Plume, *Astron. Astrophys.*, 2009, **508**, L35.
37. M.-M. Mac Low and R. S. Klessen, *Rev. Mod. Phys.*, 2004, **76**, 125.
38. M. R. Krumholz and A. Burkert, *Astrophys. J.*, 2010, **724**, 895.
39. C. Matzner, *Astrophys. J.*, 2002, **566**, 302.
40. P. Colín *et al.*, *Mon. Not. R. Astron. Soc.*, 2013, **435**, 1701.
41. T. G. Bisbas, P. P. Papadopoulos and S. Viti, *Mon. Not. R. Astron. Soc.*, 2015, **803**, 37.
42. C. Cecchi-Pestellini, in *Proceedings of the Dalgarno Celebratory Symposium*, ed. J. F. Babb, K. Kirby and H. Sadeghpour, Imperial College Press, London, 2010, p. 173.
43. D. Neufeld *et al.*, *Astron. Astrophys.*, 2010, **521**, L10.
44. B. Godard *et al.*, *Astron. Astrophys.*, 2012, **540**, 87.
45. S. von Hoerner, *Z. Astrophysik*, 1951, **30**, 17.
46. C. F. von Weizsäcker, *Astrophys. J.*, 1951, **114**, 165.
47. R. B. Larson, *Mon. Not R. Astr. Soc.*, 1981, **194**, 809.
48. B. C. Elmegreen and G. Scalo, *Annu. Rev. Astron. Astrophys.*, 2004, **42**, 211.
49. P. Hennebelle and E. Falgarone, *Astron. Astrophys. Rev.*, 2012, **20**, 55.
50. M. Heyer, C. Krawczyk, J. Duval and J. Jackson, *Astrophys. J.*, 2009, **699**, 1092.
51. *Magnetic Fields in Diffuse Media*, ed. A. Lazarian, E. M. De Gouveia Dal Pino and C. Melioli, Springer, 2015.
52. E. Falgarone and T. G. Phillips, *Astrophys. J.*, 1990, **359**, 344.
53. P. Hily-Blant and E. Falgarone, *Astron. Astrophys.*, 2009, **500**, L59.
54. H. S. Liszt and R. Lucas, *Astron. Astrophys.*, 2000, **355**, 333.
55. M. Gustafsson, A. Brandenburg, J. L. Lemaire and D. Field, *Astron. Astrophys.*, 2006, **454**, 815.
56. Z.-S. She and E. Leveque, *Phys. Rev.*, 1994, **72**, 336.
57. D. Folini and R. Walder, *Astron. Astrophys.*, 2016, **587**, 120.
58. C. Gry, F. Boulanger and C. Nehmé *et al.*, *Astron. Astrophys.*, 2002, **391**, 675.
59. E. Falgarone, L. Verstraete, G. Pineau Des Forêts and P. Hily-Blant, *Astron. Astrophys.*, 2005, **433**, 997.
60. E. Habart, A. Abergel, F. Boulanger, C. Joblin, L. Verstraete, M. Compiègne, G. Pineau des Forêts and J. Le Bourlot, *Astron. Astrophys.*, 2011, **527**, A122.

61. E. Falgarone, G. Pineau des Forêts and E. Roueff, *Astron. Astrophys.*, 1995, **300**, 870.
62. K. Joulain, E. Falgarone, G. Pineau des Forêts and D. Flower, *Astron. Astrophys.*, 1998, **340**, 241.
63. B. Godard, E. Falgarone and G. Pineau des Forêts, *Astron. Astrophys.*, 2009, **495**, 847.
64. B. Godard, E. Falgarone and G. Pineau des Forêts, *Astron. Astrophys.*, 2014, **570**, A27.
65. D. S. Nolan and B. F. Farrell, *J. Atmos. Sci.*, 1999, **56**, 2908.
66. J. Stone, E. Ostriker and C. Gammie, *Astrophys. J.*, 1998, **508**, L99.
67. D. H. Porter, T. W. Jones and D. Ryu, *Astrophys. J.*, 2015, **810**, 93.
68. A. Pon, D. Johnstone and M. J. Kaufman, *Astrophys. J.*, 2012, **748**, 25.
69. P. Lesaffre, G. Pineau des Forêts, B. Godard, P. Guillard, F. Boulanger and E. Falgarone, *Astron. Astrophys.*, 2013, **550**, 106.
70. D. Mihalas and B. Weibel Mihalas, *Foundations of Radiation Hydrodynamics*, Oxford University Press, 1984.
71. L. Monchick and J. Schaefer, *J. Chem. Phys.*, 1980, **73**, 6153.
72. A. T. Myers, C. F. McKee and P. S. Li, *Mon. Not. R. Astron. Soc.*, 2015, **453**, 2747.
73. V. Valdivia, B. Godard, P. Hennebelle, M. Gerin, P. Lesaffre and J. Le Bourlot, *Astron. Astrophys.*, 2017, **600**, 114.
74. P. S. Li, D. F. Martin, R. I. Klein and C. F. McKee, *Astrophys. J.*, 2012, **745**, 139.
75. E. G. Zweibel and A. Brandenburg, *Astrophys. J.*, 1997, **478**, 563.
76. H. S. Liszt, *Astrophys. J.*, 2015, **799**, 66.
77. D. A. Neufeld *et al.*, *Astron. Astrophys.*, 2015, **577**, 49.
78. R. Lucas and H. S. Liszt, *Astron. Astrophys.*, 2002, **384**, 1054.
79. B. D. Savage, R. C. Bohlin, J. F. Drake and W. Budich, *Astrophys. J.*, 1977, **216**, 291.
80. K. Gillmon, J. M. Shull, J. Tumlinson and C. Danforth, *Astrophys. J.*, 2006, **636**, 891.
81. B. L. Rachford *et al.*, *Astrophys. J.*, 2002, **577**, 221.
82. B. L. Rachford *et al.*, *Astrophys. J. Suppl.*, 2009, **180**, 125.
83. S. Lacour, V. Ziskin, G. Hébrard, C. Oliveira, M. K. André, R. Ferlet and A. Vidal-Madjar, *Astrophys. J.*, 2005, **627**, 251.
84. M. Micic, S. C. O. Glover, C. Federrath and R. S. Klessen, *Mon. Not. R. Astron. Soc.*, 2012, **421**, 2531.
85. W. Schmidt, W. Hillebrandt and J. C. Niemeyer, *Comp. Fluids.*, 2006, **35**, 353.
86. B. G. Elmgreen, *Astrophys. J.*, 2000, **530**, 277.
87. V. Valdivia, P. Hennebelle, M. Gérin and P. Lesaffre, *Astron. Astrophys.*, 2016, **587**, A76.
88. C. Cecchi-Pestellini, S. Casu and A. Dalgarno, *Mon. Not. R. Astron. Soc.*, 2005, **364**, 1309.
89. F. Boulanger, M. L. Prevot and C. Gry, *Astron. Astrophys.*, 1994, **284**, 956.

90. D. A. Williams and C. Cecchi-Pestellini, *The Chemistry of Cosmic Dust*, Royal Society of Chemistry, 2016.
91. S. Kwok, *Stardust: The Cosmic Seeds of Life*, Springer, 2013.
92. E. R. Micelotta, A. P. Jones and A. G. G. M. Tielens, *Astron. Astrophys.*, 2010, **510**, A36–A37.
93. G. B. Ribicki and A. P. Lightman, *Radiative Processes in Astrophysics*, John Wiley and Sons, 2002.
94. J. Cho and A. Lazarian, *Phys. Rev. Lett.*, 2002, **88**, 245001.
95. P. Goldreich and H. Sridhar, *Astrophys. J.*, 1995, **438**, 763.
96. A. Beresnyak and A. Lazarian, *Astrophys. J.*, 2008, **682**, 1070.
97. J. Y. Seok, H. Hirashita and R. S. Asano, *Mon. Not. R. Astron. Soc.*, 2014, **439**, 2196.
98. A. P. Jones, A. G. G. M. Tielens and D. J. Hollenbach, *Astrophys. J.*, 1996, **469**, 740.
99. V. Guillet, G. Pineau des Forêts and A. P. Jones, *Astron. Astrophys.*, 2011, **527**, A123.
100. H. Hirashita and H. R. Yan, *Mon. Not. R. Astron. Soc.*, 2009, **394**, 1061.
101. A. G. G. M. Tielens, C. F. McKee, C. G. Seab and D. J. Hollenbach, *Astrophys. J.*, 1994, **431**, 321.
102. C. Dominik and A. G. G. M. Tielens, *Astrophys. J.*, 1997, **480**, 647.
103. F. Matteucci, *Chemical Evolution of Galaxies*, Springer-Verlag, Berlin Heidelberg, 2012.
104. J. X. Ge, J. H. He and H. R. Yan, *Mon. Not. R. Astron. Soc.*, 2016, **455**, 3570.
105. B. T. Draine, *Astrophys. J. Suppl. Ser.*, 1987, **36**, 595.
106. G. M. Muñoz-Caro, Y.-J. Chen, S. Aparicio, A. Jiménez-Escobar, A. Rosu-Finsen, J. Lasne and M. R. S. McCoustra, *Astron. Astrophys.*, 2016, **589**, A19.
107. B. T. Draine and B. Sutin, *Astrophys. J.*, 1987, **320**, 803.
108. S. A. E. G. Falle and T. W. Hartquist, *Mon. Not. R. Astron. Soc.*, 2002, **329**, 195.
109. T. W. Hartquist, S. A. E. G. Falle and D. A. Williams, *Astrophys. Space Sci.*, 2003, **288**, 369.
110. S. Casu and C. Cecchi-Pestellini, *Astrophys. J.*, 2012, **749**, 48.
111. C. Cecchi-Pestellini, D. A. Williams, S. Viti and S. Casu, *Astrophys. J.*, 2009, **706**, 1429.

CHAPTER 4

Gas Dynamics under Gravity: Star Formation

At first sight the physical description of a cloud of gas that is contracting under its own self-gravity to form a star may seem very straightforward, especially if we make the gross simplifications of spherical symmetry, isothermality and the lack of any internal support mechanisms other than thermal pressure (as described in Section 4.1.3 below). However, it must be remembered that the dynamical range is huge; the number density changes from that appropriate to a diffuse cloud ($\sim$1–100 cm^{-3}) through dark cloud ($\sim 10^4$–$10^6\ cm^{-3}$) conditions to those appropriate at the centre of a star-forming cloud ($\sim 10^{10}$–$10^{13}\ cm^{-3}$) en route to forming an object with stellar densities ($\sim 10^{25}\ cm^{-3}$). The equation of state alone is astonishingly complex in these situations. Moreover, the details of the collapse process are critically dependent on the assumptions we make about the initial conditions. To study the process of star formation we therefore concentrate on sections of the evolution, and this chapter concentrates on the earliest stages.

The direct detection of the infall of gas that must precede star formation has been notoriously difficult to achieve. Whilst there can be little doubt that stars form from cold molecular cloud material, the dynamics of the collapse are very poorly constrained, and the differences between isolated/cluster and spontaneous/triggered star formation are not clear. Ockham's razor leads us to search for a singular mode of star formation, but observations are increasingly suggesting that star formation may proceed *via* a number of dynamical pathways.

We define a protostar as an object that is accreting matter before thermonuclear reactions start and the object joins the main sequence. Nature has conspired to make the earliest stages of protostar formation very difficult to detect; as a cloud of gas and dust contracts, it becomes highly opaque to

Dynamical Astrochemistry
By David A. Williams, Thomas W. Hartquist, Jonathan M. C. Rawlings, Cesare Cecchi-Pestellini and Serena Viti

Published by the Royal Society of Chemistry, www.rsc.org

optical light and the accretion source is hidden from sight. However, thermal emission from dust is optically thin at millimetre wavelengths for H_2 column densities up to 10^{26} cm^{-2}. This dust emission acts as a major coolant, allowing the gravitational potential energy to be radiated way, and also allows us to see deep into the star-forming region.

The process of star formation poses a number of long-standing and, as yet, still essentially unresolved issues, including:

(i) Why is it that not all clouds collapse to form stars, even those that are apparently gravitationally super-critical?
(ii) Why is the star-formation process apparently quite inefficient (only a small fraction of the protostellar cloud mass ends up in the protostar)?
(iii) What is the origin of the material from which stars form?
(iv) What are the dynamics of the initial stages of the collapse?

Importantly, it seems highly likely that the dynamics and efficiency of the earliest stages of star-formation control and determine the characteristics of the evolution that follows. As we shall argue in this chapter, astrochemical processes have particularly important roles to play in star-forming regions, not just in the (obvious) passive sense that the observations of molecular abundances and line profiles have the power to diagnose both the instantaneous physics and the physical evolution, but in a very active sense in that through their determination of ionization levels and cooling efficiency they may control the nature of the collapse process itself.

To tackle these and other problems, astrochemical modelling aims to identify the spectral signatures of collapse, clarify the dynamics and constrain key physical parameters (such as the age, temperature and number density, *etc.*).

It has been known for decades that most stars in our galaxy form in groups or clusters.[1] In nearby molecular clouds, much of the dense gas that is critical to star formation is found in filaments, which then fragment to form clumps of denser gas and eventually cores. Following Olmi *et al.* (2016),[2] we shall adopt the following definition for clumps and cores: a *clump* is defined as any compact density enhancement with sizes ranging from 0.1 to 1 pc, and densities of the order of 10^4–10^5 cm^{-3}, while *cores* are substructures within clumps of sizes <0.1 pc and densities of $\sim 10^5$ cm^{-3}. Cores are where stars will form. The determination of the properties of filaments is less straightforward; we shall adopt the definition by André *et al.* (2014)[3] whereby a *filament* is defined as any elongated interstellar medium structure that is significantly denser than its surroundings. How, from large-scale clouds (~10 pc) with moderate densities (~10^2 cm^{-3}), filaments, clumps, and dense cores form is still poorly understood and shall be the topic of this chapter.

While it is likely that low and high mass stars share a common formation path, we shall from now on describe their formation in two separate sections. For a lack of a better formalism we assume that 'low mass stars' are all those stars with mass less or equal to a solar mass, $M_\odot$, with intermediate mass stars covering masses up to 10 $M_\odot$, and 'high mass stars' defined as having masses >10 $M_\odot$.

4.1 Formation of Low Mass Stars

4.1.1 Introduction

It is often helpful to divide the star-formation process into separate phases, although the borderlines between each phase may actually be quite ill defined. The initial stage is one of fragmentation of a molecular cloud into subunits (clumps, cores, filaments, *etc.*). These sub-structures then collapse to form protostars, with accretion discs. The end of this stage is marked by the breakdown of an assumption of isothermality and spherical symmetry. As the infalling material encounters a centrifugal barrier, the next phase considers the formation of disc accretion and evolution, and is dominated by the presence of strong outflows, until the final phase when the reservoir of infalling matter is either exhausted or infall is arrested by outflows so that mass accretion is halted. In this chapter we are primarily interested in the second of these phases—the initial (approximately spherical and isothermal) collapse phase, although the all-important initial conditions will be determined by the evolution in the first phase as well.

In this section we first of all concentrate on the paradigm of low mass star formation. Low mass stars are, in many ways, ideal laboratories in which to both study astrochemical processes and to use our knowledge of astrochemistry to diagnose the physical process of infall and star formation. There are a number of reasons for this. Low mass stars are longer-lived and much more common than high mass stars and so, statistically, there is a good sample of low mass star-forming regions that are relatively close to us. Low mass stars tend to form in isolation or in binaries. The protostellar clouds from which they form are often not subject to external factors (as in triggered star formation), have relatively simple morphologies, are often nearly isothermal and—in many cases—show little evidence for non-thermal internal pressure. The energetics are relatively low so that, due to efficient cooling by molecules and dust (see below) the earliest stages are approximately isothermal ($T \approx 7$–15 K). Outflows are always present, but are usually well collimated and decoupled from the infall. Their energetics are also typically lower than in high mass star-forming regions, so that they do not completely dominate the kinematical signature of line profiles. In summary, low mass star forming regions are very well physically constrained—more so than high mass star-forming regions and, indeed, more so than molecular clouds.

4.1.2 The Evolution and Classification of Low Mass Protostars and Prestellar Cores

Historically, and prior to the advent of high spectral and spatial resolution spectroscopy, protostellar infall and evolution was inferred from mid- to far-infrared continuum spectral energy distributions (SEDs) and maps of molecular distributions. The original framework for the classification of low mass protostars was devised on the basis of the gradient of the SED in the 2–20 μm range by Adams *et al.* (1987).[4] The wavelength dependence of the luminosity yields an infrared spectral index:

$$n = \frac{d(\log \lambda L_\lambda)}{d(\log \lambda)}. \quad (4.1)$$

SEDs can be modelled on the reasonable assumption that the observed emission originates from protostellar accretion luminosity that has been absorbed and re-radiated by the static, plus infalling envelope and the central accretion disc.

In order of perceived evolution, Class I, II and III sources have $n \geq 0$, $-1.5 \leq n < 0$ and $n < -1.5$, respectively, and thus represent a progression from cooler to hotter (and therefore younger to older, more-embedded to less-embedded) sources. These sources can be identified as 'protostars' that have close associations with the molecular gas clouds, 'classical (or embedded) T-Tauri stars', and 'naked (or weak-line) T-Tauri stars', respectively.

Subsequently, millimetre and sub-millimetre observations started to reveal the presence of sources at an even earlier stage of evolution, which, although they possessed sub-millimetre point sources, had no IRAS point source. These objects are only detected at sub-millimetre or longer wavelengths and are undetectable at $\lambda < 10$ μm, so cannot be classified according to the scheme described above. To accommodate these, André *et al.* (1993)[5] extended the scheme to 'Class 0 protostars', which were originally defined, canonically, as possessing infall envelopes that are of greater or equal mass to the (hydrostatic) protostellar core. Observationally, this translates to the sources having a ratio of the bolometric to sub-millimetre (defined as $\lambda > 350$ μm) luminosities of ≤ 200. Compared to Class I sources, Class 0 sources are colder (black-body temperature $T_{BB} < 30$ K) and more isothermal (possessing a narrow, single temperature grey-body SED). They have massive envelopes ($M_{env} \geq 0.6$ M$_\odot$), are highly obscured ($A_v > 1000$ magnitudes) and have approximately spherical symmetry. They are identified with those objects that are undergoing the earliest, near-isothermal, stages of protostellar collapse. Typical Class 0 sources whose physics and chemistry have been extensively studied include B335, L1527, IRAS16293, NGC1333-IRAS4 and VLA1623. Generally, they have centrally condensed cores, of size ~0.01–0.03 pc, masses of ~0.5–20 M$_\odot$, and densities in the range 10^7–10^8 cm^{-3}.

Bridging the gap between molecular clouds and Class 0 protostars are objects that are generically described as 'starless cores'. A sub-set of gravitationally bound and centrally peaked objects is considered to be the precursor of Class 0 sources and is termed as 'pre-protostellar' cores (PPCs). PPCs lack a centrally heating protostellar source and show no evidence for infall, other than in the form of quasi-static contraction. Well-studied PPCs include L1498 and L1544.

Morphologically, the later stages are characterized by the emission from the accretion disc; an optically thin disc, plus strong IR, dust excess and Hα in the case of Class II, and an optically thick disc, no envelope IR excess or dust emission and weak Hα in the case of Class III. Class 0 sources can be thought of as early, or infalling, protostars, which, although they possess accretion discs and strong bipolar outflows, are approximately spheroidal. They have a typical lifetime of $\sim 10^4$ years. All Class 0 sources possess powerful and energy-efficient bipolar outflows that are indicative of two-dimensional (disc) geometries in the central regions. Indeed, as high resolution interferometric observations are revealing, the formation of accretion discs seems to be universal from the earliest stages of the infall process.

Figure 4.1 shows a good example of a protostellar source (the Bok globule BHR71), which shows many of the key features of Class 0 sources. The cloud is (optically) dark and the protostar itself is completely shrouded by dust obscuration. The dusty media also results in the reddening of stars seen through the cloud. This reddening can be used to determine the column density of the dust and, by inference, gas along the line of sight, and hence the radial density profile within the cloud. The only indicator of dynamical activity is the luminous and well-collimated jet (angled towards the observer) that erupts from the cloud. Outflow jets, often with bright patches, or knots,

Figure 4.1 Four colour visual image of the Bok globule BHR71, taken with the Very Large Telescope in Chile. The central source, which is not visible in this image, is probably 10 times as luminous as the Sun.
Reproduced with permission. Credit and Copyright: J. Alves, E. Tolstoy, R. Fosbury, and R. Hook.

of emission (Herbig–Haro objects, see Chapter 5) and broader (less well collimated) CO molecular outflows are common in all protostellar sources.

Class I sources are evolved protostars, and—like Class II sources—have a thick disc, and a typical lifetime of $\sim 10^5$ years, as deduced from statistical measures. They often have an internal temperature structure that is very far from isothermal and possess a central IR source. Very little of the collapse envelope remains by the Class II stage, which is dominated by emission from the accretion disc, which itself has thinned out or disappeared by the Class III stage.

4.1.3 The Paradigm of Spherically Symmetric Isothermal Collapse

It is perhaps rather surprising to discover that even the dynamics of a perfectly spherical, isothermal gravitational collapse, where the only internal pressure is isothermal, is not fully understood. The origin of the confusion mainly derives from the assumptions that are made about the physical initial conditions in a gravitationally bound cloud. To put this in context, we first of all consider the simplest possible case of the spherically symmetric collapse of a pressure-less gas—the so-called free-fall collapse.

The very earliest stages of protostar formation are nearly isothermal (~7–10 K), with efficient emission by dust and low-lying rotational transitions in ^{12}CO and optically thin isotopologues, such as ^{13}CO, efficiently radiating away much of the infall energy.

Starting from a sphere of gas with a uniform density distribution (ρ_0), the equation of motion of a thin shell, in spherical symmetry is given by

$$\frac{d^2r}{dt^2} = \frac{-4\pi G r_0^3 \rho_0}{3r^2} \tag{4.2}$$

where r is the displacement from the centre, with initial value r_0, and G is the gravitational constant. If we let $x = r/r_0$ and $\tau = t/t_{ff}$, where

$$t_{ff} = \sqrt{\frac{3\pi}{32G\rho_0}} \tag{4.3}$$

then eqn (4.2) becomes

$$\ddot{x} = \frac{d^2x}{d\tau^2} = \frac{-\pi^2}{8x^2} \tag{4.4}$$

which we can integrate, and apply boundary conditions, to find

$$\frac{\theta}{2} + \frac{1}{4}\sin 2\theta = \frac{\pi\tau}{4} \tag{4.5}$$

where $x = \cos^2\theta$. The shell reaches the centre when $x = 0$ (*i.e.* $\theta = \pi/2$). This yields a free-fall collapse time of $t = t_{ff}$, *for all shells*—*i.e.* the collapse is homologous. The entire cloud collapses on the so-called free-fall timescale $t_{ff} \approx 3.4 \times 10^7 / n_H{}^{1/2}$ years, where n_H is the gas number density (cm^{-3}).

This free-fall timescale then gives a good measure of the collapse timescale of a pressure-less sphere of gas. We can compare this to the sound-crossing time for the cloud; if the latter is smaller, then sound waves can act to smooth out pressure gradients and inhibit collapse. Equating the two timescales yields a condition of criticality and a critical mass for collapse (the so-called Jeans mass):

$$M_{crit}/M_{\odot} \approx 5.5 \times 10^4 \, (T/300)^{3/2} \, (10^6/n_H)^{1/2} \tag{4.6}$$

where T is the temperature in Kelvin and n_H is the number density of hydrogen (cm^{-3}) in both H atoms and H_2 molecules: $n_H = n(H) + 2n(H_2)$. It is evident from eqn (4.6) that, as the density rises, M_{crit} falls, so that a cloud can potentially sequentially fragment into sub-units—although such a process would be terminated when the clouds becomes optically thick to cooling radiation and the temperature rises. However, even with the gross simplification of the absence of internal non-thermal pressures and perfect spherical symmetry, the simple analysis above is entirely dependent on the assumptions concerning the initial conditions—the density, velocity and temperature structures in the initial gas cloud.

This, then, is the major source of disagreement between the various hydrodynamical models of protostellar cloud collapse and we investigate its importance in our discussion of the test case L1544, below.

In addition, there are a variety of complicating factors, including:

1. Internal turbulent pressure and deviations from isothermality. As a cloud collapses and becomes optically thick to cooling radiation, it will heat up and thermal pressure may halt the collapse.
2. Internal magnetic pressure. Magnetic fields are seen to pervade clouds in a semi-ordered fashion (as determined from polarization maps). The field is strongly coupled to the ion/electron fluids and hence to the neutral gas through collisions.
3. Angular momentum. Protostellar clouds will inevitably contain some angular momentum. Therefore, as a cloud contracts and its moment of inertia falls, it will spin up. This process leads to the formation of accretion discs, but most of the angular momentum must be dissipated, or else collapse will not proceed.

4.1.4 Hydrostatic Equilibrium of a Single, Isolated, Spherical Cloud

Returning to the case of simple spherical collapse, we consider first of all the simplest formulation of the initial conditions, on the assumption that hydrostatic equilibrium pertains.

Thus, we consider an isothermal, symmetric cloud that is in pressure balance throughout, on the assumption that gravitational forces are balanced by thermal pressure alone. We also assume that there is no external pressure acting on the cloud.

A thin shell, at radius r, has mass dM:

$$\frac{dM(r)}{dr} = 4\pi r^2 \rho(r). \tag{4.7}$$

In equilibrium, the inwards gravitational force on the shell due to the mass, $M(r)$, within it is balanced by the pressure gradient across the shell ($dP/dr < 0$).

$$\frac{dP(r)}{dr} = \frac{-GM(r)\rho(r)}{r^2}. \tag{4.8}$$

In addition we have the isothermal equation of state:

$$P(r) = c_s^2 \rho(r) \tag{4.9}$$

where c_s is the isothermal sound speed. Eqn (4.7)–(4.9) collectively form the equations of hydrostatic equilibrium. The equations are transcendental, with no analytical solution, but can be combined into a single equation

$$\frac{d^2\varphi}{d\xi^2} + \frac{2}{\xi}\frac{d\varphi}{d\xi} = e^{-\varphi} \tag{4.10}$$

with the boundary condition

$$\varphi(0) = \frac{d\varphi}{d\xi(0)} = 0 \tag{4.11}$$

and where

$$\xi = \frac{r}{c_s}\sqrt{4\pi G\rho_c} \tag{4.12}$$

and

$$\varphi = \ln\frac{\rho_c}{\rho}. \tag{4.13}$$

Eqn (4.10) is the Lane–Emden equation, which has only one free parameter, ξ_{max}, which corresponds to the cut-off radius or density contrast between the edge and the centre of the cloud. If the latter is >14.3 the cloud will spontaneously collapse and the marginal case is the so-called critical Bonnor–Ebert sphere, which has a mass given by Ebert[6] and Bonnor:[7]

$$M_{BE} = \frac{1.18c_s^4}{\sqrt{P_{ext}G^3}} \tag{4.14}$$

where c_s is the isothermal sound speed, $(kT/m)^{1/2}$, and P_{ext} is the external pressure.

Bonnor–Ebert spheres are the family of solutions to the Lane–Emden equation, the most extreme of which is the infinite, singular isothermal sphere for which $\rho \propto r^2$ throughout.

Such a solution would actually be unphysical in that the cloud would have infinite mass and is singular (*i.e.* tending to infinite density as $r \rightarrow 0$).

4.2 The Hydrodynamics of Gravitational Collapse

In the discussion that follows we limit ourselves to the cases of isolated star formation, where external influences are assumed to be minimal and, in particular, we do not address the more complex issue of induced and/or sequential star formation.

Although a multitude of modes of star formation are possible, essentially defined by the choice of initial conditions, and the characterization of any non-thermal support mechanisms that may be present, one of the simplest representations—now often referred to as the 'inside out' or 'standard model' of isolated star formation—was formulated by Shu (1977).[8] This model has largely superseded the earlier models of Larson (1969, 1972)[9,10] and Penston (1969),[11] which were criticized for their adoption of a very specific set of initial conditions, including a supersonic inflow velocity, which is probably more applicable to the physical conditions some time *after* the formation of a protostellar core. Shu's model is based on the collapse of a truncated (pressure-bounded) singular isothermal sphere (SIS, $\rho \propto r^{-2}$) supported by thermal pressure alone, and which is in critical Bonnor–Ebert equilibrium. This initial configuration is assumed to result from the quasi-static (sub-sonic) contraction of a gravitationally bound core. In this model, some (unspecified) process initiates collapse at the centre and propagates outwards (the so-called 'inside-out' collapse). Inside some boundary (the 'collapse expansion wave', CEW) material rapidly approaches free-fall—and develops a $\rho \propto r^{-3/2}$ profile—whilst outside it is static, with the SIS density profile. The CEW expands into the envelope at the local sound speed and, at any one instant, ~49% of the mass within the CEW has accreted onto the protostellar object. Thus, using the definition of a Class 0 object as one in which the mass in the infalling envelope is less than or equal to the mass of the protostar, in the standard model the boundary between Class 0 and I occurs at the time when the CEW reaches the outer boundary of the cloud (at which point the self-similar model breaks down).

There have been numerous revisions and refinements to the model, but the original formulation is still widely used. There are a number of reasons why this model is attractive:

(i) The solution is self-similar with only two free parameters; the local sound speed and the age of the collapse.
(ii) The description naturally explains the distinction between pre-protostellar, Class 0 and Class I sources.
(iii) The initial conditions are consistent with what is expected for a thermally supported, gravitationally bound sphere of gas.
(iv) The numerical solution is simple, and easy to apply/test.
(v) Many of the predictions of the model are compatible with observations.

There are however, some significant drawbacks and limitations, including:

(i) The initial SIS configuration is unphysical and highly unstable; collapse probably initiates before a cloud quasi-statically contracts to a SIS.
(ii) The self-similarity of the solution implies that all protostars should follow a similar evolutionary track. This means that one would expect the locus of points on a luminosity-age or luminosity-temperate diagram to lie on a line. We return to this subject in the context of episodic accretion below.

4.2.1 Magnetically Controlled Collapse and Ambipolar Diffusion

A possible resolution to some of the limitations of the inside-out collapse model could be that star-forming regions could be magnetically, rather than thermally, supported against collapse.

We can define a magnetic critical mass, similar to the thermal critical mass defined above, below, in which the increased pressure from magnetic field compression would resist and inhibit collapse;

$$M_\varphi = \frac{(\text{const})\varphi}{G^{1/2}} \tag{4.15}$$

where φ is the magnetic flux density.

The stability and evolution of a molecular cloud will depend on its mass (M_c) relative to M_{crit} and M_φ (Table 4.1).

We can then speculate that many clouds could be in the second of these ranges and their evolution is determined by the physics of ambipolar diffusion. The problem with this hypothesis is that there is quite a significant body of evidence that suggests that line broadening in star-forming regions

Table 4.1 The stability and evolution of a molecular cloud depending on its mass (M_c) relative to M_{crit} and M_φ.

$M_c > M_{crit}$, $M_c > M_\varphi$	The cloud is both thermally and magnetically super-critical and could collapse on a timescale that is of the order of the free-fall timescale (typically a few times 10^6 years).
$M_c > M_{crit}$, $M_c < M_\varphi$	The cloud is thermally super-critical, but magnetically sub-critical. It can evolve, quasi-statically, as ion–neutral drift occurs—on an ambipolar diffusion timescale ($>\sim 10^7$ years).
$M_c < M_{crit}$, $M_c < M_\varphi$	The cloud is both thermally and magnetically sub-critical and is unlikely to collapse.

is predominantly thermal in nature, and the extended timescales associated with ambipolar diffusion may be incompatible with the statistics of sources in different states of evolution.

Alternatively, the cores could start off as hydrostatic entities that are at least partly supported by turbulence, which would almost certainly be magnetohydrodynamic (MHD) in nature. As this turbulence slowly decays, the core contracts quasi-statically. Such a picture may, however, be at odds with the larger-scale supersonic turbulence that pervades the interstellar medium (see Chapter 3) but it is possible, for instance, that as the supersonic turbulence decays the cores sequentially fragment until sub-sonic turbulence and thermal pressure allow for the existence of the longer-lived entities that we identify as PPCs.

4.3 The Chemical Perspective

Between the stable, supported cores and those that are super-critical and will collapse, there may be metastable cores that could be oscillating or transient. These objects also have tell-tale chemical signatures; the transient structures being traced by the (unresolved) presence of 'early-time' chemical species—such as CS—but deficient in species such as NH_3 that trace denser, more evolved gas.[12]

4.3.1 Timescales and the Importance of Chemistry

As we have seen above, a useful metric of the dynamical timescale for protostellar collapse is the free-fall timescale: $\tau_{ff} \approx 3.4 \times 10^7 / n_H{}^{1/2}$ years, where n_H is the gas number density of hydrogen nucleons (cm^{-3}). The timescale for gas-phase chemical reactions (driven by cosmic-ray ionization) is $\tau_{chem} \approx 3 \times 10^5$ years, whilst that for the freeze-out of molecules onto the surface of grains is $\tau_{fo} \approx 3 \times 10^9 / n_H$ years. The timescale for cooling (*via* the collisional excitation of low-lying molecular rotational levels) is approximately given by $\tau_{cool} \approx 3 \times 10^5\ (10^4 / n_H)$ years. The timescale for ambipolar diffusion is $\tau_{AD} \approx 4 \times 10^5\ (X_i / 10^{-8})$ years, where X_i is the fractional ionization, which is itself determined by the chemistry. Significantly, for the ionization levels and densities typical for pre-protostellar cores ($n_H \approx 10^3$–10^5 cm^{-3}), these various timescales are all of the same order of magnitude.

This has a number of implications including (i) the need to study chemical and dynamical evolution together (rather than simply 'bolting on' the chemistry to the dynamics), (ii) the need to employ multi-point (Lagrangian) as opposed to single-point 'pseudo time-dependent' chemical models of the cores, and (iii) recognizing that the active interplay between the chemistry and dynamics may be important.

An obvious implication of point (i) is that the chemical evolution may be critically dependent on the assumptions that are made about the chemical initial conditions. In this respect, astrochemical models often (indeed usually) assume a 'molecular hydrogen, plus atomic gas' starting point. However, if cores form on timescales that are shorter than that for the conversion of H to H_2 this may not be correct. The significance of this was investigated[13] and it was found that—even in the case of static molecular clouds—the time-dependence of certain key species (*e.g.* OH, NH_3, CN and HNC) are dependent on the assumed initial H:H_2 ratio. This follows, quite simply, from reactions such as

$$CH + H \rightarrow C + H_2 \tag{4.16}$$

which effectively inhibits carbon chemistry, or

$$O + H \rightarrow OH + h\nu \tag{4.17}$$

which promotes oxygen chemistry (and the subsequent lead-in to nitrogen chemistry) in atomic hydrogen-rich regions.

Point (iii) can be seen when we consider that many processes in the interstellar medium, and especially star formation, are controlled by the balance between gravity and internal pressure. The pressure can be simply due to thermal motions, as used in the determination of the structure of Bonnor–Ebert spheres (above), or it can have non-thermal (magnetic, or turbulent) components (as described in Chapter 3). Chemical processes can have a controlling role in all three of these pressure sources. Collisionally excited rotational emission from molecules (together with dust) is the dominant coolant in cold molecular gas, whilst both static magnetic and turbulent support (which is of a magnetohydrodynamic nature) depend on the ionization level, which, in turn, is controlled by chemical processes. This needs a little explanation: a partially ionized magnetized cloud may be supported by magnetic pressure; ions and electrons are effectively tied to magnetic field lines, which, if ordered, resist collapse as a result of increased magnetic pressure when the flux density rises due to magnetic field constriction in a collapsing core. However, the process of *ambipolar diffusion* (actually ion–neutral drift) can allow the bulk of the (neutral) gas fluid to move relative to the ions. The efficiency of this process is, obviously, related to the ionization level—with ambipolar diffusion allowing collapse to occur for lower ionizations. Alternatively, there may be disordered, stochastic, support from turbulence. If, as is usually the case, the magnetic field strength in the core is >50 μG (5 Nanotesla), the turbulence will be magnetohydrodynamic (MHD) – Alfvénic – rather than hydrodynamic in nature (*e.g.* see Chapter 3). The survival of MHD turbulence is also dependent on the ionization level; typically, if the ionization level falls below some critical value (which has been found to correspond to extinctions $A_v > 1$)[14] then the turbulent wave motion in the ion/electron fluids becomes decoupled from the neutral fluids and the turbulence can be efficiently damped.

Therefore, the ionization level is of crucial importance and the general significance of astrochemistry in this respect originates from the fact that the dissociative recombination reaction of molecular ions with electrons is typically much faster than the radiative recombination of atomic ions with electrons. Thus a molecule-rich cloud may have a significantly lower ionization level than would otherwise be the case.

Of course, there are a number of factors to be considered, so that, for example, the dependence of ionization with visual extinction may be quite non-linear; moving in from the edge of a cloud (with a relatively high ionization level) then, as the extinction rises, photoionization becomes less efficient and the ionization fraction falls. However, as molecules freeze-out onto grains dissociative recombinations are suppressed and the ionization can rise again.

4.3.2 The Nature of Chemical Processes

The physical conditions in collapsing molecular clouds and star-forming regions are such that the whole range of gas-phase and gas–grain interactions must be considered. As described in Chapter 1, these include gas-phase chemistry, the freeze-out of atoms and molecules onto dust grain ice mantles, surface chemistry (and perhaps bulk, solid-state chemistry) between species trapped on, or in, the ices, and the desorption (or sublimation) of ice components back into the gas phase. Our understanding of these processes varies quite considerably; the gas-phase chemistry (at least for relatively small molecular species, such as CO, H_2O, NH_3, H_2CO *etc.*) is mostly very well known, with a few queries existing over some chemical pathways. The same cannot be said of the chemistry of larger species, including the so-called complex organic molecules whose gas-phase formation and destruction pathways are poorly characterized. For freeze-out, it is fairly safe to assume that atomic and molecular species that impact bare dust grains or ice mantles stick with an efficiency of order unity. There is sound empirical backing for this, both from laboratory studies, and from the efficiency with which H is converted to H_2 in diffuse clouds. For the conditions that pertain in molecular clouds, dust grains carry a small negative charge, so that the effective sticking efficiency for (positive) atomic and molecular ions will be enhanced. There is some uncertainly how this affects molecular ions, which, in the gas phase, usually dissociate on recombination with electrons—it is possible, if not likely, that the release of bond energy is sufficient to result in the immediate desorption of the dissociation products.

There have been considerable advances in our understanding of surface chemical processes in recent years, although much of this is theoretical, or informed by laboratory analogues of dust grains employing perfect, idealized surfaces. However, the simplest approximation, that surface reactions are dominated by reduction/hydrogenation reactions is a good starting point; the high surface abundance and mobility of hydrogen atoms will allow for the rapid conversion of C, N, O and S to CH_4, NH_3, H_2O and H_2S, respectively, whilst CO can be progressively hydrogenated to H_2CO (formaldehyde) and CH_3OH (methanol). At higher densities, the H to H_2 ratio is lower and surface oxidation reactions may be important (*e.g.* $O + O \rightarrow O_2$, $CO + O \rightarrow CO_2$, $CO + OH \rightarrow H_2O$, *etc.*). There is quite an extensive literature on more complex surface reactions, much of which has been deduced or verified through experiments involving laboratory analogues. This literature is described in some detail in the book by Williams and Cecchi-Pestellini.[15]

There are numerous continuous desorption processes that operate in molecular clouds and star-forming regions. The most significant are photodesorption (by the interstellar radiation field, and by cosmic-ray induced secondary photons), cosmic ray heating (either localized 'spot' heating or whole-grain heating), and enthalpy of formation driven desorption (from the energy released by the formation of H_2). Other desorption mechanisms include sudden, whole mantle sublimation as a result of chemical explosions, driven by trapped radicals, and cosmic ray sputtering of ices.

The efficiencies of these processes depend on a number of factors, such as the binding energy of the molecules and the energy spectra of the radiation field and the cosmic rays. It is also important to note that some processes (*e.g.* cosmic ray heating and mantle explosions) affect the bulk of the ices, whereas others (*e.g.* photodesorption) affect only the surface layers of the ices—in which case the desorption efficiencies very much depend on the layering structure of the ices.

Added to this list we must also include episodic desorption, which may result in the sudden, and often complete desorption of the grain ice mantles. Processes that are included in this category are rapid thermal sublimation (following, for example, the rapid turn-on of a stellar source or an episodic accretion event), grain–grain collisions and shocks. In addition, there is an extreme variant of the enthalpy-driven desorption discussed above in which sudden, catastrophic, recombination of hydrogen atoms trapped in the ice to form H_2 results in whole ice-mantle explosions and a rapid instantaneous gas-phase chemistry (so-called 'rapid radical' association). However, despite the various uncertainties in these mechanisms, unlike the continuous desorption processes, the chemical consequences of episodic desorption tend to be easier to quantify as they are generally not species-selective; all molecules are liberated with $\sim$100% efficiency. A significant source of uncertainty in all of these mechanisms is that the products of desorption are highly uncertain, both for continuous and sudden desorption mechanisms.

4.4 Chemistry Controlling Collapse

So far, in our discussion (and in most of the literature on the subject), we have limited ourselves to hydrodynamically evolving, thermally supported clouds, but we know that magnetic pressures may be very significant in potential star-forming cores. Indeed, morphological surveys of dense cores tend to show that they have an average axial ratio of ~ 0.5–0.6 and that those that are part of filamentary structures (such as L1498, B35, L1251, L1535) are reasonably (within $\sim 20^\circ$) well-aligned with each other. This could be taken as evidence for magnetic support and would be consistent with theoretical models,[16] which suggest that magnetic field strengths as low as 30 μG (*i.e.*, 3×10^{-9} T) could support clouds as large as 5×10^5 $M_\odot$ against collapse. Unfortunately, it is extremely difficult to determine the field strengths at this level with any degree of accuracy so that the hypothesis remains largely untested.

As argued above, molecular rich gas can have a relatively low ionization level as compared to atomic gas in similar physical conditions. However, the ionization structure of clouds is highly modified in regions depleted of molecules. In a region of high molecular depletion the dominant ion is probably H_3^+, or one of its isotopologues, but the recombination rate for these is low and therefore the ionization level rises.

We have already seen how the ionization level affects the efficiency of both ambipolar diffusion and the damping of MHD waves and so—in a magnetically sub-critical cloud—we have to consider the relative timescales for depletion and these two processes. Specifically, if depletion occurs more rapidly than ambipolar diffusion, then a core may be stable against collapse.

4.4.1 MHD Wave Damping

Turbulence provides support *along* magnetic field lines, and is probably composed of a superposition of circularly polarized Alfvén waves.[17] The dissipation/damping rate of these long wavelength waves varies by several orders of magnitude in a typical cloud when the ionization level changes from 10^{-5} to 10^{-7}, as the ion waves decouple from the static neutral fluid and ion–neutral friction becomes very large. As discussed above, the ionization level is sensitive to the abundance of molecular ions, such as HCO^+. Thus, above a certain level of visual extinction (yielding a lower ionization level), the rate of damping of the turbulence is such that the waves cannot be maintained at sufficient amplitudes to support the core. This interconnection between the chemistry and the dynamics of a cloud where MHD turbulent support is significant has yet to be fully investigated. However, some simple calculations were performed in an astrochemical model[14] and the critical extinction for sharp changes in the ionization level was found to be $A_v \approx 2$–3. Such values are consistent with the observations that this range of extinctions appears to mark the boundary between starless and star-forming cores.

4.4.2 Ambipolar Diffusion

If (static) magnetic pressure—which can possibly be dissipated *via* ambipolar diffusion—is important in an otherwise thermally gravitationally supercritical core, then we can envisage a situation in which a super-critical, hydrodynamically collapsing, inner core is slowly fed with material from a quasi-statically evolving magnetically sub-critical envelope. Thus, ambipolar diffusion gradually (on a timescale that can be an order of magnitude longer than the free-fall timescale) results in the increase of the mass to magnetic flux ratio in the central regions until it becomes thermally and magnetically supercritical, and hydrodynamic collapse can ensue.

There have been relatively few attempts to model the chemistry in magnetically sub-critical clouds in a self-consistent fashion. Recent studies[18,19] have used multi-point MHD models of a collapsing pre-protostellar core, including ambipolar diffusion processes, moderated by chemically controlled ionization levels. The treatment of ambipolar diffusion was somewhat simplified (to 1D geometry) but, making the realistic assumption of flux-freezing, it was found that, even for initial magnetic field strengths of $\sim 10\ \mu G$ ($\sim$ one Nanotesla), a core would go through an initial collapse phase, followed by magnetic pressure-driven re-expansion, then a terminal collapse subsequent to the relaxation of magnetic flux density in the central regions by ambipolar diffusion. These collapse dynamics are shown in Figure 4.2.[18]

The dynamics are further complicated by the fact that in each phase of the evolution there are parts of the cloud that are infalling, whilst other parts are simultaneously expanding. The studies have also indicated that the charged dust grain fluid is an important component of the ambipolar diffusion formalism and that significant dust–gas streaming may occur—resulting in reduced dust-to-gas ratios in the central regions.

Although these effects may be critically important in defining the stability and dynamics of a core, the observational tracers may be very hard to detect; the models predict small ion–neutral streaming velocities, and also shifts in the self-absorption dips in the infall line profiles. To detect and diagnose these will require the full resolution capabilities of modern facilities.

4.5 Chemistry Diagnosing Physics

As described above, an observational survey of spectral emissions lines towards a star-forming core (several species, in multiple transitions, observed at various offset lines of sight) has the power to identify and diagnose both the dynamical and chemical status of an infall source. However, to do this we need to recognise several facts and complications:

i. The low-lying rotational levels of key tracers (including species such as HCO^+, H_2CO, N_2H^+, CN, HCN, CS, C_2S, *etc.*) are usually optically thick, and observations of higher transitions (which are optically thin) give a useful measure of the rest-frame velocity.

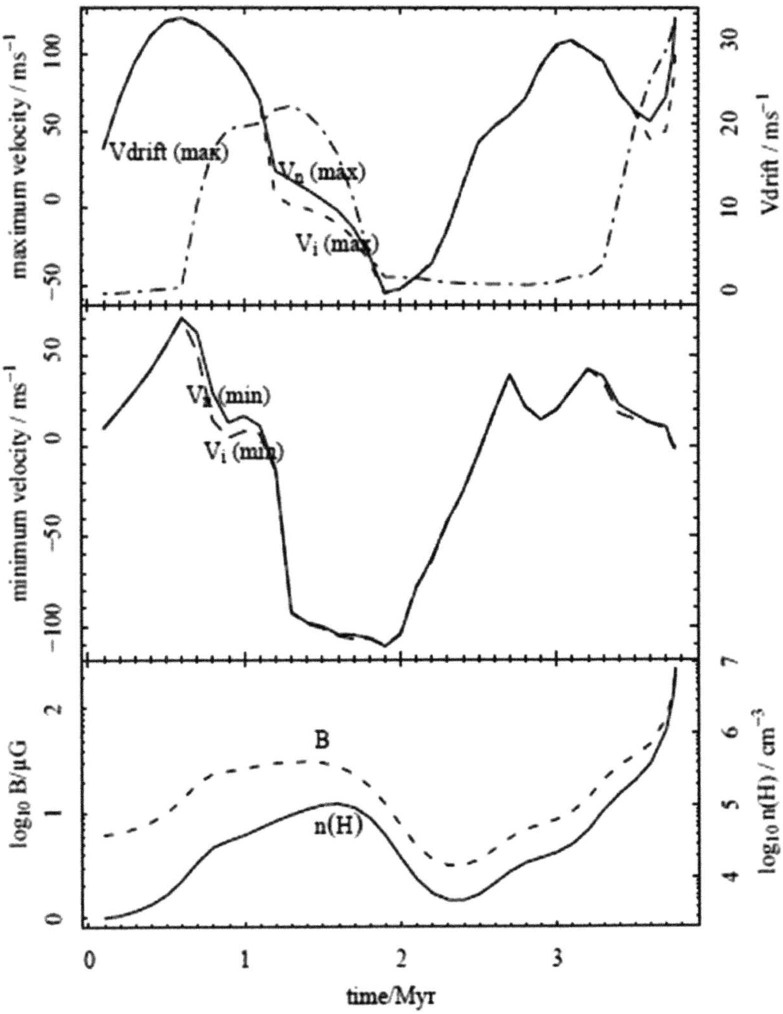

Figure 4.2 The dynamical evolution and eventual collapse of an initially magnetically subcritical core. The top panel shows the maximum ion (dashed line) and neutral (solid line) fluid velocities (scale on the left hand side) and the dot-dashed line shows the drift velocity (scale on the right hand side). The middle panel shows the minimum ion and neutral fluid velocities, and the lower panel shows the evolution of the magnetic field strength and density in the centre of the core.
Reproduced from Stace and Rawlings[18] by permission of Oxford University Press on behalf of the Royal Astronomical Society.

ii. The systemic velocities are small (~a few km s^{-1}), so that most of the cloud is well radiatively coupled, and a careful radiative transfer analysis is required.
iii. There is always the potential for confusion with other dynamical processes, such as outflows and/or rotation. Bipolar outflows are

omnipresent and can confuse the interpretations of the line profiles in the context of infall kinematics.

iv. The chemical evolution may be critically dependent on the evolutionary 'pre-history', which may, or may not, be quasi-static) prior to collapse.

In addition to this, observations of the continuum emission from the source can be used to provide a snapshot of the density and temperature profiles within the cores providing we have good knowledge of the dust opacity and make the assumption that the dust-to-gas ratio is well defined and constant.

Early modelling efforts either simply adopted diffuse cloud initial conditions, or a chemical enrichment—invoked as the cause of enhanced abundances of species such as NH_3. The enrichment is derived from a model that considers the dynamical cycling of material with collapse from an inter-core medium, chemical evolution in the core, and then ablation in turbulent boundary layers (due to the winds produced from newly forming stars). Hydrogenated molecules, such as NH_3, are formed on grains and released into the gas phase by ice mantle sputtering by shocks and thermal sublimation. This material is transported in rapid (modified free-fall) collapse to a quasi-static (isothermal sphere) configuration, followed by the 'standard' inside-out collapse. Although somewhat simplistic, and interpreted *via* a very basic optically thin large velocity gradient (LVG) radiative transfer treatment, the studies (utilizing multi-point representations of the flow) were able to identify CH, HCO, HCO^+, N_2H^+, HNO and H_2S as 'depletion-enhanced' species and that, in the case of the test object, L1498 (a starless core) the chemical signatures indicate that it is chemically 'young' and quasi-statically contracting.[20]

Rather than attempt to summarise the numerous, disparate, observational and theoretical studies of star formation, in this section we instead focus on the astrochemistry of two star-forming regions in different evolutionary phases: L1544 (a starless, pre-protostellar core) and B335 (a Class 0 protostar).

4.5.1 The Quasi-Statically Contracting Starless Core: L1544

As indicated above, there is considerable debate as to the predominant mode of star formation—even in the case of relatively isolated cores. There is a broad division between (i) those hypotheses that envisage star formation as a continuous, dynamical process—where the assembly/accretion of the protostellar envelope occurs at the same time as ongoing collapse so that there are no 'metastable' evolutionary states, and (ii) a multi-stage process, where the formation of the envelope is followed by a period of quiescence and dynamical relaxation, followed by quasi-static contraction and eventual collapse, once super-criticality is achieved. Alternatively, we can distinguish between the paradigm of dynamic/turbulent driven collapse—in which the compressible turbulence rapidly dissipates allowing prompt collapse, and the paradigm of slow collapse *via* quasi-static evolution, perhaps mediated by ambipolar diffusion processes.

There is plenty of evidence for the former, *e.g.* the presence of filamentary accretion streams, the presence of supersonic turbulence in molecular clouds, *etc.* However, the starless core L1544 is an extremely cold and very dense, isolated cloud in Taurus that seems to fall into the second category—and shows evidence for slow, quasi-static contraction. It has no internal heat/luminosity source and the only turbulence that is present is sub-sonic, resulting in only marginal additional broadening to the line profiles. In addition, it is nearly spherical (with an aspect ratio of ~ 1.5)—so that it can be represented by a cold core, well-described by the Lane–Emden equation (slightly modified for non-isothermality)—with a central density of $\sim 10^7\ \mathrm{cm}^{-3}$, externally heated by the interstellar radiation field and cosmic rays, and bounded by a photon-dominated region. Thus, modelling of the dust continuum of this, and other PPC sources,[21] using the 450/850 μm and 1300 μm intensity profiles obtained with SCUBA on the JCMT self-consistently constrained the density and temperature profiles, which were found to closely resemble those expected for a Bonnor–Ebert sphere, modified for a small temperature gradient (~7 K in the centre, $\sim$13 K at the edge). Therefore, unlike molecular clouds, whose structure is generally poorly determined and possibly dominated by a clumpy sub-structure, L1544 provides a near-unique laboratory in which to test astrochemical models.

This is important, as much of the recent progress in astrochemical star-formation studies has been driven by a paradigm shift in our understanding of fundamental chemical processes.

Traditionally, it was believed that astrochemistry was completely dominated by ion–molecule reactions, whereas the current thinking is that neutral–neutral (radical) and grain surface chemistry reactions play very significant roles, especially in the very cold and dense conditions that pertain in star-forming cores. This can be illustrated by the very simple case of H_2O whose chemistry has been studied in L1544.

The 'traditional' (cosmic-ray driven) gas-phase chemistry of H_2O formation in dark clouds is very straightforward and is essentially driven by a network of ion–neutral reactions:

$$\mathrm{H_2 + crp \rightarrow H_2^+ + e^-} \tag{4.18}$$

$$\mathrm{H_2 + H_2^+ \rightarrow H_3^+ + H} \tag{4.19}$$

$$\mathrm{H_3^+ + O \rightarrow OH^+ + H_2} \tag{4.20}$$

$$\mathrm{OH^+ + H_2 \rightarrow H_2O^+ + H} \tag{4.21}$$

$$\mathrm{H_2O^+ + H_2 \rightarrow H_3O^+ + H} \tag{4.22}$$

$$\mathrm{H_3O^+ + e^- \rightarrow OH,\ \mathbf{H_2O}} \tag{4.23}$$

(where crp represents a cosmic ray proton). However, this would suggest that in the central regions of cold cores, where the ionization level is low and the freeze-out of molecules onto cold grains is efficient, it would be expected that H_2O would be strongly depleted from the gas phase.

Recent observations have shown that this is not the case and, in particular, observations of the H_2O (1_{10}–1_{01}) line made with the HiFi instrument on Herschel (together with other molecular observations) have prompted a thorough appraisal of the chemistry and evolution of this source. This has not been easy due to the difficulty in non-LTE radiative modelling of low-level transitions in H_2O, which have sub-critical excitation rates, but some interesting results have been found—both relating to the chemical processes and also concerning the dynamics of the source.

For the reasons given above, the adopted physical model is one of slow contraction in quasi-static unstable equilibrium—starting from a critically unstable Bonnor–Ebert sphere. This has had remarkable success in the modelling of the line profiles and, moreover, it has been found that the chemistry of H_2O can be described by a remarkably simple 'lite' network of reactions:[22]

$$\mathrm{O + H_2 \rightarrow OH + H} \tag{4.24}$$

$$\mathrm{OH + H,\ H_2 \rightarrow H_2O + H} \tag{4.25}$$

$$\mathrm{OH} + h\nu \rightarrow \mathrm{O + H} \tag{4.26}$$

$$\mathrm{H_2O} + h\nu \rightarrow \mathrm{OH + H} \tag{4.27}$$

$$\mathrm{O,\ OH,\ H_2O + grain \rightarrow H_2O_{(ice)}} \tag{4.28}$$

$$\mathrm{H_2O_{(ice)}} + h\nu \rightarrow \mathrm{O,\ OH,\ H_2O.} \tag{4.29}$$

The last reaction (desorption processes) can be expanded to include cosmic-ray heating, photodesorption (direct and cosmic-ray induced), and enthalpy driven desorption. This network of reactions completely bypasses most of the gas-phase chemistry, together with the surface chemistry of oxygen-bearing molecules.

The observed line profiles and model fits are shown in Figure 4.3a[24] and b,[23] together with the results from (i) the 'lite' chemistry, and (ii) a full chemical model.[23]

As can be seen, there is an excellent fit between the modelled and observed line profiles. Equally striking is the strong similarity between the results obtained from the 'lite' and the full chemical models. The latter was a full, multi-point model, steered by the physical models of quasi-static core contraction and incorporating a comprehensive description of the chemistry, which includes gas-phase reactions, freeze-out, surface chemistry, multiple desorption mechanisms (and a description of ice-layering) with a photon-dominated region at the outer boundary.

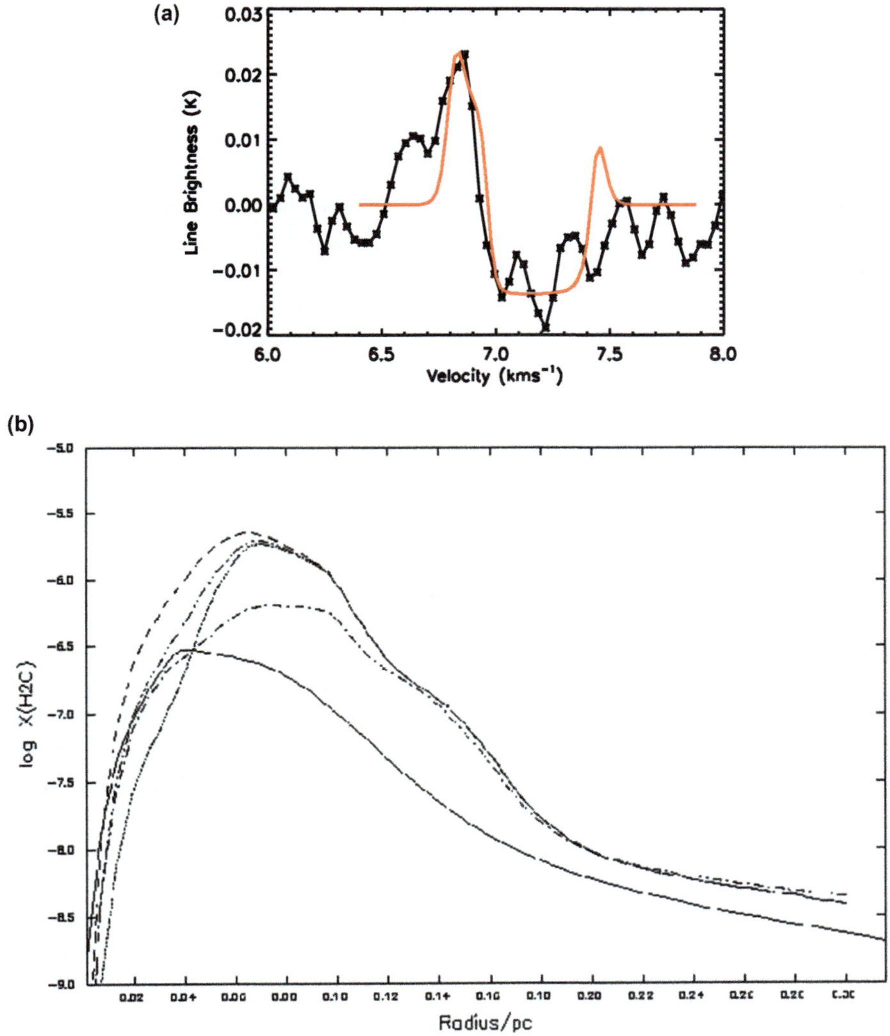

Figure 4.3 (a) The observed H_2O (1_{10}–1_{01}) line profile (black lines) towards L1544 and the modelled profile (red line) for an unstable Bonnor–Ebert sphere in quasi-static contraction. Reproduced from Figure 4.5 in The dynamics of collapsing cores and star formation, Keto, Caselli and Rawlings, *et al.*, *Mon. Not. R. Astron. Soc.*, 2015, **446**,[24] by permission of Oxford University Press on behalf of the Royal Astronomical Society. (b) The modelled abundance profile of H_2O in L1544 as a function of offset from the centre of the core. The lower (solid) line shows the results obtained from the 'lite' chemical network, whilst the dashed lines show the results obtained from a full gas–grain chemistry with several different assumptions about the cosmic ray flux, the desorption efficiencies and the assumed gas-phase rate coefficients.[23]

The chemistry of H_2O in the outer parts of the cloud tends to be dominated by the regular interstellar network given above, but since both the surface chemistry and the gas-phase chemistry have comparable efficiencies, either can give a good representation of the chemical behaviours.

Since the simulation covers a large range of physical conditions ($1 < A_v < 100$) this is a fascinating result—and implies that uncertainties in individual chemical reaction rates are effectively 'smoothed out', thus mitigating against errors in interpretation. Moreover, for the very early stages of the evolution, when the dynamical timescale is very much larger than the chemical timescale, the chemical evolution is itself quasi-static and the notion of using chemical abundance ratios as a chemical 'clock' has to be discarded.

An extension of the studies to consider the $C^{18}O$ ($1 \rightarrow 0$) line, which has a much lower critical density ($\sim 10^5$ cm^{-3}, as opposed to 10^8 cm^{-3} for the H_2O line), allows a kinematic comparison of the inner and outer parts of the envelope. Considering various scenarios, including a Larson–Penston flow, inside-out collapse and non-equilibrium Bonnor–Ebert spheres showed that only the contraction of a quasi-static critical Bonnor–Ebert sphere can explain the observations and that even a very marginal over-density of 10% would result in anomalously broad line profiles emanating from the outer envelope.[24] This is an important result as it implies that that L1544 *must* be evolving quasi-statically and is not the product of a prompt, non-equilibrium accretion process. It is very difficult indeed to see how objects like L1544 can be so close to what is expected for a thermally balanced quasi-statically contracting core unless that is indeed what they really are. This, then, poses a serious problem for star-formation theory; how is it possible for such quasi-equilibrium structures to be created in a dynamic/turbulent ISM?

An obvious possibility is that these results are consistent with the notion that the transition from a stable to an unstable contracting Bonnor–Ebert sphere is triggered by the gradual decay of turbulent support, as postulated above. However, this is something that needs further investigation before we are able to make definitive statements concerning the mode of collapse.

4.5.2 Detection and Characterizing Infall Associated with Star Formation

We have discussed how SEDs can be used to establish an evolutionary sequence in the early stages of star formation, but their diagnostic power is limited and they are strongly dependent on the assumed dust opacity law, and the source geometry/size. To gain further insight we need to understand the structure of individual cores.

Before it was possible to obtain, and analyse, high resolution molecular spectra, the main source of information came from molecular maps of potential star-forming clouds. This morphological approach—which is typically applied to the transitions of species such as CO (and its isotopologues),

NH_3, N_2H^+, CS, *etc.*—relies on the transitions of these species having different 'critical densities' (that is to say, the density above which the level populations are thermalized). Thus, for example, the critical densities for the $C^{18}O$ 2.7 mm ($J=1\rightarrow0$), NH_3 1.3 cm (J, K)=1,1, CS 3.0 mm ($J=2\rightarrow1$) and N_2H^+ ($J=3\rightarrow2$) transitions are $1.9\times10^3\ cm^{-3}$, $1\times10^4\ cm^{-3}$, $5.8\times10^5\ cm^{-3}$ and $3\times10^6\ cm^{-3}$, respectively. So, in principle, different transitions can be used to trace regions of different density and molecular 'onion skin' structure may be observed. However, the procedure is complicated by excitation/optical depth effects, abundance variations, molecular freeze-out onto cold dust grains, and the presence of a (unresolved) clumpy structure.

In fact, only spectral line profiles can reveal the kinematic information, and only through multiple observations of different transitions along different lines of sight is it possible to break the degeneracies in the free parameters.

One of the great challenges in modern star-formation studies has been the unambiguous, direct, detection—and characterization—of the infall associated with star formation.

As discussed above, the earliest stages of the star-formation process occur in very cold (~7–10 K) and dense ($n_H > 10^4\ cm^{-3}$) gas, so that the only reliable dynamical tracers are the low-lying rotational transitions in molecular species. As molecular clouds are chemically rich and have high densities, the emission lines from many of these molecular species are often very optically thick. If certain conditions are met, the spectral line profiles may show a characteristic signature or 'fingerprint' of infall—an example of which is shown in Figure 4.5.[30]

As can be seen in this figure, the line profile is self-reversed, with strong absorption at the systemic rest velocity, and possesses asymmetric emission wings, which (somewhat counter-intuitively) are stronger in the "blue" (shorter wavelength) than in the "red" (longer wavelength). The theory of the formation of this infall signature has been described in Zhou/Zhou *et al.* (1992,1993)[25,26] and can be summarized, briefly, as follows.

Consider a cloud that possesses a static envelope, surrounding spherically symmetric infalling gas, that has a velocity profile that increases towards the centre of the core, a temperature profile that also increases from the edge towards the central region, which is heated by the luminous protostellar source, and an abundance of a molecular species that has a population in rotationally excited states. If a given rotational transition is so optically thick, and the velocity gradient sufficient, then the infalling and outflowing hemispheres (*i.e.* those parts that are furthest from and closest to a distant observer, respectively) can be radiatively decoupled. The observer then sees, preferentially, emission coming from that part of each hemisphere that is closest to her/him. For the blue-shifted hemisphere that will be the regions closest to the core, which are warmest and more highly excited, whereas for the red hemisphere that will be the outer parts that are cooler and less well-excited. Hence the result shown in Figure 4.5 that the "blue" peak is stronger than the "red". The self-absorption originates from the cool, static outer envelope.

This, rather simple, analysis of the line profile makes the implicit assumption that the molecular species in question is abundant and uniformly distributed throughout the protostellar cloud. However, we know that molecular abundances vary both in time and space—partly due to time-dependences in the chemistry, and partly due to the responses to changes in physical conditions. Moreover, these changes can be very far from linear. However, this additional, complicating, factor can be turned to our advantage; although the chemical behaviours are non-linear, they are certainly not random and different species/transitions can have different behavioural trends. Thus, one species may be suppressed in regions of high density (perhaps as a result of freeze-out onto grains) whereas another may be enhanced.

As an example of how this works, consider the case of HCO^+: a major channel for the formation of this species in dark cloud environments is the reaction

$$H_3^+ + CO \rightarrow HCO^+ + H_2. \quad (4.30)$$

Whilst the abundance of CO falls due to freeze-out/depletion, this reaction can be the main loss route for both CO and H_3^+, so that, at the same time, the abundance of H_3^+ rises and the net formation rate of HCO^+ is relatively constant. HCO^+ is lost by dissociative recombination with electrons. Thus, in denser/darker regions of lower ionization, the loss rate for HCO^+ falls. In addition, in gas where H_2O is abundant, reactions of molecular ions with H_2O tend to dominate over dissociative recombination as the main loss channels for molecular ions. Thus, for HCO^+:

$$HCO^+ + H_2O \rightarrow H_3O^+ + CO. \quad (4.31)$$

As the abundance of H_2O declines with increasing depletion, so the abundance of HCO^+ *increases* until it too suffers depletion due to freeze-out. The overall effect of this is a period of enhanced abundance, a so-called *depletion enhancement*. The observational manifestation of this enhancement will be preferential emission from the inner, depleted, regions of the flow, which are relatively fast-moving, yielding significant line profile broadening.

This was investigated by Rawlings and Yates (2001)[27] and an example of this is shown in Figure 4.4.[27]

This shows the results obtained from a full chemical model as described below that mimics the conditions in the Class 0 source B335 in the context of the inside-out collapse model, preceded by quasi-static contraction from diffuse cloud conditions. The features described above are clearly evident, so that small variations in the values of the free parameters (*e.g.* the net sticking coefficient for freeze-out) can have dramatic effects on the shapes of the spectral line profiles.

We can therefore identify three epochs of chemical activity: (i) early chemistry, mostly dominated by gas-phase reactions, (ii) the early depletion phase—as described above—when differential depletion drives extreme

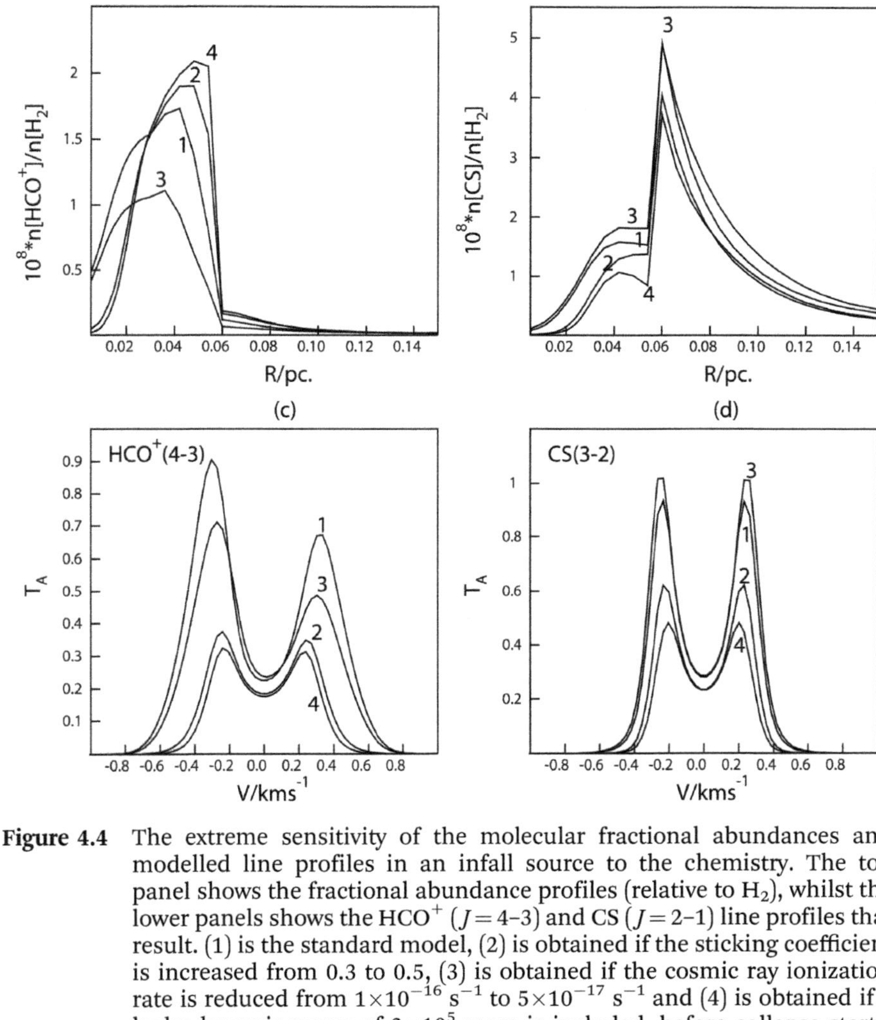

Figure 4.4 The extreme sensitivity of the molecular fractional abundances and modelled line profiles in an infall source to the chemistry. The top panel shows the fractional abundance profiles (relative to H_2), whilst the lower panels shows the HCO^+ ($J=4$–3) and CS ($J=2$–1) line profiles that result. (1) is the standard model, (2) is obtained if the sticking coefficient is increased from 0.3 to 0.5, (3) is obtained if the cosmic ray ionization rate is reduced from 1×10^{-16} s^{-1} to 5×10^{-17} s^{-1} and (4) is obtained if a hydrodynamic pause of 2×10^5 years is included, before collapse starts. Reproduced from Figure 4.3 in Modelling line profiles in infalling cores, Rawlings and Yates, *Mon. Not. R. Astron. Soc.*, 2001, **326**,[27] by permission of Oxford University Press on behalf of the Royal Astronomical Society.

variations in gas-phase abundances, and (iii) the late depletion phase, when abundances are largely determined by the balance between freeze-out and desorption processes. Obviously, the duration of the chemically interesting second epoch will be limited. Moreover, the expected physical evolution of Class 0 sources only allows a relatively narrow window of opportunity for the detection and diagnosis of infall; the earliest stages are very close to isothermal and so the line profiles will lack much of the infall asymmetry, whilst in the later stages the infalling gas will be warm enough to allow ice mantle sublimation to occur, further confusing the interpretation of the observations.

In principle, therefore, it is possible to plan observational programmes to map the line profiles of different transitions of different molecular species at different offsets from the source centre (and possibly with different telescope resolutions) to map the dynamics and chemical evolution of a star-forming core. In effect, the varying chemical behaviours delineate the dynamical equivalents of 'critical densities'.

4.5.3 The Class 0 Source B335

There are, of course, a number of caveats to this approach; as discussed above, due to the similarities of the chemical and dynamical timescales, the precise behaviours will depend on the assumptions made about the initial conditions, and the early dynamical evolution. A sophisticated computational approach is also required; the simple one-point pseudo-time-dependent models that are applied to many astrochemical problems are not suitable and, instead, a multi-point Lagrangian flow model is necessary. Moreover, high spatial resolution is necessary to elucidate the density and dynamical profiles. As a result, this potentially very powerful procedure has yet to be fully utilized other than in a few, specific sources.

A notable example, and very much the prototype for astrochemical studies of protostars, is B335. The dark (Bok) globule, B335, which harbours the low mass protostar IRAS 19349 + 0727, was chosen as the test case for a 'classic' Class 0 infall source for a variety of reasons; it is a fairly spherical core that, until recently, was believed to be located $\sim$250 pc away, which is apparently evolving towards a solar-type star. At the assumed distance of 250 pc its mass is 0.42 $M_\odot$ and its luminosity is 3 $L_\odot$ (Zhou *et al.*, 1993,1994)[26,28]—although see below. Extensive observations (reddening maps, continuum dust emission maps) and modelling have yielded well-defined density and temperature structures that are consistent with the spherically symmetric thermally supported model of core formation. In addition, there is very little evidence for non-thermal line broadening (other than that caused by the systemic dynamics) so that the turbulent velocity field is sub-sonic.

As with all Class 0 sources, bipolar outflows are present and dominate the kinematics. However, the outflow jets are almost in the plane of the sky. So, most significantly, it was the first source for which the envelope components were disentangled and found to show line profiles (in H_2CO $J=2_{12}-1_{11}$, $3_{12}-2_{11}$ and CS $J=2-1$, 3–2, 5–4) that unambiguously exhibit the 'infall fingerprint' and that are consistent with the standard (inside-out) collapse model.

The models were further constrained by (JCMT beam-averaged) fractional abundance determinations for H_2CO, CS and NH_3 (4.6×10^{-9}, 5.5×10^{-9} and $0.5-2\times10^{-9}$, respectively) and initially analysed in a scenario similar to that applied to L1498, as described above. With a limited set of observational constraints it is inevitable that there are some degeneracies in constraints on the values of the (many) free parameters, but two solutions were found: (i) prompt collapse of a cloud embedded in a larger cloud complex with an extinction of $A_v \approx 5.5$, or (ii) delayed collapse (delay $\sim 1/2$ Myr), with an external extinction of $A_v \approx 7.0$. Although not conclusive, the results from this modelling seem to prefer fairly prompt, but slowly evolving collapse modes.

The conversion of molecular abundance profiles (coupled to the physical structures of the infall) to spectral line profiles that can be compared directly to observations requires a careful approach that recognises that much of the flow inside young protostellar clouds is sub-sonic and that simple LVG techniques are invalid. Instead, a variety of Monte Carlo and accelerated lambda iteration (ALI) codes have been developed and applied in the synthesis of both single dish and interferometric observations.

Using this combined astrochemical, dynamical and radiative transfer approach it was possible to constrain key parameters. The earliest studies made no attempt to model the chemical evolution and adopted spatially invariant abundances for the molecular species, which were left as free parameters and constrained by the modelling. In addition, it was assumed that the gas and dust are well coupled (so that $T_{gas} = T_{dust}$ throughout). Despite these major simplifications, the models had some success and derived the location of the 'infall radius' (the location of the collapse expansion wave) in the context of the 'standard' model. Nevertheless, the simplifications and assumptions meant that this could only be regarded as a result that is consistent with, rather than validating, the infall model.

The next step was to use B335 as a test bed for full astrochemical modelling of an infall source (Evans *et al.*, 2005).[29] This model still assumed the validity of the Shu (1977)[8] model, whose free parameters were constrained by the earlier modelling. The physical framework was further constrained by their modelling of the dust continuum emission. Dust emission has the advantage of being optically thin, not subject to 'depletion' and the only free parameters are the temperature and opacity law. There are, however, some assumptions—notably that the opacity law and dust-to-gas ratios are well defined and spatially/temporally invariant.

These assumptions would not hold (for example) if there were significant aggregation of dust grains, strong variations in the amount of ice coverage, or gas–dust streaming—as would be the case if strong magnetic fields were present.

However, the observational evidence suggests that—for this source at least—the idealized situation is reasonably accurate. With these assumptions it was then possible to determine the radial profiles of the density and (dust) temperature. This was then used to model the gas temperature profile. A model of the time dependence of the source luminosity was then used to determine the time dependence of the temperature profiles.

With this 'preliminary' modelling completed, the only free parameters are then the abundance profiles for the various molecular species. A large number of molecular observations were modelled with a Monte Carlo radiative transfer model, and represented by two different chemical models: (i) where the abundance profiles are described by two-value step functions, and (ii) where the chemistry is modelled self-consistently, starting from a sequence of contracting Bonnor–Ebert spheres (similar to that described for L1544 above) followed by the inside-out collapse. Obviously, (i) lacks any self-consistency and has a large number of free parameters (three per species), whilst (ii) only has a few free parameters. However, both models had some degree of success and reasonably good fits to the observed line profiles were obtained.

The chemical modelling was somewhat crude (and, for example, did not include any surface chemistry or detailed desorption mechanisms)—and was not informed by the sort of advances that were applied to the modelling of L1544, but it provided a proof of concept that will, no doubt, be expanded upon and utilized as high resolution observations and advances in our understanding of astrochemical processes progress. However, there were some very definite results: (i) the observations cannot be fit by the Larson–Penston series of infall models, which predict line profiles that are too broad, (ii) the adopted quasi-static evolution, plus inside-out collapse seems to work well for this source, with an 'infall radius' of $\sim$0.03 pc (at the assumed distance of 250 pc), (iii) there is some possible evidence for thermal sublimation of species such as H_2CO in the innermost regions by the central source, (iv) there are some kinematic features that are unexpected from the standard model, which suggests that it may not be wholly accurate. This is supported by the SED and continuum mapping, which suggest a slightly different (flatter) density profile than would be expected from the standard inside-out collapse model.

Most recently, B335 has been observed at high resolution with ALMA and the results analysed in the context of models similar to the ones described above (although not utilizing a full model of the position- and time-dependent chemistry).[30] The interpretation of the data had to be renormalized in the light of recent observations,[31] indicating that B335 is significantly closer ($\sim$100 pc) than had been previously believed. Following this revision, an infall radius of 0.012 pc, a collapse age of 5×10^4 years (consistent with the inferred dynamical age of the outflow) and a source luminosity of 0.72 $L_{\odot}$ were deduced. At high resolutions, the observations are effectively a 'pencil beam' and reveal some subtle additional features, such as the presence of slightly redshifted *absorption* against the continuum source. Despite the simplicity of the chemical modelling and assumption of purely spherical symmetry, pretty good agreement with the observations of HCO^+ and HCN (observed on-source and at offset positions) was obtained—see Figure 4.5. Other interesting results were the apparent absence of any significant rotation and the fact that nearly all the matter within the inner 25 au has accreted onto the central protostar. In addition the presence of complex organic molecules, such as CH_3CHO (acetaldehyde), $HCOOCH_3$ (methyl formate), NH_2CHO (formamide), CH_3OCH_3 (dimethyl ether), C_2H_5OH (ethanol), C_2H_5CN (ethyl cyanide) and CH_3COCH_3 (acetone) in the inner few tens of au indicate the presence of a 'hot corino'.[32] The fact that the line profiles are all broad ($\Delta v > 5$ km s^{-1}) is indicative of them originating from inflowing material.

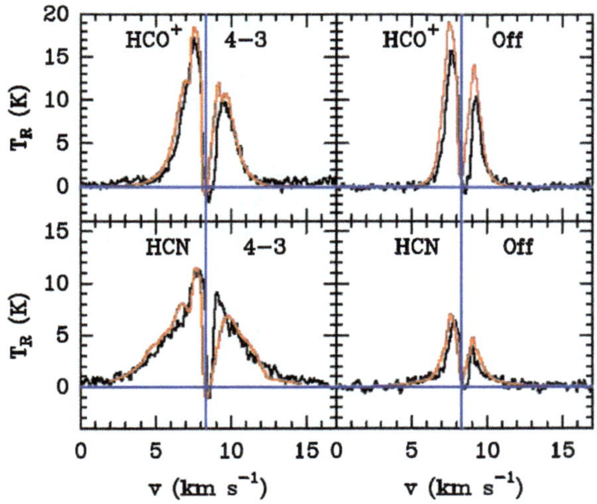

Figure 4.5 The observed (black lines) and modelled (red lines) line profiles for HCO^+ and HCN towards the Class 0 source, B335. The profiles on the left hand side are for centred spectra, whilst those on the right are averaged spectra for offsets of 0.5'' north and south of the peak. The blue vertical lines indicate the systemic velocity.
Reproduced by permission from the AAS from Evans *et al.*[30]

Overall, these models all indicate the relative importance and duration of the different stages of chemical and dynamical evolution:

(1) The ambient cloud material from which star-forming cores form.
(2) Evolution to a bound 'pre-protostellar' state.
(3) The possibility of a period of static chemical evolution.
(4) The isothermal collapse phase ($n_H \approx 10^4$–$10^8\,cm^{-3}$) when the gas-phase, freeze-out and continuous desorption processes occur.
(5) Non-isothermal, late-stage, collapse, where the chemistry is dominated by sudden ice mantle evaporation.

Each of these phases must be carefully considered and incorporated into any full astrochemical description of the early phases of star formation.

Astrochemical models are now entering a phase of unprecedented complexity, with adaptive mesh refinement techniques being used to describe systems of arbitrary complexity and geometry, coupled to chemical models that include a range of gas-phase, gas–grain and surface processes and have the ability to discriminate between the various components of star-forming regions (*e.g.* the central core, the outflow, the accretion disc and the envelope).[33] However, as the complexity increases, so does the number of free parameters and the potential for degeneracy. Thus, the advances in modelling complexity and the constraints imposed by many high-resolution observations must go hand in hand.

4.6 Later Stages—'Hot Corinos'

Once a protostellar accretion source has been formed, the inner parts of the flow are heated and then ice mantles are sublimated, resulting in the so-called 'hot corinos'.

The popular image of star formation is one of a dimly luminous collapsing cloud that suddenly 'flashes into life' as a bright main-sequence star is formed. This is, of course, quite incorrect. The luminosity of a protostar is driven solely by accretion, so that

$$L = \frac{GM_*(t)\dot{M}}{R_*} \tag{4.32}$$

where M_* and R_* are the mass and radius of the protostar, and $\dot{M}$ is the infall rate. Obviously, as M_* rises so does the luminosity, but the most significant term is $\dot{M}$, which declines sharply as the protostellar envelope is depleted. Consequently, a typical protostar starts its life with a luminosity that may be orders of magnitude greater than the star that it will become. It then moves *down* the Hayashi track on the Hertzsprung–Russell diagram (declining luminosity at near constant temperature) and then on the Henyey track (rising temperature at near-constant luminosity) until thermonuclear reactions start and the newly formed star joins the main sequence. Thus the birth of a Sun-like star is almost at the lowest point on the luminosity track.

The significance of this is that the high luminosity of the accretion source can warm the infalling gas and dust (to temperatures that are sufficiently high for ice mantle sublimation to occur) out to distances that are spatially resolvable with observational facilities.

As material flows into regions that are denser and hotter, then rapid ice mantle sublimation can occur, leading to a dramatic (but short-lived) chemical enrichment that is akin to what is seen in the 'hot cores' associated with high mass star-forming regions. The significant feature of these sources is that the line profiles are always strongly broadened (by several km s^{-1}) so that, unlike the physical model that is usually adopted for high mass 'hot cores', the sublimation phase is triggered by the flow of the gas, rather than temporal changes in the source luminosity. This, once again, implies that there are a number of competing timescales; in addition to those that we have already discussed, there are characteristic radii of sublimation and the dynamical timescale of the inflow.

4.6.1 The Case of IRAS 16293-2422

One of the most studied hot corino environments to date is that found around IRAS 16293-2422, a Class 0 object, still embedded in a large gas and dust envelope (~3000 au). In fact, two main sources are clearly visible: IRAS 16293-2422 A and B (IRAS A and IRAS B from now on) separated by ~600 au, both emitting in the centimetre waveband. The presence of quadruple outflows is an indication that this source is indeed a binary. IRAS A is much more complex than IRAS B but both sources exhibit a rich chemistry and are populated with complex organic molecules, the latter arising from the most compact central regions of both sources. There are many chemical survey programs dedicated to IRAS 16293-2422, with both sub-millimetre single dish telescopes[34,35] and interferometers.[36] From a dynamical point of view both IRAS A and B are very interesting sources: IRAS A exhibits broader line profiles (up to 8 km s^{-1}) than IRAS B (up to 2 km s^{-1}). It is generally accepted that IRAS A is protostellar in nature, although there is still some debate as to the nature of IRAS B. Pineda and collaborators[37] observed P-Cygni profiles in complex organic molecules, such as $HCOOCH_3$ (methyl formate) and H_2CCO (ketene), towards IRAS B, and associated such profiles with infall motions from the inner envelope. IRAS A is a more complex source and high spatial resolution observations of different molecules allow us to trace and study the several gas components found towards it; for example, the kinematics of OCS suggests the presence of an infalling-rotating envelope, while the compact emission of CH_3OH (methanol) and $HCOOCH_3$ (methyl formate) trace a ring-like structure at about 80 au from the source.[38]

4.7 Other Collapse/Evolution Scenarios and Protoplanetary Disks

Future studies will address much more complicated dynamical and chemical scenarios that may be common. A few of these are mentioned, very briefly, below:

4.7.1 Episodic Accretion

In recent years it has become apparent that the simple monotonic evolution of a protostar on the temperature–luminosity plane is not compatible with observations. In fact there is a huge scatter in luminosity for any given bolometric luminosity, L_{bol}, (and inferred evolutionary status) and both the mean and the median value of L_{bol} are lower (by a factor 10 of more) than expected. This has been termed the 'luminosity problem'[39] and a possible, if not likely, resolution may be that the accretion process is not smooth and continuous, but subject to quasi-periodic enhancements ($\geq 10^{-4}$ $M_\odot$ $year^{-1}$) due to gravitational instabilities in the accretion disc. This 'episodic accretion', of which FU Orionis is regarded as the prototype source, is then characterized by a low-level luminosity due to disc accretion and short periods of strong enhancement due to (partial) disc collapse onto the protostar.

4.7.2 Filamentary Collapse

Recent observations, and particularly those made using the Herschel Space Observatory, have revealed that much of the Galactic star formation occurs in filamentary networks of gas and dust. Indeed, the well-known source TMC-1 is now considered to be a filamentary structure, with the various clumps perhaps exhibiting different chemical signatures due to them being in different stages (or even modes of) dynamical evolution.[40] These filamentary networks appear to be ubiquitous (and are also present in regions where no star formation is occurring) and have some rather puzzling properties; in the Gould Belt it is found that the mean (projected) width of the filaments is $\sim$0.1 pc, although the amount of material that they contains varies very significantly. This length scale is hard to explain either in terms of gravitationally self-supporting objects, or hydrodynamical collapse.

One possibility is that these filamentary structures are magnetically supported structures, perhaps connected to the MHD-controlled collapse scenarios described above. If so, then their formation may follow from those models that describe the cascade and decay of MHD turbulence and the formation of oscillating and/or metastable cores. This idea has yet to be investigated properly with the inclusion of astrochemical processes.

4.7.3 Protoplanetary Discs

All molecular clouds possess some angular momentum, so that, as a cloud collapses, significant spin-up and deviation from spherical symmetry are to be expected. As a result, the final stages of protostellar evolution are characterized by the presence of first an accretion disc and then a remnant protoplanetary disc (PPD), within which planets may, or may not, form. There has been considerable recent interest, both observational and theoretical, in these PPDs that surround (low mass) T-Tauri stars. This is partly driven by the realization that a high proportion of protostars harbour extra-solar planets and, by inference, PPDs, and is partly due to the fact that interferometric facilities such as ALMA have the capability to provide direct imaging of these systems (*e.g.* see Figure 4.6a below).

There is some debate as to whether or not the chemistry of these systems is fed by the parent collapsing cloud, but in any case astrochemical studies can be used to (a) provide diagnostics of the discs, and (b) help answer outstanding questions concerning the delivery of water and pre-biogenic molecules to the early Solar System and the Earth. For example, interferometric observations of $H_2{}^{18}O$ $3_{1,3}$–$2_{2,0}$ towards the Class 0 protostar NGC1333-IRAS4B have revealed the presence of water in a 25 M_{Earth} disc of radius $\sim$25 au.[41]

An increasingly rich inventory of molecules has been detected in PPDs. For example, CO, HCO^+, CN, HCN, CS, C_2H, N_2H^+ and H_2CO (plus isotopologues) are seen in the sub-millimetre, and CO, HCN, OH, H_2O, CO_2 and C_2H_2 infrared emission features are detected from the inner disc.[42] To try and understand/interpret these observations, various dynamical/chemical models have been developed. The physical and chemical conditions in PPDs are most unlike those that have been discussed in this chapter so far. The gas is very much more dense (10^{10}–10^{14} cm^{-3}), hotter (in parts >350 K) and

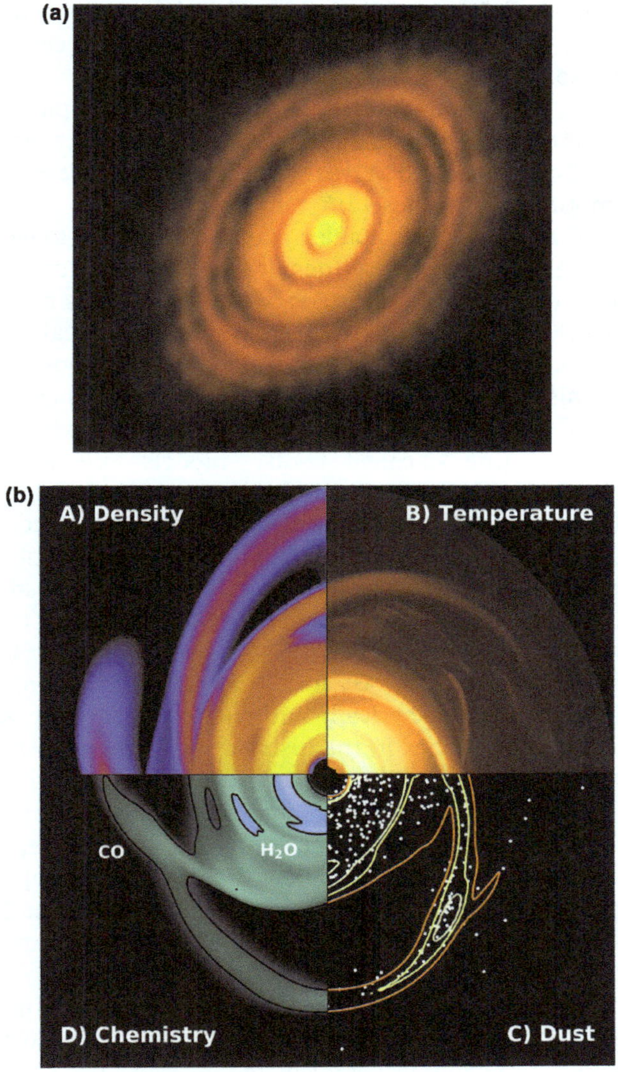

Figure 4.6 (a) Sub-millimetre images of the protoplanetary disc of HL Tau, obtained with ALMA, showing rings and gaps, perhaps indicative of the presence of planets. Image courtesy of ALMA (ESO/NAOJ/NRAO), NSF. (b) Results from a full hydrodynamical/chemical simulation of a gravitationally unstable disc, showing the spiral structure in the density, temperature, dust and chemical distributions. The latter shows the presence of the H_2O 'snow line'. Image courtesy of John Ilee, Institute of Astronomy, University of Cambridge (Ilee *et al.*, 2011).

the disc is strongly illuminated by the central source, in both UV and X-rays. In addition, there is a possible additional source of luminosity and heating deriving from the radiogenic decay of heavy elements within the disc. Moreover the column densities through the plane of the disc are so high that the attenuation of all forms of radiation, and also cosmic rays, has to be included. Strong and ordered magnetic fields are almost certainly present and this will also affect the transport of cosmic rays.

Added to this, the dynamics of PPDs can be very complex, and careful consideration of the transport of angular momentum, viscosity, turbulence and self-gravitation is required. A wealth of literature now exists that makes various assumptions about these different factors.

In the case of well-developed, stable systems (as appears to be the case for HL Tau) the structure of the disc (density, temperature, *etc.*) can be considered to be smooth and axisymmetric, with fluid parcels moving in near-perfect circular (Keplerian) orbits. The physical and chemical conditions are then just functions of the radial position (r), the height above the plane (z), and the time. This has led to the formulation of a series of successful models (*e.g.* Walsh *et al.*, 2010[42]) based on the analytical 'α-prescription' for the kinematic viscosity (the so-called 'α-disc' models—see Section 2.7.5.2), which usually assume that there is no fluid motion in the radial direction. Physically, the models include the effects of heating by the (accretion shock-generated) UV and X-rays, and solve the equation of hydrostatic equilibrium (in the z-direction) to obtain the density and temperature structure. The models have included a fairly comprehensive description of the chemistry (including three-body gas-phase reactions, gas–grain processes, surface chemistry and the effects of the UV and X-ray radiation fields). The results show how the disc is both physically and chemically layered, with distinct vertical and radial boundaries in the distribution of molecules—largely determined by the positions at which the sublimation rate for a species is balanced by freeze-out (the so-called 'snow lines').

A next level of refinement would be to include turbulence, microscopic mixing, and an analytical representation of (non-axisymmetric) motions, but in the youngest (Class 0 and Class I) systems, the mass of the PPD may be close to that of the protostar and hence be subject to gravitational instabilities. The additional factor of self-gravity in young, massive, discs leads to the formation of spiral waves, shocks, mass and momentum transport, turbulence and, possibly, protoplanetary fragments. In these circumstances a fully coupled hydrodynamic–chemical model must be employed. Some recent studies[43–45] have used 3D hydrodynamical codes to couple the motion of some 1000 Lagrangian fluid elements to a chemical model of gravitationally unstable PPDs (0.17–0.39 $M_\odot$) surrounding low mass (0.8–1 $M_\odot$) stars. There are many dynamical implications of the unstable, non-equilibrium, nature of the disc and one of these is that the fluid parcels are often seen to make very significant deviations in motion out of the plane, into regions of higher z, where very different physical conditions prevail. This, obviously, will have a significant implication for the chemical evolution.

The chemistry in these models is, so far, less well developed as compared to some of the α-disc models and, for example, doesn't include any surface reactions. Nevertheless, the results from the model (an example of which is shown in Figure 4.6b) show how the shocks strongly affect the chemistry.

Most significantly, the balance between the freeze-out and desorption of species it crucial so that shock-induced ice evaporation drives up abundances significantly.[43] By generating synthesized ALMA maps from the model results it was possible to identify that transitions in $C^{17}O$, HCO^+, OCS, and H_2CO are probably the best tracers of the disc kinematics and structure and can clearly delineate the shocks.[44] This is probably not very dependent on the disc mass, although the relative weakness of the shocks in the lower mass discs may make it more difficult to detect the spiral structure.[45] Significantly, the spiral shock waves can also act to trap dust and promote the growth of grains and planetesimals. However this structure develops, it clearly defines the initial conditions out of which planets form.

4.8 Formation of High Mass Stars

In the previous sections we described the dynamics involved in the formation of low mass stars and how the chemistry of the different stages of low mass star formation are affected by the changes in density, temperature and pressure the gas undergoes during the multi-million year cycle. The formation process for high mass stars differs from that outlined above in several significant ways. For example, the Kelvin–Helmholtz timescale (*i.e.*, the time over which the collapsing object can be powered by the energy supplied by the gravitational contraction) of low mass stars is much longer than the time required to make them; in other words, the free fall collapse time is much shorter than the thermal timescale and the centre of the collapsing core will therefore heat up almost adiabatically. This is not true for high mass stars; above a certain critical mass accretion will occur even once the star has reached the main sequence, yielding a considerable radiation pressure that should act on the grains in the inflow and hence halt the collapse. In principle, stars of mass >10–40 $M_\odot$ (depending on the grain properties) should not exist. In fact the existence of stars with masses of 100 $M_\odot$ was already known in the early 1980s (see the review by Zinnecker and Yorke 2007[46]). Different theories for the formation of massive stars to circumvent the radiation pressure problem have been put forward and we shall review them in Section 4.9. Continuum and molecular observations, especially in the last decade, have helped enormously in discriminating between different theories, and in determining a somewhat empirical classification of the early stages of high mass star formation. We shall review the evolutionary sequence as determined by observations in Sections 4.10 and 4.11.

4.9 Theories of Massive Star Formation

High mass stars, usually found in multiple systems or clusters, are the result of gravitational collapse of massive, dense clumps with masses ≥100 $M_\odot$ inside large molecular clouds. The collapse is believed to occur either through direct accretion of material in massive cores within the clump without further fragmentation (*e.g.* McKee and Tan 2003; Tan *et al.* 2013),[47,48] or *via* competitive accretion, where several low mass cores accrete in a highly fragmented clump (Bonnell *et al.* 2004).[49] While the first scenario can account for single as well as multiple systems, the latter will yield multiple systems of both high and low mass stars. As Fontani *et al.* (2016)[50] point out, the rate and occurrence of fragmentation seems to be key in discriminating between these two scenarios.

For massive stars to form *via* accretion a number of momentum boundary conditions must be satisfied (Larson and Starrfield 1971).[51] Let us first discuss the outward boundary conditions; for infall to occur the outward radiative acceleration must be less than the inward gravitational acceleration:

$$\Gamma = \frac{kL}{4\pi GM(r)} \tag{4.33}$$

where Γ is the ratio of the above accelerations, k is the flux mean radiation pressure coefficient, L is the luminosity, and r is the outer boundary of the flow. k is a function of the properties of the grains, and for a standard MRN grain size distribution (see Chapter 1) it can be shown that infall of a massive protostellar cloud (*i.e.*, ending in a star with a mass larger than 10 $M_\odot$) does not occur. Modifications to the MRN distribution do not lead to a scenario where accretion for more massive stars occurs. Hence, if Γ has to be less than unity, core masses cannot exceed 15–30 $M_\odot$.

The inner boundary conditions are determined by the inward and outward momentum fluxes. Because the visible and UV radiation from the central gas is absorbed in a thin layer called the 'momentum deposition region', for infall to be maintained the inward momentum flux needs to exceed the outward momentum flux, or in other words:

$$\dot{M}_\mathrm{i} > \frac{L}{u\,c} \tag{4.34}$$

where u is the velocity just beyond the deposition zone and $\dot{M}_\mathrm{i}$ is the mass inflow rate. Solutions to this equation for various conditions show that the collapse of a massive star requires a very turbulent or warm region to produce high enough accretion rates to circumvent the radiation pressure.

One should note that the limitations of the accretion theory as described above are based on the assumption that accretion occurs spherically. In fact, the angular momentum of the infalling gas will lead to the formation of a disc, which leads to accretion in the equatorial plane at higher rates per unit area, while radiation will escape at the poles. Simulations of the collapse of slowly rotating non-magnetic massive clumps show that, in principle, massive stars can indeed form *via* accretion through a disc. A consequence of this formation scenario would be the presence of a powerful radiation-driven outflow in the polar directions and a thick disc. As we shall see in Section 4.10 such signatures are now observed. We discuss discs and outflows in detail in Chapter 5.

The above accretion scenario is not the only one that can circumvent the problem of the radiation pressure. Spherical accretion can in fact be rescued by the 'turbulent core accretion model', sometimes also called 'monolithic collapse' model, which incorporates the effects of supersonic turbulence and high pressures observed in massive-star-forming regions (see Section 4.10). In this model, the inclusion (i) of turbulent motions, which increase with radius, and (ii) of higher external pressures and densities, due to the fact that massive stars form in more 'crowded' regions, leads to shorter free-fall times and an increase in the accretion rates with time. The critical mass, M_{crit}, above which clouds will not be in hydrostatic equilibrium under this scenario, can be equated to the Bonnor–Ebert mass, and Tan (2014)[52] reviews the expressions for M_{crit} depending on whether the cloud is magnetized or not. This model does, however, make some assumptions such as spherical accretion and a steep density distribution with radius. One consequence of this model is that, as in the case of a low mass star, the final mass of the star only depends on the mass of the collapsing core and it is therefore independent of the rest of the cloud.

An alternative theory to that described above is the so-called 'competitive accretion'; here star formation is considered as a function of the characteristics of the cluster where it occurs. In this model both the gas and stellar motions are governed by the cluster's gravitational potential. The premise of this approach is that, observationally, young clusters are seen to be associated with massive clumps of molecular gas, which should therefore play a role in the formation of the stars in the clusters. The gas is accreted at a rate

$$\dot{m} = \pi \rho_{\rm cloud}\, v r_{\rm acc}^2 \tag{4.35}$$

where ρ_{cloud} is the local density and v is the relative velocity of the stars with respect to the gas in the clump. The accretion rate can then either depend on the tidal radius or on the Bondi–Hoyle accretion radius (Bondi–Hoyle accretion assumes that the flow is gravitationally focused behind the massive star and forms a shock in the downstream direction). The latter is the most likely case for massive star formation; hence, if one assumes that all prestellar clumps have the same initial mass, and that their mass is increased *via* Bondi–Hoyle accretion then the accretion rate must be a function of their location in the cluster potential. If the clump is near a gravitational well, where more mass is available, it will accrete more than if it were close to the edge of the cluster. The young stars can also interact among themselves, quenching the accretion. Hence, there is no connection between the mass of the birth core and the final stellar mass.

Fragmentation is key in determining the most likely scenario for massive star formation. Since the conception of these two theories many advances in the dynamics and physics of the models have been achieved. For example feedback mechanisms, opposing gravity, are now routinely included in most models; it is shown that the number, the mass, and the spatial distribution of the fragments strongly depend on whether intrinsic turbulence, radiation feedback (including thermal feedback and ionization),[53] or magnetic pressure[54] are the main competitors of gravity. Mechanical feedback includes the effects of outflows and supernovae. Outflows help by sustaining turbulence against infall by injecting energy and momentum into the system, while supernova feedback works at larger scales and can terminate any star formation in its vicinity. Ionization of hydrogen *via* extreme UV photons can halt further formation of stars but only if accretion is completely spherical. If discs form then most of the ionization will occur in the discs. Recent work by Palau and collaborators[55] seems to show that mechanical feedback due to outflows is not dominant.

If radiation feedback is ignored, the collapsing clump can fragment, in principle, into as many cores as the total mass, divided by the Jeans mass, allows. However, the temperature rise due to the gravitational energy radiated away from the densest parts of the clump can suppress fragmentation. Models that include strong magnetic support, such as those by Hennebelle and collaborators,[54] predict fragments more massive and less numerous than those predicted by the models with weak magnetic support. What mainly determines the effect of the magnetic support is the ratio between the total mass of the clump and the magnetic flux. Other important factors are the initial conditions of the clump (temperature, density and angular momentum). Once magnetic fields have been included in the models, fragmentation of the original cloud, included in the original accretion models either *via* adding turbulence (in the 'monolithic collapse' model) or by setting the initial mass of the cores to be very small (in the 'competitive accretion' model), would lead to fewer fragments with a higher mass. Feedback needs to be taken into consideration in any model of massive star formation because massive stars do form in associations, and the presence of existing massive stars will affect the birth and formation of new stars. Radiation from already born stars will heat the neighbourhood, raising the Jeans mass but reducing the number of clumps. This can lead to more massive, but fewer, stars in the cluster.

4.10 Characterization of the Early Stages of High Mass Star Formation

One of the reasons why the process that leads to the formation of a massive star is not well understood is that observationally it is still rather difficult to classify the early stages of massive star formation as we routinely do for low mass stars; *e.g.*, there is not yet a definite equivalent of a Class 0 or a Class I for massive stars. Nevertheless, recently, with the advent of far-infrared telescopes such as Herschel, several groups have categorized the high mass star formation sequence. For example, Battersby and collaborators[56] organized the evolution in four stages: their first stage corresponds to infrared dark clouds (IRDCs), which we defined in Section 3.2.2. These clouds are regions of high extinction viewed against the bright diffuse mid-infrared galactic background: the molecular material in these regions is cold (<25 K), dense ($>10^5\ \mathrm{cm^{-3}}$) and has high column densities (10^{23}–$10^{25}\ \mathrm{cm^{-2}}$). The second

Table 4.2 Possible evolutionary sequences for high mass star formation. In column 2 we give the most used names of each stage; in column 3 the typical column densities and temperatures can be seen; in column 4 we list the molecules most used to trace each stage; in column 5 we list the type of chemistry that drives each stage; finally in column 6 we list rough lifetime estimates. We note that the lifetime of filaments is unknown but simulations show that fragmentation may occur within the timescales reported in our table.

Stage	Name	Physical characteristics $N(H_2)(cm^{-2})$, T (K)	Example of molecular tracers	Dominant chemistry	Lifetime (years)
1	Filaments	10^{21-23}, 15–50	CO, CS, N_2H^+	Ion–neutral	10^5
2	IRDCs	10^{5-6}, <25	NH_3, N_2H^+, HCN, HNC	Ion–neutral, freeze-out	10^4
3	HMPOs	10^{24}, 50–100	HCN, C_2H, N_2H^+, H_2CO, SO	Ion–neutral, neutral–neutral, thermal processing	6×10^4
4	HCs	10^{24-25}, 100–300	CH_3CN, CH_3OH, NH_3	Cosmic rays, surface processes, thermal processing	4×10^4
5	UCHII	$>3\times10^{21}$, 20–75	Thermal CH_3OH, CS, H_2O, CO	Ionization	10^4

stage is still IR dark but already contains signatures of star formation, such as shocks or masers. This stage may be also referred to as a high mass protostellar object (HMPO) (*e.g.* Beuther *et al.* 2007).[57] Their third stage sees the massive star already formed; this stage may be what in the past decades we have referred to the 'hot core phase'; hot cores are small (<0.1 pc), dense ($n_H \approx 10^7$ cm^{-3}) and warm ($T \approx 200$–300 K) objects believed to be fragments of infalling material very close to the protostar but not incorporated into it. The material is, however, strongly irradiated and warmed by the nearby protostar. The final stage is characterized by ultra compact HII regions (hereafter referred to as UCHII). Less than 10^{17} cm in diameter, these are ionized nebulae surrounded by a high column density of gas and dust. However, the Herschel observatory has shown that filaments are in fact the birthplace of massive stars and hence we shall consider them here as the first stage of the massive star formation process. This sequence, including filaments, is summarized in Table 4.2, where we list their physical and chemical characteristics together with possible lifetimes as determined by chemical considerations by Gerner *et al.*[58] We stress, however, that the stages are not as yet well defined and the defining line between them is blurred.

4.11 Chemistry of the Early Stages of Star Formation

Up to now we have described the formation of massive stars *via* their physical characteristics. In fact, chemistry is key in the determination of the very early stages of star formation. The chemistry of star forming gas, regardless of the final mass of the star, is described at length in Section 4.3. In this section we simply highlight the characteristic chemical processes for each evolutionary stage. These processes can now be identified thanks to the recent observational studies of large samples of objects differing in ages, such as those performed by Gerner *et al.* (2014),[58] which have allowed us to draw a consistent picture of the chemical evolution of massive star forming regions.

Despite such advances, there is still not much known of the chemistry that drives the formation and evolution of filaments; observations are revealing a wide range of filament widths that may indicate different environmental conditions under which filaments form. For example, variations in cosmic ray ionization rates in our Galaxy (from 10^{-17} to 10^{-15} s $^{-1}$) or on the magnetic field structure inside the filament can significantly affect the resulting width of the filament.

The second stage of the formation of massive stars is already characterized by a much more complex chemistry. As the temperatures and number densities in IRDCs are low ($T<25$ K) and high ($n(H_2)\approx10^3$–10^4 cm^{-3}), respectively, atoms and molecules 'freeze-out' onto the surfaces of dust grains, forming icy mantles. CO is, in fact, observed to be heavily depleted in IRDCs.[59] Other species, such as NH_3, N_2H^+, HCN, HNC and deuterated species are therefore better tracers of IRDCs, just as they are of prestellar cores. However, in contrast to their lower mass counterpart, IRDCs are also rich in complex species such as CH_3OH, and shock tracers such as SiO.

During the HMPO phase, the temperature rises and radicals become more mobile on the dust grains leading, possibly, to a more effective and complex surface chemistry. A temperature increase will at the same time lead to partially evaporated ices, with the newly formed complex species returning therefore to the gas phase.

Hot core chemistry is driven by cosmic rays and dust grain processes that lead to the formation of ices and their subsequent thermal processing. Thermal desorption experiments and chemical models of hot cores[60–62] show that thermal mantle desorption occurs selectively and is a strong function of the dust temperature, which increases as a function of the age of the core, but it also decreases with distance from the source. Distinct chemical events will therefore occur at specific grain temperatures. The observed chemistry will be rich in hydrogenated species, as well as complex organic molecules, believed to be formed, at least partially, on the surface of dust grains. Species such as H_2O, NH_3, H_2S, CH_3OH, CH_3CN, CH_2CHCN (vinyl cyanide), CH_3CH_2CN and C_2H_5OH can be enhanced by over two orders of magnitude compared to standard cold dense core chemistry. The study of multiple species may therefore reveal a record of the physics of the collapse process and selected molecules could be used to trace the ignition of the nearby star. However, for molecules to be viable 'clocks' the temperature structure of the hot core must be known. Methanol, with its large number of observable transitions in the sub-millimetre, may allow us to determine this temperature structure. It is also important to be able to disentangle material thermally evaporated from the icy mantles from that sputtered by the presence of shocks. The high dipole moment methyl cyanide molecule can be an excellent tracer of the kinematical structures, while sulfur-bearing species can be a useful tracer of the presence of shocks. As in the case of low mass cores, ices can be observed towards hot cores, despite being in lower abundances.

In UCHII regions the gas is more ionized, possibly due to their more extended morphologies; UCHII have a much poorer chemistry compared to that found in hot cores; however they are found to have stronger thermal CH_3OH. These regions have been mapped in the molecules of CS, in several transitions, which allows the determination of their density structure. High spectral resolutions of CO and H_2O observations can be used to determine their expanding velocity structure, and spatial variations.

4.12 Observational Tracers of the Massive-Star-Formation Process

Observational studies of massive star formation have come a long way in the last decade, especially due to the advent of the Herschel Observatory and sub-millimetre interferometers such as ALMA.

The Herschel infrared galactic plane survey (HiGal)[63] has revolutionized our observational picture of filaments with their dust continuum observations, which revealed, without doubt, that the ISM is pervaded by filaments. Recent molecular observations of CO (1–0) and N_2H^+ (1–0) at high spatial resolution have revealed the density and velocity structure of such filaments; for example, molecular observations of one of the most massive and dense star-forming regions in the Galaxy, the filament containing the well-known sources DR21 and DR21(OH), at a distance of 3 kpc, have revealed infall signatures (*via* HCO^+ and ^{12}CO line profiles) as well as a large-scale network of filamentary structures (revealed by the maps of ^{13}CO (1–0), CS (2–1), N_2H^+ (1–0)).[64] Such structures can be seen in Figure 4.7.[64]

The second stage of the formation of massive stars is already characterized by a much more complex chemistry. One of the most studied IRDCs is G035.39 − 00.33, which resides at a distance of 2.9 ± 0.5 kpc. In particular, Jimenez-Serra and collaborators carried out a comprehensive study of the molecular line emission and chemical composition of the gas toward this region.[59,65] Among their findings, of particular interest is the large-scale shocks traced by SiO emission. This emission presents two very different components: the first SiO component is compact, with a large fractional abundance and broad line profiles, characteristics of shocked gas in outflows associated with high mass star formation. The more extended second component has instead very narrow line profiles, with low SiO abundances possibly consistent with large-scale shocks, remnant of the IRDC formation processes. Despite the presence of shocks, the dense gas in this IRDC is

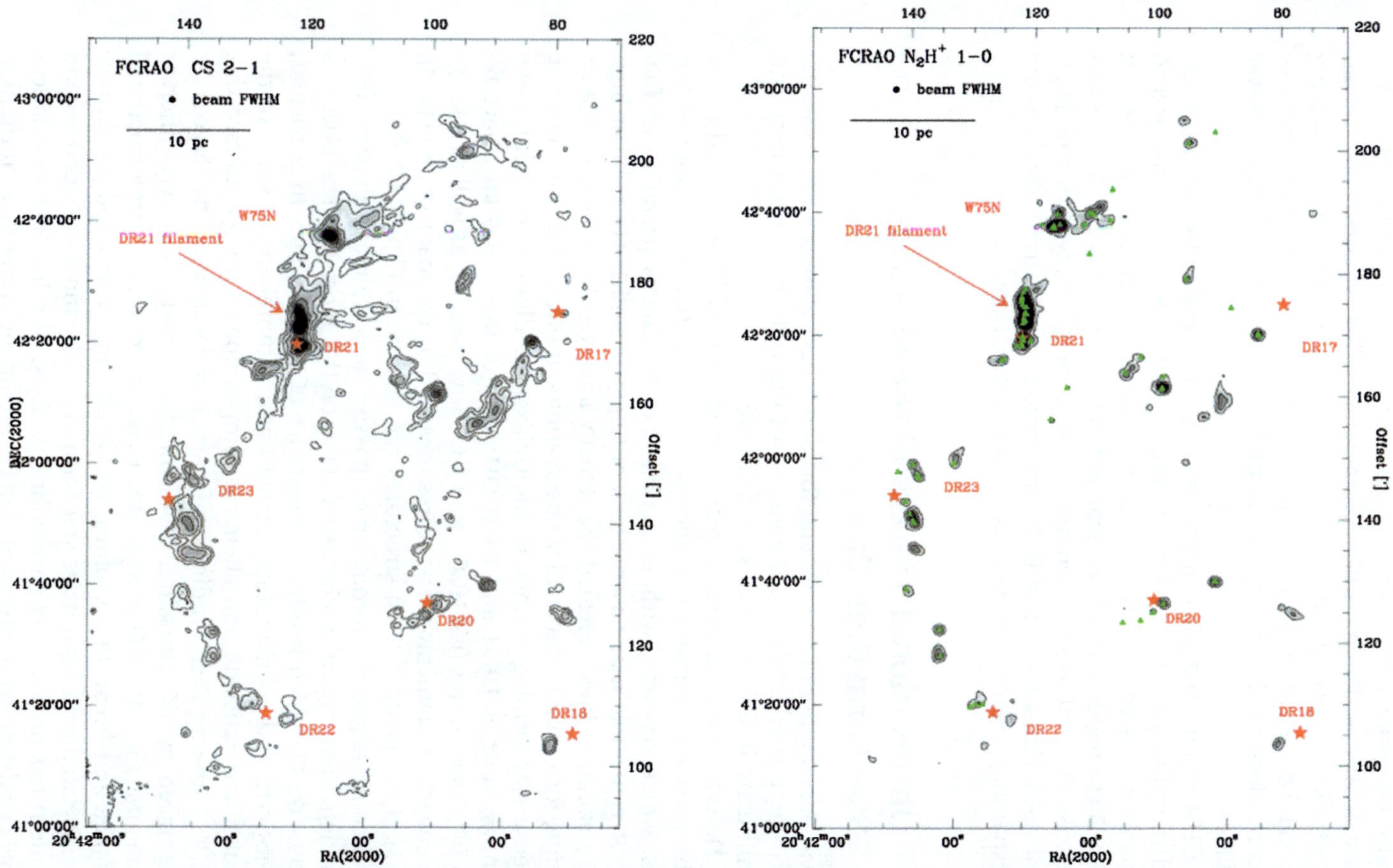

Figure 4.7 The 0th moment maps of the CS 2–1 transition (left) and the N_2H^+ 1–0 transition (right) emission in Cygnus X North. The red stars mark the thermal HII regions, while the millimetre continuum sources are marked by green triangles. The filamentary structures are more pronounced in CS but they are still visible in the weaker tracer N_2H^+.
Schneider *et al.*, A&A, **520**, 49 (2010), reproduced with permission © ESO.[64]

found to be heavily depleted in CO; by relating the depletion timescale to the dynamical timescale this study concludes that the initial conditions for star formation are created from environments where approximate pressure equilibrium has been established. N_2H^+ observations depict a scenario for G035.39 − 00.33 where a 'gentle'[66] merging of filaments occurs and may be responsible for the large-scale SiO (2–1) emission seen. A shift in velocity between the N_2H^+ and the CO molecules (see Figure 4.8[66,78]), as well as the supersonic gas motions, is consistent with the proposed merging. Massive dense cores within this region are preferentially found in the locations of the merging. Finally, this cloud also exhibits widespread deuteration.[67] N_2D^+ and DN_2H^+ towards G035.39 − 00.33 yield an average deuterium fraction of ∼0.04, three orders of magnitude higher than the interstellar [D/H] ratio and significantly lower than the values found in quiescent IRDCs, but consistent with those to be expected in chemical equilibrium. Such an equilibrium would have taken at least ∼3 Myr to be established, providing us with a lower limit dynamical timescale of ∼8 free-fall times.

HMPOs are generally not very luminous compared to the objects characteristic of the later stages, with a luminosity that is usually lower than 10^5 $L_\odot$. An exception is IRAS 23151 + 5912, which, despite being very luminous (at least 10^5 $L_\odot$), it is also very young. This object has been the subject of several studies of continuum and line emission at sub-arcsecond resolutions.[57,68,69] This source, at a distance of 5.7 kpc, shows H_2O maser emission but has no associated hypercompact region. It has been detected in several molecular species, including ground-state and vibrationally torsionally excited CH_3OH lines, sulfur-bearing species (^{34}SO, SO_2, $C^{34}S$), SiO and even few complex molecules such as CH_3CH_2CN and CH_3OCH_3 (usually typical of hot cores) (see Figure 4.9[57]). The forest of CH_3OH lines observed can be used to determine the kinetic temperature of the gas. Beuther and collaborators[57] found that the two main sub-millimetre continuum peaks range from 80 to 150 K. Observations of the CO (2–1) and SiO (2–1) lines are of particular importance because they reveal a bipolar outflow. The derived outflow rate of 10^{-3} $M_\odot$ per year yields an estimate of the accretion rate of the same order of magnitude. Two sub-sources are detected, each driving one of the molecular outflows observed in SiO. The detection of an intermediate dense spectral line forest with 27 lines within the 4 GHz bandpass allows the investigation of additional physical and chemical properties. The 19 CH_3OH lines provide temperature estimates toward the two main sub-millimetre continuum peaks to $T(\text{Peak1}) \approx 150 \pm 50$ K and $T(\text{Peak2}) \approx 80 \pm 30$ K, respectively.

Until a decade or so ago, the only way of tracking and studying the early stages of a massive star, when still embedded, was by studying hot cores. As already mentioned, the abundances of species detected in hot cores can be very large and different molecular species appear to probe different parts of the hot core, and possibly different evolutionary stages. In fact, by the 1990s,[70] sub-millimetre spectral surveys of samples of hot cores were revealing large ranges of abundances for almost every species across hot cores; these differences have been attributed to different luminosities as well as

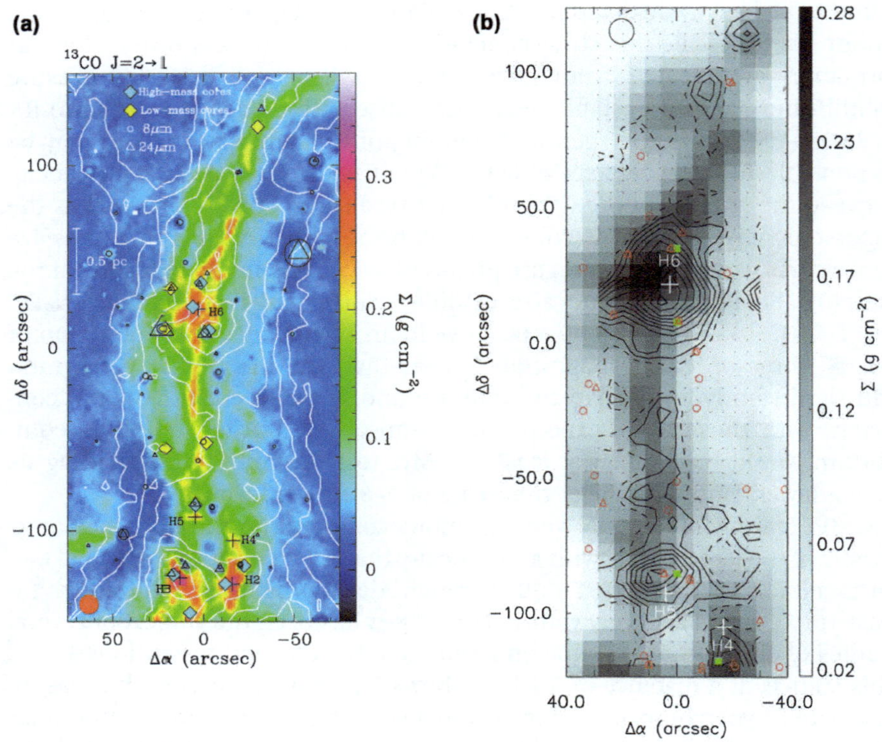

Figure 4.8 (a) The integrated intensity of the ^{13}CO $J=2$–1 line emission observed toward the IRDC G035.39 − 00.33 (white contours) for the velocity range 40–50 km s^{-1}, superimposed on the mass surface density map of the cloud (in colour). The white crosses mark the positions of massive cores within the region. The black symbols are the 8 μm (open circles) and 24 μm (open triangles) sources detected with Spitzer toward G035.39 − 00.33. The beam size of the 13CO $J=2$–1 observations is shown in the lower left corner of the panel. Reproduced from Figure 4.1 in Gas kinematics and excitation in the filamentary IRDC G035.39 − 00.33, Jimenez-Serra *et al.*, *Mon. Not. R. Astron. Soc.*, 2014, **439**, by permission of Oxford University Press. (b) The integrated intensity of N_2H^+ (3–2) towards the IRDC G035.39 − 00.33 (contour) overlaid over the mass surface density. Contours are 3σ to 5.5 K km s^{-1}, by 1 σ, where $\sigma = 0.46$ K km s^{-1}. The white crosses mark the positions of massive cores within the region. The coloured symbols are 4.5 μm (green squares), 8 μm (red circles) and 24 μm (red triangles) sources found in the direction of this this infrared dark cloud. Analysis of the comparison of the emission of the CO with that of N_2H^+ in terms of velocities indicates signs of merging. Reproduced from Figure 4.3 in Complex, quiescent kinematics in a highly filamentary infrared dark cloud, Henshaw *et al.*, *Mon. Not. R. Astron. Soc.*, 2013, **428**,[78] by permission of Oxford University Press.

different evolutionary stages. Disentangling the two effects remain a challenge, especially as shocks in the form of molecular outflows appear at an early stage of high mass protostellar evolution and hence they affect the

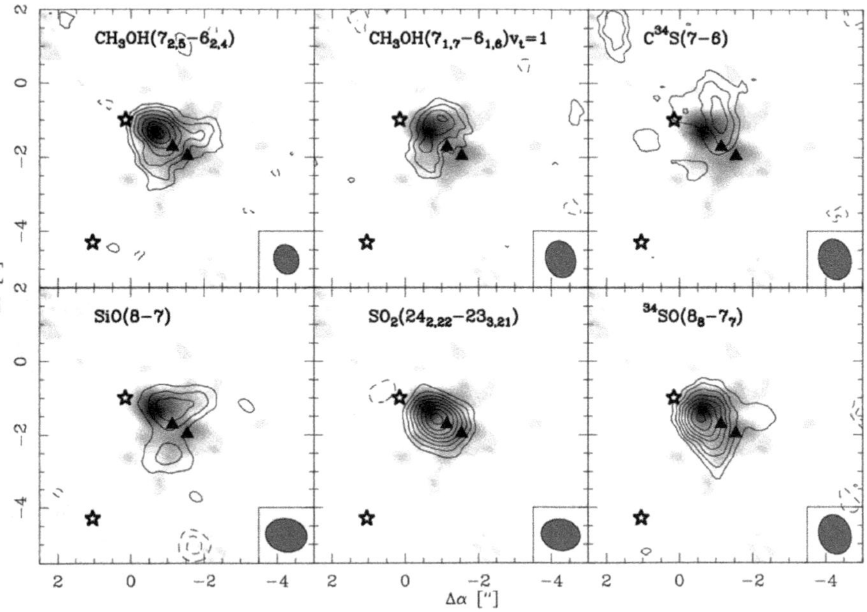

Figure 4.9 Images of selected molecules in the HMPO IRAS 23151 + 5912. The synthesized beams are shown at the bottom right of each panel. The grey-scale shows the sub-millimetre continuum emission. Triangles and stars mark the H_2O maser and the near infrared peaks in the region.
Beuther *et al.*, A&A, **473**, 493, (2007), reproduced with permission © ESO.[57]

chemistry of a hot core. It is therefore clear that the dynamics of the hot core is affected and in turn affects the chemistry observed. An interesting example of the interplay of infall and outflow material in hot cores is provided by the G31.41 + 0.31 object; located at a distance of 7.9 kpc it is one of the hottest molecular cores discovered so far and probably one of the richest in terms of chemical variety. It exhibits strong ammonia emission of high excitation transitions. In fact recent high angular resolution VLA observations[71] of the (2,2) to (6,6) ammonia transitions reveal differences between the red- and blue-shifted channel maps, a clear signature of gravitational infall (see Figure 4.10[71]). On the other hand velocity gradients observed in methyl cyanide (CH_3CN) reveal the presence of deeply embedded YSOs and the possibility of the co-existence of infall and outflow material has been investigated[72] but not yet confirmed, while rotation is for sure contributing to the large velocity gradients found in all molecular species. A way forward in the disentangling of the different dynamical components could be the study of more complex molecules, as chemical models show that they may be ideal evolutionary tracers due to their compactness. G31 was the first star-forming region where the complex organic molecule of glycolaldehyde was detected (by Beltran and collaborators[73]) and it was indeed found to be very compact. The very recent work by Rivilla and collaborators[74] corroborated this result;

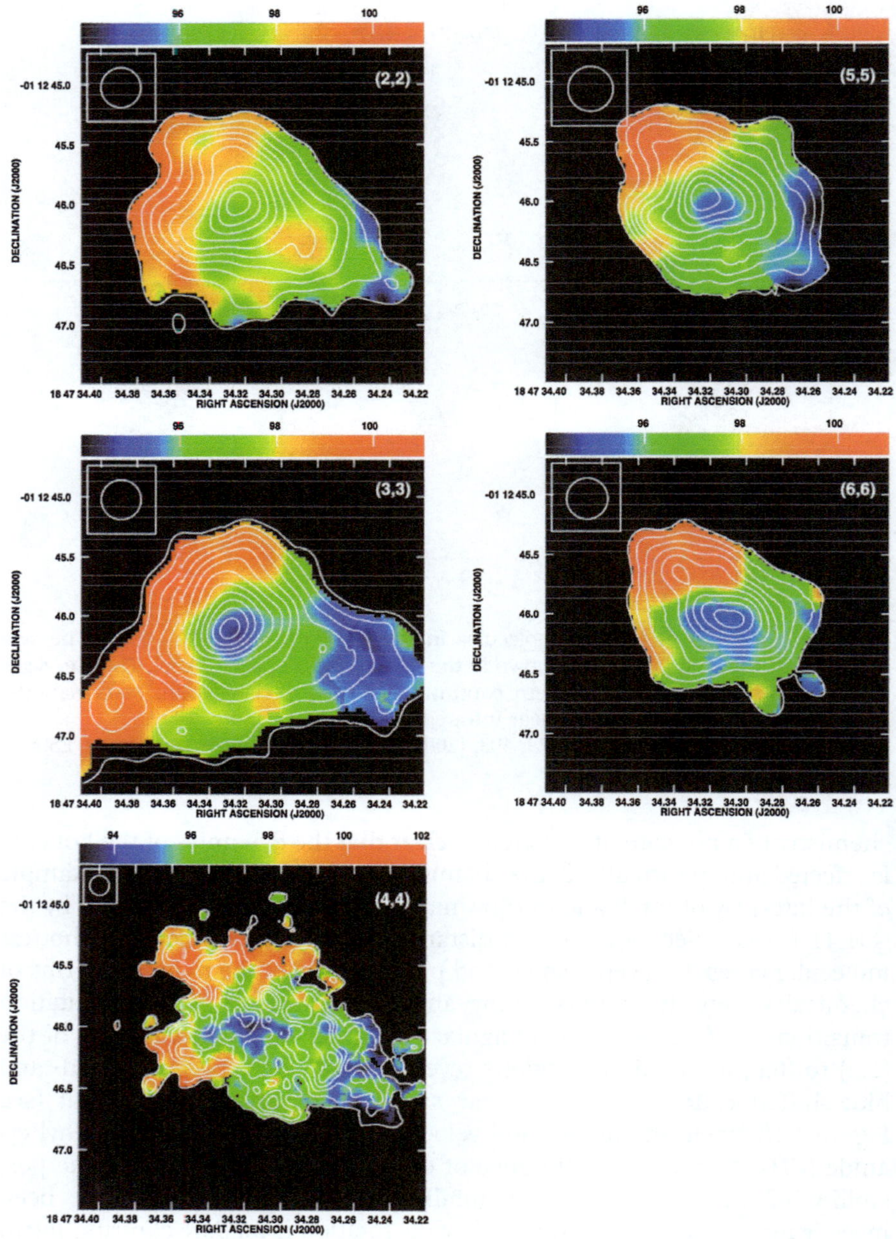

Figure 4.10 The overlay of the integrated intensity in contour and the intensity-weighted mean, in colour, velocity maps of the ammonia transitions (2,2), (3,3), (4,4), (5,5) and (6,6) towards the hot molecular core G31.41 + 0.31. The synthesized beam is shown in the upper left hand corner of each map. These transitions reveal differences between the red- and blueshifted channel maps, indicating infall (see Section 4.5). Reproduced from Figure 4.2 in Signatures of infall motions in the images of the molecular emission of G31.41 + 0.31 hot molecular core, Mayen-Gijon *et al.*, *Mon. Not. R. Astron. Soc*, 2014, **437**,[71] by permission of Oxford University Press on behalf of the Royal Astronomical Society.

they detected multiple transitions of other complex organic molecules such as CH_2OHCHO (glycolaldehyde), CH_3OCHO, CH_3OCH_3 and C_2H_5OH, although the interpretation of the differences in the abundances of these species across cores is hampered by the fact that little is known about their formation and destruction routes.

By the time high mass star-forming cores host UCHII regions, outflows can be visible. We discuss the chemistry of outflows in Chapter 5 but here we review the observations of an interesting UCHII region, W28 A2, also or better known as G5.89 − 0.39 (see Figure 4.11[76]). This UCHII region has been associated with a near-infrared source believed to an O5 (or earlier) star and is

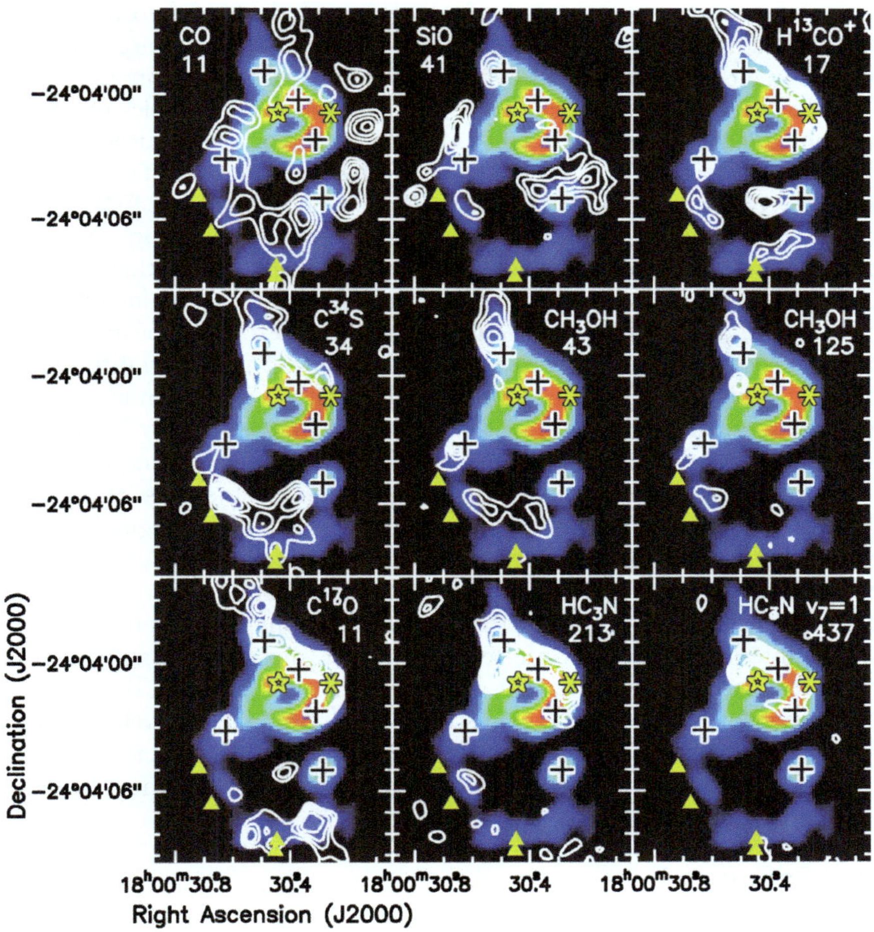

Figure 4.11 The integrated intensity maps of selected molecular emission from the UCHII region G5.89 − 0.39. The line intensity maps are overplotted over the 875 μm continuum image. The black crosses mark the position of sub-millimetre sources. The yellow triangles mark the position of the near infrared H_2 knots, while the yellow asterisk is where an outflow originates. Below each species name one can find the lower energy level in units of cm^{-1}.
Reproduced by permission from the AAS from Hunter *et al.*[76]

associated with a strong bipolar outflow identified in CO.[75] The outflow emission has been also observed in HCO^+ and SiO. Sub-arcsecond submillimetre studies by Hunter and collaborators[76] in several molecular tracers, including HC_3N, SO and SO_2, have in fact determined that the IR source is an intermediate mass star that harbours a second outflow, as well as possibly a cold dusty object just outside this region. A third outflow was discovered a year later.[77] G5.89 − 0.39 is therefore an ideal astrochemical laboratory to study multiple star-formation processes at different evolutionary stages.

4.13 Conclusions

As we have discussed in this chapter, there is a plethora of theories of infall and protostar formation and it may seem that many of the arguments go around in circles, with various hypotheses seemingly at the whim of fashion.

There are some major uncertainties in the fundamental physics of star formation. The debate as to the relative importance and origins of turbulence and magnetic fields continues, and will do so as long as the magnetic field strengths are still very poorly quantified. If, as would seem likely, magnetic fields play an important role in the early stages of collapse, then—either through static support or MHD turbulence—they could well regulate the star-formation efficiency.

There are, however, reasons to be optimistic, and these are largely driven by the huge advances in the spatial and spectral resolution of modern observational facilities. To interpret these new data it is essential to employ descriptive dynamical, astrochemical and radiative transfer models. These are now at such a high level of sophistication that we are on the verge of being able to produce and utilize combined modelling techniques to give a comprehensive analysis and diagnosis of the early stages of the star-formation process in both high and low mass star-forming regions.

However, these models must be informed by accurate and unambiguous chemical data. There has been a notable paradigm shift in the last few decades as it has become apparent that surface reactions have a very significant role in interstellar media. Unfortunately, whilst (most) important gas-phase reactions are well-characterized, the same cannot be said of gas–grain interactions and surface chemistry, which are most likely critically dependent on the properties of the (grain/ice) surface. It is therefore probably in the realm of laboratory astrophysics that the next big step must be taken.

References

1. C. J. Lada and E. A. Lada, *Annu. Rev. Astron. Astrophys.*, 2003, **41**, 57.
2. L. Olmi, M. Cunningham, D. Elia and P. Jones, *Astron. Astrophys.*, 2016, **594**, 58.
3. P. André, J. Di Francesco and D. Ward-Thompson, *et al.*, *Protostars Planets VI*, 2014, **914**, 27.
4. F. C. Adams, C. J. Lada and F. H. Shu, *Astrophys. J.*, 1987, **312**, 788.

5. P. André, D. Ward-Thompson and M. Barsony, *Astrophys. J.*, 1993, **406**, 122.
6. R. Ebert, *Z. Astrophys.*, 1955, **37**, 217.
7. W. B. Bonnor, *Mon. Not. R. Astron. Soc.*, 1956, **116**, 351.
8. F. H. Shu, *Astrophys. J.*, 1977, **214**, 488.
9. R. B. Larson, *Mon. Not. R. Astron. Soc.*, 1969, **145**, 271.
10. R. B. Larson, *Mon. Not. R. Astron. Soc.*, 1972, **157**, 121.
11. M. V. Penston, *Mon. Not. R. Astron. Soc.*, 1969, **144**, 425.
12. S. D. Taylor, O. Morata and D. A. Williams, *Astron. Astrophys.*, 1996, **313**, 269.
13. J. M. C. Rawlings, T. W. Hartquist, D. A. Williams and S. A. E. G. Falle, *Astron. Astrophys.*, 2002, **391**, 681.
14. D. P. Ruffle, T. W. Hartquist, J. M. C. Rawlings and D. A. Williams, *Astron. Astrophys.*, 1998, **334**, 678.
15. D. A. Williams and C. Cecchi-Pestellini, *The Chemistry of Cosmic Dust*, RSC Publishing, 2016.
16. G. E. Ciolek and T. Ch. Mouschovias, *Astrophys. J.*, 1994, **425**, 142.
17. J. Arons and C. E. Max, *Astrophys. J. Lett.*, 1975, **196**, 77.
18. H. A. Stace and J. M. C. Rawlings, *Mon. Not. R. Astron. Soc.*, 2017.
19. K. Tassis, K. Willacy, H. W. Yorke and N. J. Turner, *Astrophys. J.*, 2012, **753**, 29.
20. J. M. C. Rawlings, T. W. Hartquist, K. M. Menten and D. A. Williams, *Mon. Not. R. Astron. Soc.*, 1992, **255**, 471.
21. N. J. Evans II, J. M. C. Rawlings, Y. L. Shirley and L. G. Mundy, *Astrophys. J.*, 2001, **557**, 193.
22. E. Keto, J. M. C. Rawlings and P. Caselli, *Mon. Not. R. Astron. Soc.*, 2014, **440**, 2616.
23. J. M. C. Rawlings, E. Keto and P. Caselli, *Mon. Not. R. Astron. Soc.*, 2017.
24. E. Keto, P. Caselli and J. M. C. Rawlings, *Mon. Not. R. Astron. Soc.*, 2015, **446**, 3731.
25. S. Zhou, *Astrophys. J.*, 1992, **394**, 204.
26. S. Zhou, N. J. Evans II, C. Kömpe and C. M. Walmsley, *Astrophys. J.*, 1993, **404**, 232.
27. J. M. C. Rawlings and J. A. Yates, *Mon. Not. R. Astron. Soc.*, 2001, **326**, 1423.
28. S. Zhou, N. J. Evans II, Y. Wang, R. Peng and K. Y. Lo, *Astrophys. J.*, 1994, **433**, 131.
29. N. J. Evans II, J.-E. Lee, J. M. C. Rawlings and M. Choi, *Astrophys. J.*, 2005, **626**, 919.
30. N. J. Evans II, J. Di Francesco, J.-E. Lee, J. K. Jorgensen, M. Choi, P. C. Myers and D. Mardones, *Astrophys. J.*, 2015, **814**, 22.
31. S. Olofsson and G. Olofsson, *Astron. Astrophys.*, 2009, **498**, 455.
32. M. Imai, N. Sakai and Y. Oya *et al.*, *Astrophys. J. Lett.*, 2016, **830**, L37.
33. U. Hincelin, B. Commerçon, V. Wakelam, F. Hersant, S. Guilloteau and E. Herbst, *Astrophys. J.*, 2016, **822**, 12.
34. E. Caux, C. Kahane and A. Castets, *Astron. Astrophys.*, 2011, **532**, 23.

35. C. Ceccarelli, A. Bacmann and A. Boogert *et al.*, *Astron. Astrophys*, 2010, **521**, L22.
36. J. K. Jørgensen, M. H. D. van der Wiel and A. Coutens *et al.*, *Astron. Astrophys.*, 2016, **595**, 117.
37. J. E. Pineda, A. J. Maury and G. A. Fuller *et al.*, *Astron. Astrophys.*, 2012, **544**, 7.
38. Y. Oya, N. Sakai, A. López-Sepulcre, Y. Watanabe, C. Ceccarelli, B. Lefloch, C. Favre and S. Yamamoto, *Astrophys. J.*, 2016, **824**, 88.
39. M. M. Dunham, N. J. Evans II and S. Tereby *et al.*, *Astrophys. J.*, 2010, **710**, 470.
40. Y. Choi, J.-E. Lee, T. L. Bourke and N. J. Evans II, *Astrophys. J.*, 2017, **229**, 38.
41. J. K. Jorgensen and E. F. Van Dishoeck, *Astrophys. J. Lett.*, 2010, **710**, L72.
42. C. Walsh, T. J. Millar and H. Nomura, *Astrophys. J.*, 2010, **722**, 1607.
43. J. D. Ilee, A. C. Boley and P. Caselli *et al.*, *Mon. Not. R. Astr. Soc.*, 2011, **417**, 2950.
44. T. A. Douglas, P. Caselli and J. D. Ilee *et al.*, *Mon. Not. R. Astr. Soc.*, 2013, **433**, 2064.
45. M. G. Evans, J. D. Ilee and A. C. Boley *et al.*, *Mon. Not. R. Astr. Soc.*, 2015, **453**, 1147.
46. H. Zinnecker and H. W. Yorke, *Annual Rev. Astron. Astrophys.*, 2007, **45**, 481.
47. C. F. McKee and J. C. Tan, *Astrophys. J.*, 2003, **585**, 850.
48. J. C. Tan, S. Kong, M. J. Butler, P. Caselli and F. Fontani, *ASPC*, 2013, **476**, 123.
49. I. A. Bonnell, S. G. Vine and M. R. Bate, *Mon. Not. R. Astron. Soc.*, 2004, **349**, 735.
50. F. Fontani, B. Commercon and A. Giannetti *et al.*, *Astron. Astrophys*, 2016, **593**, L14.
51. R. B. Larsson and S. Starrfield, *Astron. Astrophys.*, 1971, **13**, 190.
52. J. C. Tan, Proceedings IAU Symposium No. 315, 2015.
53. M. R. Krumholz, R. I. Klein and C. F. McKee, *Astrophys. J.*, 2007, **656**, 959.
54. P. Hennebelle, B. Commerçon, M. Joos, R. S. Klessen, M. R. Krumholz, J. C. Tan and R. Teyssier, *Astron. Astrophys.*, 2011, **528**, 72.
55. A. Palau, A. Fuente and J. M. Girart *et al.*, *Astrophys. J.*, 2013, **762**, 120.
56. C. Battersby, A. Ginsburg, J. Bally, S. Longmore, M. Dunham and J. Darling, *Astrophys. J.*, 2014, **787**, 113.
57. H. Beuther, Q. Zhang, T. R. Hunter, T. K. Sridharan and E. A. Bergin, *Astron. Astrophys.*, 2007, **473**, 493.
58. T. Gerner, H. Beuther, D. Semenov, H. Linz, T. Vasyunina, S. Bihr, Y. L. Shirley and Th. Henning, *Astron. Astrophys.*, 2014, **563**, 97.
59. A. K. Hernandez, J. C. Tan, P. Caselli, M. J. Butler, I. Jimenez-Serra, F. Fontani and P. Barnes, *Astrophys. J.*, 2011, **738**, 11.
60. M. P. Collings, M. A. Anderson, R. Chen, J. W. Dever, S. Viti, D. A. Williams and M. R. S. McCoustra, *Mon. Not. R. Astron. Soc.*, 2004, **354**, 1133.

61. S. Viti, M. P. Collings, J. W. Dever, M. R. S. McCoustra and D. A. Williams, *Mon. Not. R. Astron. Soc.*, 2004, **354**, 1141.
62. D. J. Burke, F. Puletti, W. A. Brown, P. M. Woods, S. Viti and B. Slater, *Mon. Not. R. Astron. Soc.*, 2015, **447**, 1444.
63. S. Molinari, B. Swinyard and J. Bally, *Astron. Astrophys.*, 2010, **518**, 100.
64. N. Schneider, T. Csengeri, S. Bontemps, F. Motte, R. Simon, P. Hennebelle, C. Federrath and R. Klessen, *Astron. Astrophys.*, 2010, **520**, 49.
65. I. Jiménez-Serra, P. Caselli, J. C. Tan, A. K. Hernandez, F. Fontani, M. J. Butler and S. van Loo, *Mon. Not. R. Astron. Soc.*, 2010, **406**, 187.
66. I. Jiménez-Serra, P. Caselli, F. Fontani, J. C. Tan, J. D. Henshaw, J. Kainulainen and A. K. Hernandez, *Mon. Not. R. Astron. Soc.*, 2014, **439**, 1996.
67. A. T. Barnes, S. Kong, J. C. Tan, J. D. Henshaw, P. Caselli, I. Jiménez-Serra and F. Fontani, *Mon. Not. R. Astron. Soc.*, 2016, **458**, 1990.
68. T. Rodríguez-Esnard, V. Migenes and M. A. Trinidad, *Astrophys. J.*, 2014, **788**, 176.
69. H. Beuther, Q. Zhang, E. A. Bergin and T. K. Sridharan, *Astron. J.*, 2009, **137**, 406.
70. J. Hatchell, M. A. Thompson, T. J. Millar and G. H. MacDonald, *Astron. Astrophys. Suppl.*, 1998, **133**, 29.
71. J. M. Mayen-Gijon, G. Anglada, M. Osorio, L. F. Rodríguez, S. Lizano, J. F. Gómez and C. Carrasco-González, *Mon. Not. R. Astron. Soc.*, 2014, **437**, 3766.
72. R. Cesaroni, M. T. Beltrán, Q. Zhang, H. Beuther and C. Fallscheer, *Astron. Astrophys.*, 2011, **533**, 73.
73. M. T. Beltrán, C. Codella, S. Viti, R. Neri and R. Cesaroni, *Astrophys. J.*, 2009, **690**, 93.
74. V. M. Rivilla, M. T. Beltrán, R. Cesaroni, F. Fontani, C. Codella and Q. Zhang, *Astron. Astrophys.*, 2017, **598**, 59.
75. P. M. Harvey and T. Forveille, *Astron. Astrophys.*, 1988, **197**, 19.
76. T. R. Hunter, C. L. Brogan, R. Indebetouw and C. J. Cyganowski, *Astrophys. J.*, 2008, **680**, 1271.
77. Y.-N. Su, S.-Y. Liu, K.-S. Wang, Y.-H. Chen and H.-R. Chen, *Astrophys. J.*, 2009, **704**, 5.
78. J. D. Henshaw, P. Caselli, F. Fontani, I. Jiménez-Serra, J. C. Tan and A. K. Hernandez, *Mon. Not. R. Astron. Soc.*, 2013, **428**, 3425.

CHAPTER 5

Stellar Jets and Outflows

5.1 Introduction

In the preceding chapter, we saw how the gravitational collapse of a cloud of interstellar gas may lead to the formation of a new star. We described, in particular, how chemistry is intimately involved in the formation of stars, by providing molecules that act both as tracers of the physical processes occurring and also as active participants in those processes. We saw that molecules control star formation in two important ways: firstly, molecules exert control through the cooling they provide and, therefore, by moderating the gas kinetic pressure that opposes collapse; and, secondly, the chemistry forming the molecules also determines the electron abundance in the gas, and consequently the level of magnetic support within the cloud. Obviously, magnetic pressure, like kinetic pressure, must be controlled if gravitational collapse is to be sustained.

In this chapter, we shall describe the consequences following the appearance of new stars in the star-forming region. The situation around new stars is very dramatic, with an intense interplay between magneto-hydrodynamics and chemistry. Infall onto new stars of material from circumstellar discs continues during the early evolutionary phases, from the deeply embedded Class 0 objects (in which most of the mass remains in the infalling envelope), through Class I (in which most of the envelope mass has been transferred to the protostar), to the optically revealed Class II objects (or T Tauri stars) in which the envelope has been removed but accretion continues from the circumstellar disc.

As described in Section 5.2, new stars are usually associated with very large velocity ($\sim10^2$ to $\sim10^3$ km s^{-1}) collimated jets that may extend for large distances (~parsecs) and may possibly escape entirely from the star-forming region that generates the jets. These jets (and accompanying counter-jets)

Dynamical Astrochemistry
By David A. Williams, Thomas W. Hartquist, Jonathan M. C. Rawlings, Cesare Cecchi-Pestellini and Serena Viti

Published by the Royal Society of Chemistry, www.rsc.org

show evidence of internal shocks; their main working surfaces at the heads of the jets excite and irradiate the ambient cloud in a characteristic manner, creating the so-called Herbig–Haro (HH) objects. Star formation is also typically associated with bipolar outflows of rather smaller velocities (up to $\sim$30 km s^{-1}) than the jets, but with the symmetry axis defined by the jets. The opening angles of these outflows appear to increase with time; this process may ultimately terminate the infall of material onto the star from directions perpendicular to the symmetry axis. The interfaces between the outflows and the ambient molecular cloud gas may be mixing zones for two types of gas: molecular cloud gas and stellar wind gas; such zones may generate a characteristic chemistry.

The study of jets and outflows is developing very rapidly, both observationally and theoretically. A useful review of both aspects is given by Frank *et al.* (2014).[1] It is generally agreed that fast protostellar jets are driven and collimated by rotating magnetic fields anchored in the protostellar disc (Pudritz *et al.* 2007;[2] Shang *et al.* 2007[3]), although the precise location of the anchoring is still under discussion. This picture predicts that the jet and outflow should rotate, as has been observed (*e.g.* Zapata *et al.* 2015[4]).

Observations of star-forming regions have been transformed by modern high-resolution, high sensitivity facilities. In the next section we present a survey of the observational situation. We discuss stellar jets and their chemistry in Section 5.3 and stellar outflows and their chemistry in Section 5.4, and make some conclusions in Section 5.5.

5.2 Observations of Jets and Outflows

5.2.1 Observations of Jets

Jets originate very close to the young stars, where densities and temperatures are at the highest values. As the dynamic of the jet's structure changes along the jet, different observational techniques are employed to study the spatial variation of jet material.

The location where the jet originates is generally probed by radio continuum centimetric emissions and by water masers. Radio continuum observations allow us to determine the direction and collimation of the jet ejected by the young stars in recent times; at very high angular resolution, observations at centimetre wavelengths can trace the base of a jet down to the injection radius, at a scale of a few au, where the ionized jet is expected to begin. Comparing these observations with wind models provides a description of the jet physical properties. For example, combining the flux densities and sizes from the observations with free-free models, one can estimate the ionized mass loss rate in the jet, $\dot{M}$, as well as the radius where ionization starts, r_0. From a sample of the most well known jets it is found that the mass loss rates range from 10^{-8}–10^{-6} $M_\odot$ y^{-1} while the ionization radius is always less than 60 au, and in some cases (Shang *et al.* 2004)[5] as close to the star as 10 au (see Table 5.1).

Table 5.1 The ranges of some physical and chemical parameters associated with jets from protostars.

n_e (10^3 cm^{-3})	n_H (10^4 cm^{-3})	T_{eff} (10^3 K)	$\dot{M}$ ($M_\odot$ y^{-1})	Radius	Depletion (*cf.* solar)	Ionization fraction
0.05–5.7	0.01–6	0.5–10	10^{-8}–10^{-6}	3–5 $R_\star$	0.21–1	0.01–0.7

Water maser observations at 22.2 GHz also allow us to explore the kinematics of the jet gas in the vicinity, down to sub-au scales, of the accretion disc. Water masers are well known to be associated with outflow phenomena and with the earliest evolutionary phases of star formation. In fact, using a number of telescopes such as the Nobeyama 45-m telescope with Very Long Baseline Interferometry (VLBI), multi-epoch H_2O maser surveys have been performed, revealing a high maser activity during the earliest phases of low mass star formation, decreasing towards the end of the main accretion and outflow phase, and disappearing after the pre-main sequence evolution, when the star is optically visible and no significant accretion is occurring, implying that the water maser structure is correlated with the age of the Young Stellar Object (YSO). Finally, observations of the water masers also allow the determination of proper motions and line of sight velocities with very fine time sampling.

The regions downstream from the jet launch region are usually observed *via* infrared (IR) lines (Nisini *et al.* 2009).[6] Diagnostics based on IR spectroscopic techniques have been widely used in the last fifteen years to probe the jet streams, in particular, lines of [FeII], [SII] and H_2 rovibrational and rotational lines, in the 0.8–2.5μm spectral range. [FeII] lines will trace gas at low ionization with temperatures in the range 8000–20 000 K and densities of 10^3–10^5 cm^{-3}. Combining lines of [FeII] in the J band (1.1–1.4 μm) and H band (1.5–1.8 μm) give measurements of the reddening, while ratios of [FeII] lines in the H band, such as 1.533 μm/1.644 μm, or 1.60 μm/1.644 μm are used to measure the electron density. In the jet regions where the extinction is low, such as when the jet has emerged from the dense circumstellar envelope, optical and IR lines can be used together to get additional information on the physical conditions. For example the [FeII] 1.64 μm/0.862 μm ratio is a sensitive temperature indicator without having to make assumptions about the abundance or ionization fraction.

Observations of the H_2 rovibrational lines reveal temperature stratification along the jets, ranging from 100 to 4000 K near the YSOs; such stratification can only be reproduced by time-dependent shock models. It is indeed possible to see time-dependent effects in young flows and measure the age of the flow if the H_2 excitation diagram is sufficiently sampled over a large energy range (*e.g.* Bourlot *et al.* 2002).[7] H_2 is most abundant at the interfaces between the bow shock regions and the interstellar medium, where cooling is most efficient. Finally, mass flux rates can be also derived from H_2 line luminosity if the molecular jet velocity is known. A schematic diagram showing the interaction of a jet with a gaseous environment may be viewed in Figure 5.12a. The bow shock structure can be quite complex.

During the very early stages of star formation, when the central object is deeply embedded, the highly collimated jets will interact at high speed with nearby gas and form the so-called HH objects. These objects can be observed in H_2 infrared lines (see, *e.g.*, Schwarz and Greene 2003[8]) and also in CO lines (see, *e.g.*, Arce *et al.* 2013[9]) in the millimetre waveband. The morphology and geometry of the HH objects can be determined with wide-field narrowband Hα and S[II] optical images, which can reveal the number of HH knots present, as well as their separation (see, *e.g.*, Podio *et al.* 2009[10]).

An important component of jets is dust, which until recently was not routinely observed. Gas-phase abundances of elements such as iron (Fe), magnesium (Mg), silicon (Si), and calcium (Ca) are depleted by factors of 10^2–10^4 with respect to solar abundances as they become 'locked' in dust grains in the form of silicates and oxides. However, during the jet evolution, refractory material from dust grains can be returned to the gas phase during the passage of shock waves. As described in Sections 2.5.3.4 and 2.6.3, gas–grain and grain–grain collisions can lead to partial or total sputtering and vaporization of the grains. The sputtering may be total if the shock velocity exceeds ~ 200 km s^{-1}. The fact that along the jet we detect bright [FeII] lines in dense jets is clear evidence that most of the dust grains have been destroyed by the passage of the shocks, releasing refractory elements into the gas phase (Podio *et al.*[10]) Along the jets, the gas-phase abundances of Fe are typically found to be $\sim$ 80% of solar. However, in the inner jet regions, the abundance is as low as 20–30% of the solar value. This is a strong indication that a significant fraction of dust grains is still present in the initial jet beam. Table 5.1 summarizes the range of physical and chemical characteristics of jets.

The UV radiation generated by the HH objects is intense enough to drive photochemistry in nearby cold dense gas, leading to characteristic tracer species, such as NH_3 and HCO^+ observed in the gas around a large sample of HH objects (Girart *et al.* 1994).[11] The formation of these molecular species has been attributed to the release of NH_3 molecules enhanced by surface hydrogenation on icy mantles of dust, and of the parent species, H_2O, which then react with C^+ ions (also enhanced by the UV radiation) to form HCO^+ (Taylor and Williams 1996).[12] The latter species is found to be enhanced by 1–2 orders of magnitude compared to its typical dark dense cloud abundances. CH_3OH and H_2CO are also observed to be strongly enhanced compared to normal dark cloud chemistry and, in fact, molecular surveys ahead of several HH objects show that the emission of most species seems to be enhanced with respect to that of a typical dense clump, probably due to the exposure to a high UV radiation field from the HH objects (Viti *et al.* 2006).[13] Chemical differentiation among the positions is also observed. We list in Table 5.2 the range of values of column densities (cm^{-2}) observed for each molecule, detected by single-dish telescopes, towards up to ten individual HH objects. Chemistry at the terminal working surfaces of jets is discussed further in Section 5.3.6.

Many HH objects have been studied in detail. As an example, we report on the observational characteristics of one pair of HH objects, HH46/47, in the

Table 5.2 The ranges of column density for selected molecular species detected in molecular clouds ahead of, and illuminated by, HH objects. The observations from which these data are taken were made by single-dish telescopes. The detections were made towards up to ten distinct HH objects.

Molecule	Column density (cm^{-2})	Molecule	Column density (cm^{-2})
CO	3.3×10^{17}–3.5×10^{18}	CH_3OH	2.9×10^{13}–4.1×10^{14}
^{13}CO	6.5×10^{15}–5.5×10^{16}	C_3H_2	8.5×10^{11}–8.9×10^{12}
HCO^+	5.0×10^{12}–4.0×10^{13}	HCN	$<6\times10^{11}$–2.3×10^{13}
$H^{13}CO^+$	6.0×10^{11}–4.2×10^{12}	H_2S	6.0×10^{12}–5.0×10^{13}
DCO^+	1.9×10^{11}–5.0×10^{12}	CS	2.0×10^{12}–6.2×10^{13}
H_2CO	1.0×10^{13}–9.6×10^{13}	SO	8.2×10^{12}–8.6×10^{13}

Gum nebula, at a distance of 450 pc. The HH 46/47 objects have been studied through extensive optical and IR observations over the past few decades. The blue (northern) lobe of HH 46/47 is detected at optical wavelengths well outside the Bok globule where it originates, while the red lobe mostly resides inside the globule and is best seen in IR images (Zhang *et al.* 2016).[14] HH 46/47 is a complex object; wide-field, narrowband Hα and [SII] optical images of the region reveal that HH 46/47 is in fact in the innermost part of a giant HH flow that extends 2.6 pc in the plane of the sky. Several parameters are well known (Arce *et al.*[9]), such as the inclination between the jet and the plane of the sky ($\sim$30°) and its average jet velocity ($\sim$300 km s^{-1}). This object has been fully mapped at low resolution in CO and targeted by most ground (*e.g.* APEX) and space telescopes (Herschel) in CO, OH and H_2O lines. The molecular emission traces the colder, slower gas along the flow and arises from both non-dissociative C-type shocks (for example much of the hot H_2O) as well as from dissociative J-type shocks (*e.g.* OH). Molecular observations are therefore an excellent tool for studying the shock structure of jets. Recently, high spectral (0.08 km s^{-1}) and spatial resolution ($\sim$3″) maps (Arce *et al.*[9]) have revealed remarkable details of the flow (see Figure 5.1), in particular they find for the first time evidence of much higher velocities than usually associated with the molecular component of these jets (30 and 40 km s^{-1} in the blue and red lobes, respectively), indicating that at least this particular HH object is more energetic and carries more momentum than previously thought. The origin of the wide range of velocities observed is unknown but it is inconsistent with rotation. In fact, the ALMA maps revealed a clumpy nature for HH 46/47, and the clumps' spatial and velocity structures indicate that the molecular component of the jet is formed by the entrainment of an underlying wide-angle protostellar wind. We shall discuss this wind component in the next section.

5.2.2 Observations of Molecular Outflows from Low Mass Protostars

As we saw in the previous section, jets from young stars also give rise to classical bipolar outflows, probably as a result of erosion of the ambient

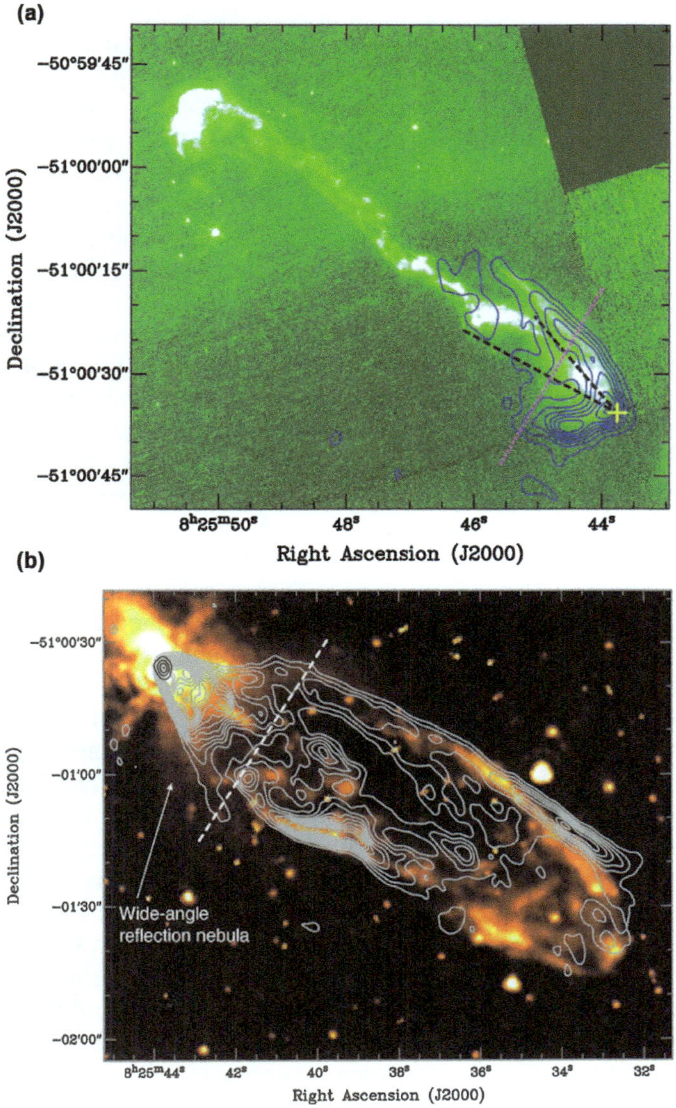

Figure 5.1 Observations of the red- and blue-shifted CO lobes of the HH46/47 outflows. (a) In the blue-shifted lobe, the HST optical image of the jet is shown in white, and is superimposed by blue contours of CO emission obtained in ALMA observations; the yellow cross is the position of the source driving the outflows, HH47 IRS. The dashed lines indicate the range of jet orientations producing the "wiggles" in the jet. (b) The red-shifted lobe is traced in grey contours of CO emission superimposed on the Spitzer 4.5 μm false-colour image. The source driving the outflow (HH47 IRS) is within the black contours tracing 3 mm continuum emission, at the upper left.
Reproduced by permission from the AAS from Arce *et al.* (2013).[9]

medium through shocks and entrainment. Such outflows are observed mainly in CO and H_2 emission lines. However, several other molecular species may be good tracers of these molecular components, including HCO^+ and SiO. The former is very abundant (and enhanced with respect to dark cloud abundances) in the outflow walls; HCO^+ is believed to be enhanced here due to the release of the icy mantles induced by the photons, which easily penetrate in the interface between the outflow and the cold ISM (Rawlings *et al.* 2004;[15] Viti *et al.* 2002[16]). HCO^+ observations imply hydrogen number densities along these interfaces of about $\sim 10^6$ cm^{-3}, with very low visual extinctions (~0.5 mag) and radiation field intensities of the order of 10 times the mean interstellar radiation field intensity. SiO on the other hand is observed to be abundant both in the cavity walls as well as within the outflow, and it is certainly tracing the presence of strong shocks. In the outflow, SiO indicates the presence of density structures larger than 10^6 cm^{-3}. The most interesting types of outflow from a chemical point of view are molecular outflows that originate from much younger protostars, where infall may still be occurring.

The most interesting chemically rich outflow of this type is the outflow originating from the Class 0 protostar L1157-mm (see Figure 5.2), at a distance of 250 pc. On its discovery it was clear that strong shocks (implied by the temperature derived from NH_3 and SiO observations) were present (Umemoto *et al.* 1992[17]). Several bow shocks along the outflow have been identified and studied through their rich chemical content. The brightest of these, L1157-B1, has a complex, clumpy structure (Benedettini *et al.* 2007[18]) and an age of the order of 1000 years (Gueth *et al.* 1996[19]). Several molecules have been observed in emissions throughout the 1990s and early 2000s and their abundances were a clear sign of mantle evaporation or sputtering. With the advent of high spectral resolution infrared space missions and high spatial resolution ground based single dish and interferometric telescopes, L1157 has been amply surveyed in most frequency ranges where molecules emit. Such surveys not only confirmed the chemical richness of this object but also discovered and characterized a complex structure and the presence of both J- and C-type shocks. For example, using CO, the presence of three velocity components was revealed, tracing three different kinematic components of the gas. The velocities are 12.5 km s^{-1}, 4.4 km s^{-1} and 2.5 km s^{-1}, which are respectively associated with a J-type shock where the protostellar jet impacts the B1 cavity, the walls of the B1 cavity and an older cavity, B2, (Lefloch *et al.* 2012[20]). Other molecules would not necessarily fit all three components, so, for example, line emissions from CS, SO^+, HCS^+ and H_2S could only be fitted by two lowest velocity components, which are also likely to have gas shocked by C-type shocks (Holdship *et al.* 2016[21]). Water emission, on the other hand, seems to come mainly from one of these components. However, the water transitions with the highest upper energy levels seem to require a fourth, hot component. Clearly, all these velocity components represent distinct physical conditions in the gas, and a comparison among the profiles of different molecules allows the derivation of the properties of the shocked and non-shocked gas in L1157. For

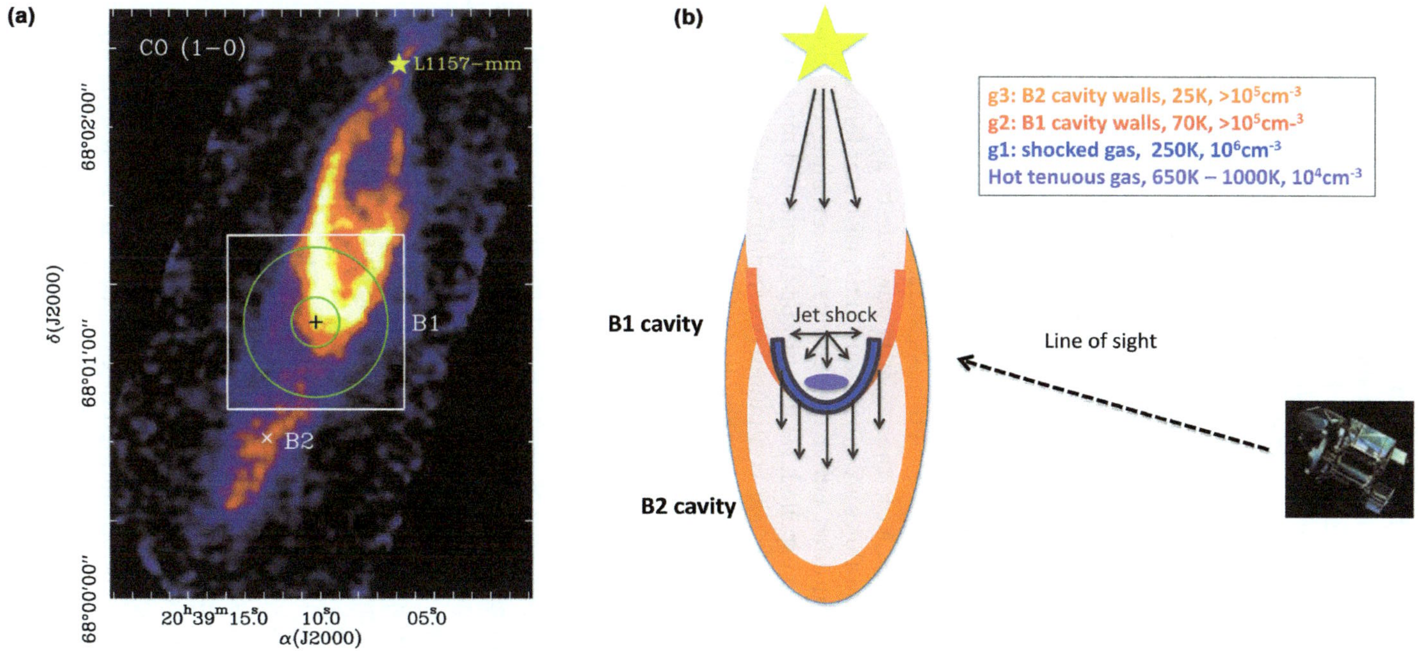

Figure 5.2 The outflows from the Class 0 protostar, L1157-mm. (a) The protostar is indicated by a star symbol and powers the outflow lobe observed in CO (1–0). The black cross marks the nominal position of the bow shock (L1157-B1) established by the outflow, and the white cross indicates the position of the bow shock (B2) associated with an earlier outflow. Studies were made within the region indicated by the white square in several lines of H_2O for which the green circles indicate the largest and smallest beam sizes. (b) The sketch shows the inferred L1157 blue lobe system. The B1 and B2 cavities are indicated in red and orange, respectively. The two shock components are indicated in blue for the warm shocked gas and violet for the hot tenuous gas. The physical conditions deduced for each component are shown in the box above, and the line of sight inferred for the observations is indicated on the right.
Busquet *et al.*, A&A, **561**, 120 (2014), reproduced with permission © ESO.[23]

example, the NH_3/H_2O intensity ratio decreases by a factor of ~5 moving towards higher velocities, which has been interpreted as a decrease in the abundance ratios. Other tracers of shocked material such as CH_3OH and H_2CO show the same profile as that of NH_3. Shock models suggest that the differences observed in the profiles of the different molecular tracers are due mainly to the temperature of the gas: if the latter undergoes a period at a temperature close to 4000 K, then NH_3 is easily destroyed by the reaction with hydrogen, which leads to $NH_2 + H_2$ (this reaction has a high barrier of ~5000 K), while H_2O remains high in abundance. The observations point to the presence of a C-type shock across which the temperature varies and reaches 4000 K in the region of the B1 shock features of L1157 (Viti *et al.* 2011[22]). Observational data for L1157 are shown in Figure 5.2a and an explanatory sketch is in Figure 5.2b.

5.2.3 Observations of Outflows from High Mass Stars

Observations of outflows from young high mass stars are sparser than for their lower mass counterparts. Nevertheless, it is now safe to assume that deeply embedded massive outflows are ubiquitous in massive star formation regions. Observations of outflows either concentrate on revealing the kinematics of the material being expelled, *via* line profile atomic and molecular observations, masers and ionized gas continuum observations, or they reveal the shocks along the flows, *via* radio continuum and line emission of shock tracers, and finally they can reveal the outflow morphology (*i.e.*, its cavities, bow shocks, and collimation).

Physically, outflows from early B-type stars have mass outflow rates 10^{-5}–10^{-3} $M_\odot$ $year^{-1}$, momentum rates 10^{-4}–10^{-2} $M_\odot$ km s^{-1} $year^{-1}$, and mechanical luminosities of 10^{-1}–10^{2} $L_\odot$, while outflows from O stars can reach mechanical luminosities of more than 10^4 $L_\odot$ (Shepherd 2005[24]). Assuming that outflows around massive stars are generated in a similar manner to low mass outflows (see Section 5.4.1) then similar observations to those described in the previous sections should provide us with information on the short-lived early stages of massive star formation. However, one must not forget that massive stars form in clusters, and the clustering renders the observations of individual outflows a challenge due to crowding and source confusion. Moreover, due to the high gas densities involved in the very early stages of high mass star formation, the extinction is so high that near infrared and optical observations are impossible. For example, some of the most studied high mass star-forming regions (*e.g.* AFGL 5142) show ample evidence of multiple outflows (Zhang *et al.* 2007[25]). Finally, one must not forget that high mass stars are found further afield and their distance means that even the highest resolution interferometric observations will hardly resolve the density and temperature stratification of massive outflows. Nevertheless, in order to trace the low velocity, extended, entrained gas component of the outflow, observations of CO and HCO^+, and their isotopologues, are routinely performed, while observations of SiO emission will trace the hotter, more collimated jets. In particular, surveys of the SiO (2–1) and (3–2) line emission in large samples of high mass YSOs found that the intensity of the SiO line becomes fainter as the luminosity-to-mass ratio increases, and this may be an indicator of the evolutionary stage of YSOs. The variation of the SiO line intensity can either be due to a decrease in the SiO abundance with time or to a decrease in the jet/outflow mass with time, or both. When mapping of a sub-sample of these objects was performed in both SiO, HCO^+, N_2H^+, and sometimes larger species such as CH_3CN, it confirmed that the SiO abundance seems in fact to decrease with time (from 10^{-8} to 10^{-9}), possibly indicating that early stages are associated with stronger shocks (Sánchez-Monge *et al.* 2013[26]). It also shows, consistently, that the HCO^+ abundance seems instead to increase with time, due to the increase of material being swept by the jet and hence more material being entrained. The densities derived from molecular observations yield gas number densities quite similar to their low mass counterparts (10^5–$10^6\,cm^{-3}$) but with temperatures as high as 500 K.

These observations seem to point to a common evolution of low mass and high mass outflows; both are collimated when young, and their collimation decreases with age. This conclusion seems to be corroborated by monitoring of water masers in a handful of high mass objects. There are still some questions that are highly controversial concerning outflows in high mass star-forming regions; in particular, we do not know whether massive outflows are mainly driven by radiation pressure or stellar winds, or whether they are equally driven by both. However, recent observations (Feng *et al.* 2016[27]) of collimated SiO (2–1) line emission from high mass, low luminosity, dark clouds clearly indicate the presence of strong bipolar outflows from heavily embedded sources, which still have an SED consistent with starless cores (see Figure 5.3). SiO is an excellent tracer of very early outflows and may yield support for a radiation driven origin.

One of the most chemically interesting outflow regions associated with massive (B) stars is Cepheus A. Observations of typical outflows and shock tracers show multiple outflows in different directions, all originating from the group of B-type stars located in Cep A East (Codella *et al.* 2005[28]). Interestingly, different molecular species seem to trace different outflows within this region, or at least different gas components within each outflow, and at the interface between outflows and ambient environment. In particular it is found that $HC^{18}O^+$ and CH_3C_2H only trace the hot core component around the B cluster, while standard shock tracers and most sulfur-bearing species (*e.g.* CS, SiO, H_2S, SO_2, and S) trace the bulk of the outflow at different velocities. There is then a class of species (OCS, H_2CS, HDO, and CH_3OH) that are also associated with an additional component believed to be a turbulent interface between the outflow gas and the dense cold gas, and where dynamical mixing and diffusion are occurring (Codella *et al.* 2006[29]). This region is a prime example of how outflows in massive star-forming regions can provide us with a rich laboratory where one may study co-existent multi-phase interstellar media.

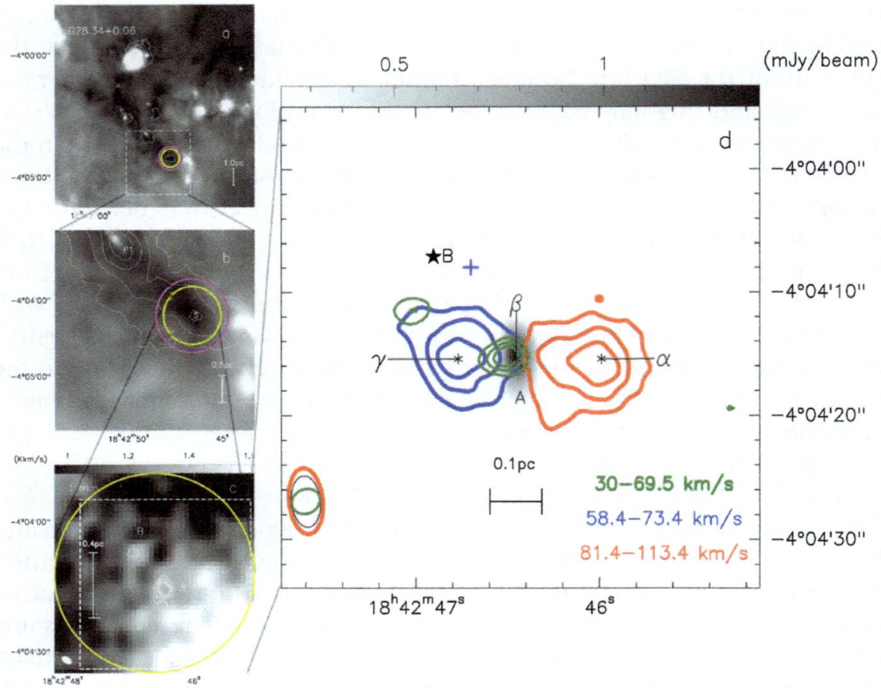

Figure 5.3 A compilation of observations of increasing angular resolution of the dark clump G28.34S, previously considered to be a candidate for a starless core. In (a) and (b), a greyscale map of dust emission at 70 μm from Herschel data is overlaid with contours of 870 μm emission. In (c), contours of 1.1 mm emission are overlaid on a greyscale map of $C^{18}O$ (2–1) emission, while in (d) contours of red- and blue-shifted SiO (2–1) emission from outflows in G28.34S-A are shown, along with blue-shifted SiO (5–4) emission, and are overlaid on a greyscale map of 3.4 mm continuum emission. This source appears to be at an early evolutionary stage, and shows some indications of a second outflow.
Reproduced by permission from the AAS from Feng *et al.* (2016).[27]

5.3 Chemistry in Jets

As we discussed in Section 5.2, the early phases of star formation are normally accompanied by the launch of hypersonic jets, with velocities up to about 10^3 km s^{-1}. These jets are very highly collimated on size scales of less than about 50 au. They carry a significant amount of kinetic energy (equivalent to about 10% of the accretion mechanical luminosity in sources of low stellar mass). Jets may be important in the removal of angular momentum from the accreting system. Consequently, jets may even affect planet formation and planet migration within a planetary system (see reviews by Turner *et al.* 2014;[30] Baruteau *et al.* 2014[31]). The detailed origin of jets remains a topic of vigorous discussion, but it is widely accepted (see, *e.g.*, Frank *et al.* 2014[1]) that the jets are driven by MHD processes in which accreted matter interacts with the magnetic field of the rotating stellar disc or the star (or both). Thus, material in the jets has come from relatively dense infalling matter, and—constrained by the rotating magnetic field—is dramatically accelerated from the disc and collimated in the jet-launch process. A schematic diagram representing this process, on a scale of a few au, is shown in Figure 5.4.

In principle, the nature of the material in the jet, and hence its dynamical history, may be approached through its emitted radiation. In the following sections we shall discuss the chemistry that may generate suitable diagnostics. Our ideas concerning the physical nature of the jets have developed rapidly following studies at high sensitivity and high angular resolution using the Spitzer, Herschel and ALMA facilities. As discussed in the previous section, jets are observed to be partially atomic and partially molecular, and they are also partially ionized. Densities and temperatures are typically high compared to those of cold dark interstellar clouds. There is good evidence from depletion studies that at least some jets are dusty. The material in the jets is accelerated to very high velocities of hundreds of km s^{-1} in a launch process that allows these important properties—*i.e.*, partially neutral, partially molecular, somewhat dusty—to survive. Direct evidence of a magnetic field is difficult to obtain; however, evidence for a magnetic field is seen in the HH80/81 jet (originating in a massive star, rather than a more typical low mass star) through its detected synchrotron emission. The structure of this field is revealed in the polarization of the emission (Carrasco-Gonzalez *et al.* 2010[32]), and appears to form a helix around the jet axis. Similar observations for jets around low mass stars have not yet been reported, probably because they do not emit synchrotron radiation.

Class II objects may be observed optically, and their jets are seen to have wide opening angles close to the source but collimated within ~ 100 au of the

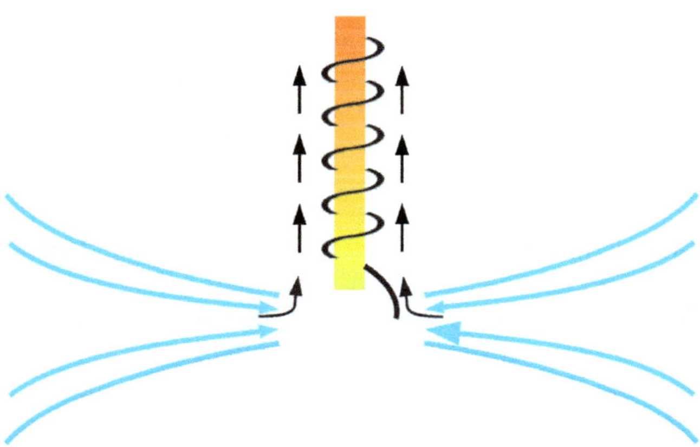

Figure 5.4 A schematic diagram of jet formation on a scale of a few au. The entire system of accretion disc (blue lines), wind (black arrows), and magnetic field lines (continuous black lines) is rotating. The field lines are rooted in the rotating disc and tend to constrain the wind into a narrow jet (yellow/orange).

star (Ray *et al.* 2007[33]). Optically obscured Class 0 and Class I objects may be observed in molecular lines (such as SiO; Cabrit *et al.* 2007[34]) at millimetre wavelengths or in the infrared (*e.g.* through [FeII], [SI], [SiII], and H_2 transitions; Dionatos *et al.* 2009,[35] 2010;[36] Davis *et al.* 2001[37]). Such observations show that very effective focusing of jets in these early objects also occurs, just as in the later objects, and on a similar size scale. This strongly indicates that the infalling envelope is not a major influence in focusing a jet, and that a general collimation mechanism is required that operates in all classes. Such a mechanism is provided by magnetic fields anchored in the rotating disc.

In the following section, we shall describe briefly some launch processes that are under discussion and their implications for jet chemistry. We shall then consider the observed structure in jets, its possible causes and chemical consequences. Finally, we shall consider processes at the main working surfaces of the jets, which may be outside the molecular cloud in which the jets have originated.

5.3.1 Jet Launch Processes

The launch process is clearly an MHD process, but the precise nature of the launch region remains unclear. It could be the stellar wind itself (a region of a few stellar radii), the magnetosphere–disc interface (a region of less than 1 au), or the disc wind itself (up to about 30 au). It is evident that these regions are quite distinct, and the chemistry of jets away from the launch region may provide useful diagnostics to distinguish between these models. Of course, it is possible that the launch process involves more than one of these regions.

Many theoretical and computational studies have been devoted to understanding the process in which the accretion of matter on to the protostellar disc results in the ejection of jets and winds. Casse and Keppens (2002,[38] 2004[39]) used numerical simulations to make the first demonstration that a collimated outflow could be launched at high velocity from an accretion disc. Zanni *et al.* (2007),[40] Tzeferacos *et al.* (2009),[41] Murphy *et al.* (2010),[42] and Sheikhnezami *et al.* (2012)[43] showed that the disc magnetization played a controlling part in the launch mechanism, its ejection efficiency and its energetics. More recently, Stepanovs and Fendt (2014)[44] explored the evolution of the accretion–ejection process over long periods of time. Stepanovs *et al.* (2014)[45] considered the effect of a large-scale magnetic field generated by the disc dynamo. In recent simulations, Stepanovs and Fendt (2016)[46] explored the relation between the local disc magnetization and astrophysically important quantities such as the outflow velocity, the jet's mass loading, the jet's angular momentum, and the local mass accretion rate. Their results led to the conclusion that discs that are strongly magnetized should launch jets that are more energetic, and that such jets should extract more angular momentum from the disc.

These and similar studies are helping to improve very greatly our understanding of how jets are launched from the discs during the accretion–ejection process. However, from the perspective of this book we wish to be able to explore the chemical evolution of material entrained in the accretion–ejection process, so as to be able to identify molecular species that may be suitable as potential tracers of the physical events described in the simulations listed above. Such chemical studies have not been performed in those works. In the next section we focus on numerical studies of the accretion–ejection process, which include in a self-consistent way of evolving the chemical description.

5.3.2 A Coupled Chemical-dynamical Model for Protostellar Disc Winds

The most detailed models, so far, of the chemistry coupled with the dynamics occurring during the launch of jets in protostellar disc winds have been explored by Panoglou *et al.* (2012)[47] and Yvart *et al.* (2016).[48] These models require consideration of many issues not normally required in astrochemistry, and represent some of the most complex astrochemical models ever constructed. We describe here the various aspects of these models.

Panoglou *et al.*[47] adopted a steady-state, axisymmetric, dynamical MHD model (an extension of the MHD jet model of Blandford and Payne 1982[49]) to determine the 3D total density, velocity, magnetic field, and current. Disc compression and sub-Keplerian rotation (Shu *et al.* 2008[50]) are included. A restriction on the ratio of the mass outflow rate to the accretion rate is imposed to ensure that predicted atomic line centroids are in harmony with observations of T Tauri atomic jets; this is called the "slow" disc wind solution (Casse and Ferreira 2000[51]). The geometry and density for this solution are illustrated in Figure 5.5, for an accretion rate of 10^{-6} $M_\odot$ $year^{-1}$ and a protostellar mass of 0.5 $M_\odot$.

The thermal, ionization, and chemical evolution occurring along dusty flow streamlines in this model may then be explored self-consistently as a function of three parameters: the protostellar mass, the mass accretion rate, and the mid-plane radius of the magnetic surface along which matter flows. However, the location of the material close to a newly forming star means that the conventional computations appropriate for the types of chemistry described in Chapters 3 and 4 need to be extended in various ways.

For example, heating sources in disc wind models may include not only the conventional sources of the external radiation field and the flux of cosmic rays. They also include the stellar radiation field appropriate for a low mass star, (taken by Panoglou *et al.*[47] to be a black body with effective temperature ~4000 K) and hot accretion spots on the stellar surface created when infalling material impacts directly on the surface (treated as a black body with effective temperature ~10000 K). Hard coronal X-rays from the star are powerful sources of ionization at the base of the dusty disc wind, and secondary electrons generated by these ionizations also dissociate and heat the gas. X-ray ionization rates are computed following Glassgold *et al.* (1997)[52] and Shang *et al.* (2002).[53] In the Panoglou *et al.*[47] model, the rate of H_2

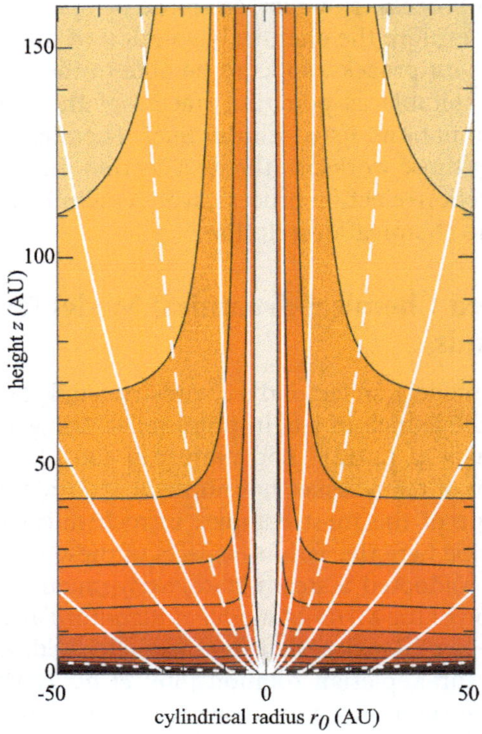

Figure 5.5 The "slow" MHD disc wind solution of Casse and Ferreira (2000)[51] in modelling of disc wind chemistry. Solid white curves represent magnetic flow surfaces, and the dashed line represents that magnetic surface anchored at a radius of 1 au in the disc. The density in the wind is determined by the accretion mass rate and the protostellar mass, and is shown for values of 10^{-6} $M_{\odot}$ $year^{-1}$ and 0.5 $M_{\odot}$, respectively; the number density increases from the lowest value of 8×10^{4} cm^{-3} in steps of 2.
Panoglou *et al.*, A&A, **538**, A2 (2012), reproduced with permission © ESO.[47]

dissociation by X-rays, and the heating efficiency of cosmic rays and X-rays are computed according to the treatment of Dalgarno *et al.* (1999).[54]

Where present, dust grains may play several important roles. In particular, dust accreted with the infalling gas absorbs and scatters radiation at all wavelength regimes. The photoelectric effect on dust grains is an important heating mechanism for the gas. Collisions with the dust may affect the gas temperature at the base of the wind; high dust temperatures may heat the gas, whereas low temperatures will tend to cool it. A suitable dust model needs to be adopted to account for all these processes. If the dust from the natal cloud is unmodified during the cloud collapse and subsequent star formation, then a conventional interstellar model may be appropriate. In the accretion jet launch process dust grains that are sufficiently close to the star may be heated to temperatures sufficiently high for sublimation to occur. Adopting a sublimation temperature of 1500 K, Panoglou *et al.* found that sublimation occurs in models (with the various radiation fields described in the previous paragraph) up to radii of about 0.2 au for Class 0 and Class I objects, and about 0.1 au for Class II objects. This suggests that while the roles of dust grains may be entirely suppressed on the innermost streamlines, grains may survive and have the active roles described above on streamlines more distant from the star. The view that sublimation of grains may be occurring in some circumstances in jets is supported by a study by Podio *et al.* (2009)[55] of the depletion of calcium in a number of jets. Their conclusion was that depletion is high (*i.e.* calcium is retained in dust grains) in low excitation jets, while depletion is low (*i.e.* calcium has a near-solar abundance, implying that grains have been destroyed) in high excitation jets. In this study, calcium may be regarded as an analogue of silicon, one of the main constituents of conventional dust grains.

Ambipolar diffusion will be an important source of heating in MHD flows. The elastic scattering between neutral atoms and molecules with ions and charged grains that are constrained to move with the magnetic field lines deposits energy in the neutral gas. Ohmic heating occurs as electrons drift through other fluids. Momentum transfer rates for these processes are computed by Panoglou *et al.*[47] according to the methods of Pinto and Galli (2008)[56] and Schunk (1975).[57]

Cooling is provided by conventional radiative processes of collisionally excited molecules, atoms, and ions. Panoglou *et al.*[47] include the following coolants: the molecules H_2, CO, H_2O, ^{13}CO, OH and NH_3 and atoms/ions C, N, O, Si, C^+, N^+, O^+, S^+, and Si^+. Abundances of these species may be determined along wind streamlines under the constraints of the MHD model, by a conventional astrochemical network of gas phase reactions, together with the formation of molecular hydrogen at grain surfaces and the possible formation or removal of ice mantles on grain surfaces. Panoglou *et al.*[47] also include a representative PAH molecule of 54 carbon atoms; with a fractional abundance relative to total hydrogen of one part in a million. The network adopted by Panoglou *et al.* describes the chemistry of 134 species interacting in 1143 reactions, with data for most of these reactions taken from familiar astrochemistry databases. Thus, the chemistry occurring in the launch regions is essentially conventional astrochemistry.

5.3.3 Chemical Results from a Protostellar Disc Wind Model

Results from the disc wind model obtained by Panoglou *et al.*[47] and Yvart *et al.*[48] are very encouraging. They adopt accretion rates of 5×10^{-6}, 10^{-6}, and 10^{-7} $M_\odot\,y^{-1}$ to represent Class 0, Class I, and Class II objects, respectively. The general structure of the predicted disc wind is that the ionization fraction and kinetic temperature decline with increasing radii at which streamlines are launched, from some critical small inner radius (~0.1–0.2 au) at which the dust grains are destroyed by sublimation. The ion–neutral coupling is sufficiently strong that neutrals are ejected along streamlines out to a radius of at least 9 AU. The kinetic temperature is controlled mainly through heating by ambipolar diffusion and cooling by molecular hydrogen. The computed temperatures in the disc wind are in the range 700–3000 K.

The chemistry determining the abundances of possible disc wind diagnostics is essentially controlled by reactions with molecular hydrogen. The disc wind model makes strong predictions about the survival of H_2 during the jet launch. For Class 0 objects, densities in the wind are high ($n_H \approx 10^{10}$ cm^{-3} just beyond the sublimation radius (~0.2 au) and about 0.1 au above the plane of the disc) and temperatures are sufficiently low (~300 K) that almost all of the initial molecular hydrogen (from the infalling gas) survives. For Class I objects, the lower accretion rate implies a lower effectiveness of dust shielding so that the far UV radiation field is much greater than in the Class 0 case. Just outside the sublimation radius (~0.2 au) photodestruction of H_2 at 1 au above the disc swamps formation on dust, and the fractional abundance is small ($n(H_2)/n_H \approx 10^{-4}$). However, on streamlines at greater radii (~1 au), most of the H_2 survives. The situation for Class II objects is similar to that for Class I objects. The survival of H_2 in the wind is a crucial effect, and arises partly from the geometry of the jet, which provides shielding for outer streamlines by dense material on inner streamlines. It also arises from the importance of timescales; in the wind, the flow timescale is short compared to the destruction timescale, so H_2 can survive in this dynamical situation. By contrast, H_2 cannot survive in static models, such as those of Rawlings *et al.* (1988),[58] Glassgold *et al.* (1989,[59]2004[60]), and Nomura and Millar (2005)[61] in which a steady-state chemistry has been assumed.

Astrochemistry is normally driven by reactions with H_2, and chemistry in disc winds is no exception. The high abundance of H_2 on streamlines at radii greater than ~1 au, together with shielding by dust grains (and by self-shielding in the case of CO and H_2) drives a conventional astrochemistry, generating substantial fractional abundances of simple molecules that have the potential to be tracers of the winds. For Class 0 and Class I objects on streamlines launched at 1 au, much of the carbon is tied up in CO so that its fractional abundance in these winds is $\sim 10^{-4}$ for very large heights (~10^3 au) above the disc plane. In Class II objects CO is predicted to have a smaller fractional abundance, at $\sim 10^{-5}$. Much of the available oxygen is tied up in H_2O formed in endothermic reactions of O with H_2 in the warm disc winds. The fractional abundance of H_2O approaches $\sim 10^{-4}$ in objects of Classes 0, I, and II. Two-dimensional cuts of H_2O fractional abundances in Class 0 and Class I objects computed by Yvart *et al.*[48] are shown in Figure 5.6.

Photodissociation of H_2O ensures that the fractional abundance of OH is also high in Classes I and II. Sulfur is mainly ionized in objects of Classes I and II by the stronger far UV radiation fields, and is mainly present as neutral atoms in Class 0 objects. A few per cent of sulfur atoms are converted to SO molecules at large heights (~10^3 au) above the disc in Class 0 objects. Note that SO (and SO_2) are very effective tracers of the HH 212 jet up to a few hundred au from the protostar (Podio *et al.* 2015),[62] a source also traced in lines of CO isotopologues, HCO^+, and SiO. Fractional abundances computed

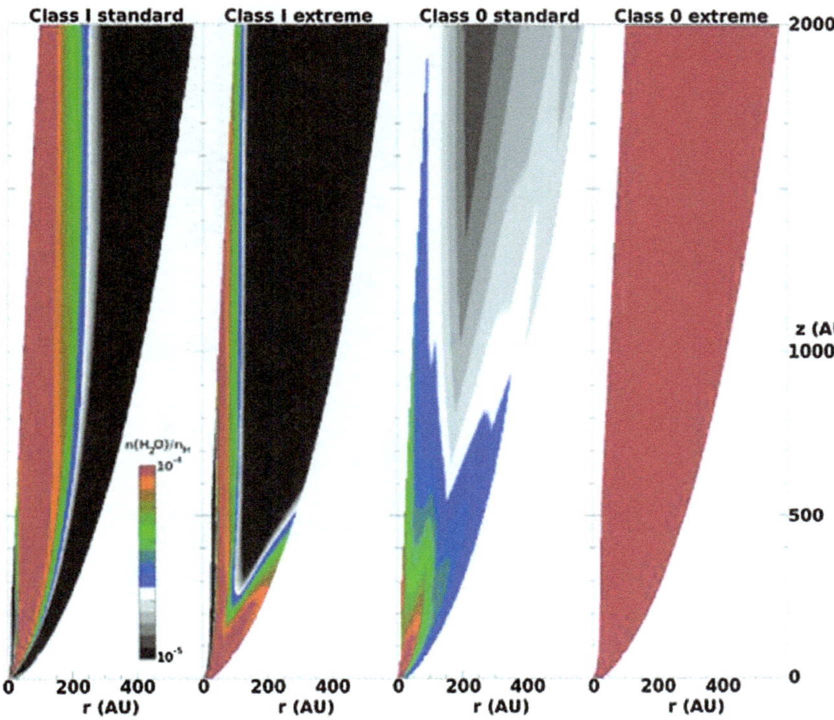

Figure 5.6 Two-dimensional cuts of the H_2O gas-phase fractional abundance for two Class 0 and two Class 1 models. The calculations are carried out for distances up to 2000 au from the star. The launch radii range from the sublimation radius out to a disc radius of 25 au. The colour scale represents fractional abundances ranging from 10^{-6} (black) to 10^{-4} (red). The models are defined by stellar mass (0.1 and 0.5 $M_\odot$ for Class 0 and I, respectively) and mass accretion rate (2×10^{-5} and 5×10^{-6} $M_\odot$ year^{-1} for extreme and standard Class 0, respectively; 5×10^{-6} and 1×10^{-6} $M_\odot$ year^{-1} for extreme and standard Class I, respectively).
Yvart *et al.*, A&A, **585**, A73 (2016), reproduced with permission © ESO.[48]

by Panoglou *et al.*[47] of some relevant atomic and molecular species on a streamline launched at 1 au are shown in Figure 5.7.

Yvart *et al.*[48] computed synthetic H_2O line profiles from the high abundances of H_2O predicted by the Panoglou *et al.* model of a dusty disc. The shapes and intensities of these profiles appear to be in harmony with observations of Class 0 and Class I objects made using the HIFI spectrograph on the Herschel satellite observatory (van Dishoeck *et al.* 2011[63]).

5.3.4 Implications for Jet Chemistry

The disc wind models discussed in Sections 5.3.1 and 5.3.2 predict that a large variety of atomic and molecular species with high fractional abundances may be considered as potential tracers of the jet launch

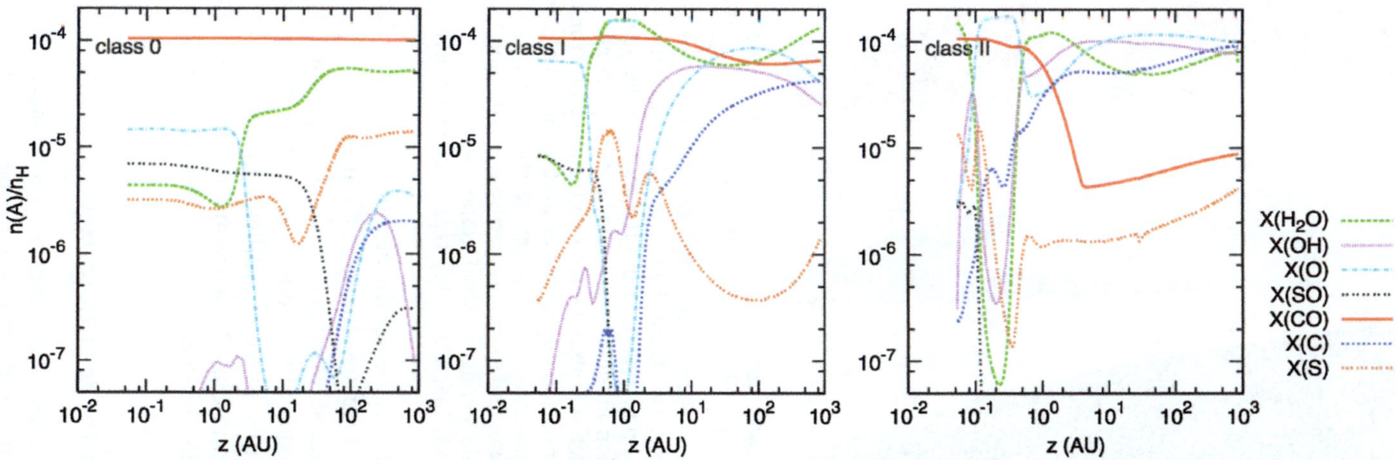

Figure 5.7 Disc wind chemistry for models of Class 0, Class I, and Class II sources; in all cases along the disc wind streamlines are launched from a radius of 1 au in the disc. The results are given as a function of *z*, the height above the disc. The fractional abundances of important O-, C-, and S-bearing species are shown for each Class model.
Panoglou *et al.*, A&A, **538**, A2 (2012), reproduced with permission © ESO.[47]

process. In particular, the detailed calculations of Yvart *et al.*[48] of H_2O line profiles from disc winds, using the Panoglou *et al.*[47] model, compare very satisfactorily with the Herschel/HIFI observations of H_2O lines (Kristensen *et al.* 2012).[64]

However, the actual launch site remains a matter for discussion. Matt and Pudritz (2008)[65] considered a launch from a stellar wind, and Zanni and Ferreira (2013)[66] discussed a launch from the zone where the stellar magnetosphere interacts with the inner disc. Such regions are likely to be more highly excited than the launch zones considered in the disc wind models and therefore more hostile environments for molecular tracers of jets.

Turbulent mixing of ambient gas with a highly supersonic jet may entrain molecules (White *et al.* 2016)[67] but the implied chemistry has yet to be explored in detail. Shock entrainment would require that the molecular gas rapidly refill the space behind preceding shocks (Davis *et al.* 2011).[68]

If the launch process occurs close to the star, then the molecular content will be small and the dust content destroyed by sublimation. A significant chemistry *internal* to the jet may then only be established through the re-formation of H_2 and subsequent gas-phase chemistry. The re-formation of H_2 in gas-phase processes is inefficient compared to H_2 formation on the surfaces of dust grains, but can be enhanced in low velocity (less than about 15 km s^{-1}) internal shocks in jets with variable velocity amplitude (see Section 5.3.5.1; Raga *et al.* 2005[69]).

5.3.5 Downstream Jet Chemistry and Structure

It is well known that Herbig–Haro jets (and, indeed, extragalactic jets as well) may show chains of high velocity emission knots in various tracers (see Figure 1.2; and the discussion of observations in Section 5.2). For example, the HH 1 jet shows a knotty structure in emissions from H_2, [FeII], [SII], and Hα (Reipurth *et al.* 2000;[70] Giannini *et al.* 2015[71]). A variety of processes have been invoked to account for the presence of knots within a jet. Recent ideas have included oblique shock focusing, possibly affected by magnetic fields, precession in the jet source, or interactions with the jet environment. In this Section we discuss briefly two models that have been put forward to account for the observed knotty structure in HH jets, and which have received much attention.

5.3.5.1 Variations in Jet Velocity

Raga *et al.* (1990[72]) were the first to explore models of HH jets, in which the ejection velocity varied with time, in an attempt to account for high proper motions in the chain of knots in the HH 34 jet (Eislöffel and Mundt 1992;[73] Heathcote and Reipurth 1992[74]). The variable jet models were also favoured by the observations (Reipurth and Heathcote 1992;[75] Reipurth *et al.* 2002[76]) that at least some of the knots were small-scale versions of the "heads" of the jets (the original HH objects themselves). Thus, the knots appeared to be a shock phenomenon, such as would occur if slightly slower elements of the jet were overtaken by faster components. The model based on this interpretation of knots has been used widely to interpret observations on chains of knots.

A recent multi-epoch study of a chain of well-defined knots traced by the Submillimeter Array in the SiO (8–7) line in the jet HH 211 powered by a Class 0 object confirms that the knots in this jet are moving with a high transverse velocity of $\sim$114 km s^{-1} (Jhan and Lee 2016[77]), as required in this model of knots. These observations of the knots are consistent with small periodical variations in the (high) jet velocity, as described by Raga *et al.* (1990).[72] The HH34 jet also shows a "wiggle", affecting symmetrically the jet and counterjet on both sides of the protostar and at the same epochs (Jhan and Lee 2016[77]). It seems likely that these "wiggles" are caused by the orbital motion of the jet source and that they propagate away from the central source. They are not caused by jet instabilities.

In this model of variability in the jet velocity, emission from knots is enhanced in the internal shocks in the jet. If emitting molecules survive the launch of the jet, then these are the molecules that are excited in the shocks (if the shock velocity is sufficiently low that dissociation does not occur—see Section 2.3.4). Another possibility is that chemistry may occur *in situ*. Raga *et al.* (2005)[69] considered the possibility that jet gas is dust free and atomic, a situation that should occur for jet launch on streamlines anchored sufficiently close to the protostar. They showed that a significant amount of H_2, $\sim$1–10%, could be formed in the jet purely by gas-phase reactions, and that emission in the H_2 1–0 S(1) line generated in the shocks could be more intense than that in Hα. However, the implications for chemistry in dust-free jets, based on *in situ* formation of H_2, remain to be explored, but products should include simple hydrides and oxides of carbon, oxygen, nitrogen, silicon, sulfur, *etc.* Figure 5.8 shows intensity maps for Hα and H_2 1–0 S(1) emission for a model of a dust-free variable jet after a time-integration of 120 years. In this model, the mean jet velocity is 200 km s^{-1} and has a sinusoidal variation of amplitude 10 km s^{-1} and a period of 10 years; the jet number density is 2×10^5 cm^{-3} and the total mass loss in the jet is 1.1×10^{-7} M$_\odot$ y^{-1}.

Perfect symmetry of knot positions in jets and counterjets is not seen in all sources. Spitzer/IRAC images of the HH 34 jet and counterjet are found close to (<0.1 pc) the protostar, but on the large scale ($\sim$1 pc) such symmetry does not occur, see Figure 5.9.

Departures from perfect symmetry of the knot positions between jet and counterjet can be attributed to the finite size of the jet production region. For HH 34, Raga *et al.* (2011)[78] found that this source size is less than 2.8 AU. This is a severe constraint on the size of the region in which the HH jet is produced.

The structures of HH jets can be regarded as a kind of "archaeological record" of the variable ejection of material from protostars (Raga *et al.*

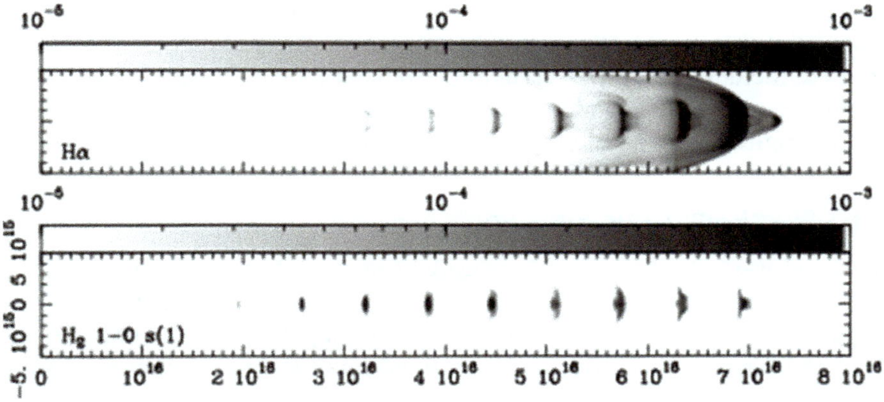

Figure 5.8 Intensity maps of Hα and H_2 1–0 S(1) emission computed for a model HH jet in which the jet number density is $2\times10^5\,cm^{-3}$, the half-amplitude of the assumed sinusoidal velocity variability is 10 $km\,s^{-1}$, and the time-averaged mass-loss rate is $1.1\times10^{-7}\,M_\odot\,year^{-1}$. The intensities are indicated in logarithmic greyscale shown on each graph, in units of $erg\,cm^{-1}\,s^{-1}\,sterad^{-1}$, and the axes are shown in cm.
Reproduced with permission from Raga *et al.*, *Rev. Mex. Astron. Astrofís.*, 2005, **41**, 137.[69]

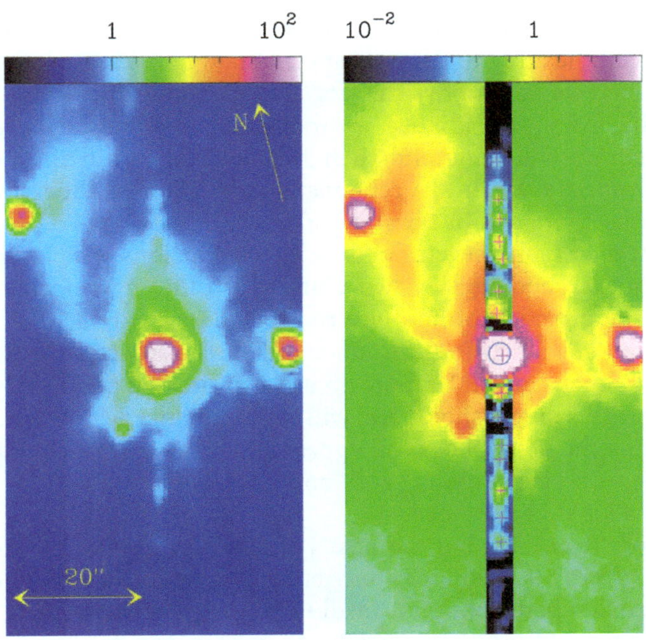

Figure 5.9 Jet/counterjet symmetry in HH 34 determined from infrared images from the Spitzer Space Telescope. The left hand panel is the image of HH 34 created by adding emission in the 4.5 and 5.8 μm channels; the right hand image has background subtraction in a strip along the jet/counterjet system, revealing the positions of the knots, which are marked by red crosses.
Reproduced by permission from the AAS from Raga *et al.* (2011).[78]

2015).[79] If this record is to include not only the aligned knots close to the protostar but also the "large head" at greater distances, then the model needs to include multimode variability; such a model requires at least a low-amplitude, short period mode to produce the near-zone knots, and a long period, high-amplitude mode for the more distant working surface.

5.3.5.2 *"Hypersonic Buckshot"*

Hartigan *et al.* (2005)[80] (see also Frank *et al.* 2009;[81] Hansen *et al.* 2016[82]) noted that there are jets with non-axial morphology in which knots are displaced to one side of the jet axis; for example, in HH47 this morphology takes the form of an apparent helical bending of the jet.[80] High resolution multi-epoch observations reveal time-dependent structures that imply the presence of heterogeneous density distributions within the jet beams (Hartigan *et al.* 2011).[83] For example, some observations of Hα emission appear to be from separate bow shocks within the beam and also from the intersection of such bow shocks.[83]

Such behaviour seems inconsistent with the model based on variations in jet velocity, described in Section 5.3.5.1, and so an alternative model has been developed in which the jets are essentially clumps embedded in an otherwise smooth beam. The clumps are of diameters less than that of the jet, and have a range of velocities (parallel to the jet velocity) so that clump collisions occur, leading to the formation of bow shocks. Such a system has been described as "astrophysical jets as hypersonic buckshot" by Frank *et al.*[81] Using fully 3D radiative hydrodynamic methods, Hansen *et al.*[82] simulated clump collisions and their bow shocks within the jet, and included appropriate cooling mechanisms that allowed them to generate synthetic emission maps in Hα and [SII] lines.

Hansen *et al.*[82] found several interesting conclusions. The intersection of two bow shocks generated by two clumps of differing velocity will move laterally to the direction of flow (see the sketch in Figure 5.10).

As Hartigan *et al.*[83] point out, since gas in the intersection has passed through two shocks, it will be more greatly compressed and heated than gas passing through a single shock, and will show enhanced emission. Thus, the intersection of the bow shocks will appear as a transversely moving object in a time-series of observations, and one that is hotter and denser than material in a single bow shock.

Regions of "froth" behind the heads of fast bow shocks are predicted to occur. This "froth" is strongly heterogeneous emission. Faster shocks show stronger density perturbations, leading to brighter and more extensive "froth". The term "froth" was introduced by Hartigan *et al.*[83] in describing emission regions in HH jets observed by HST.

The effects of several clumps acting closely together can create the appearance of a "sheet" of emission in Hα. To create such a "sheet" the clumps are required to have a spacing that is smaller than the average bow shock width. An example taken from Hansen *et al.*[82] is shown in

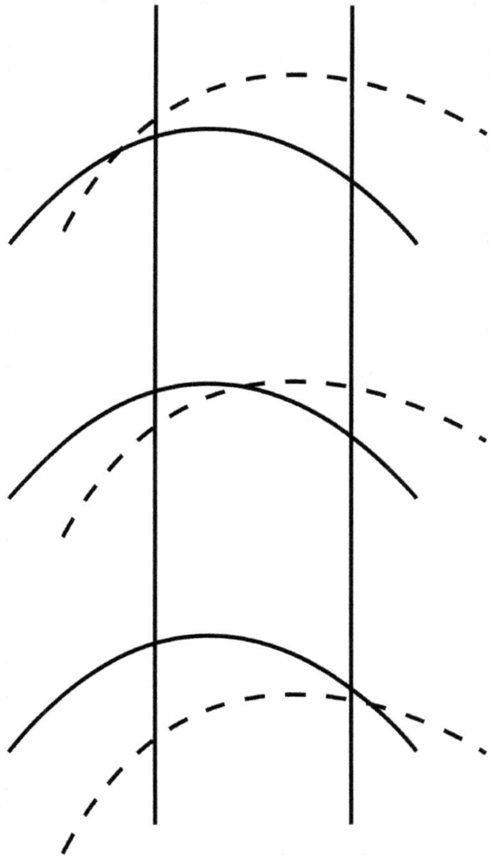

Figure 5.10 A sketch to show the apparent lateral motion of emission from the intersection of two bow shocks set up by independent clumps with differing velocities. The slower bow shock (solid line) is overtaken by the faster bow shock (dashed line). In the diagram, time increases upwards, and the intersection of the two bow shocks moves towards the left.

Figure 5.11. Ten clumps with differing speeds combine to create a transient object that dissipates as the clumps separate. Hansen *et al.* calculated the time evolution of the emission line structure from bow shocks associated with these clumps.

Hansen *et al.*[82] compared their model results with observations obtained by the *Hubble Space Telescope* over several epochs and found substantial qualitative agreement. Lateral motions of emission clumps have, for example, been found in HH 34. Evidently, a region of intense Hα emission should be regarded as a shock feature, not a clump. Hα "froth" has been widely found in observations (for example, in HH 1; Hartigan *et al.*[83]). Sheets of Hα emission have been found in HST observations. The compelling agreement with predicted and observed features in HH jets provides strong

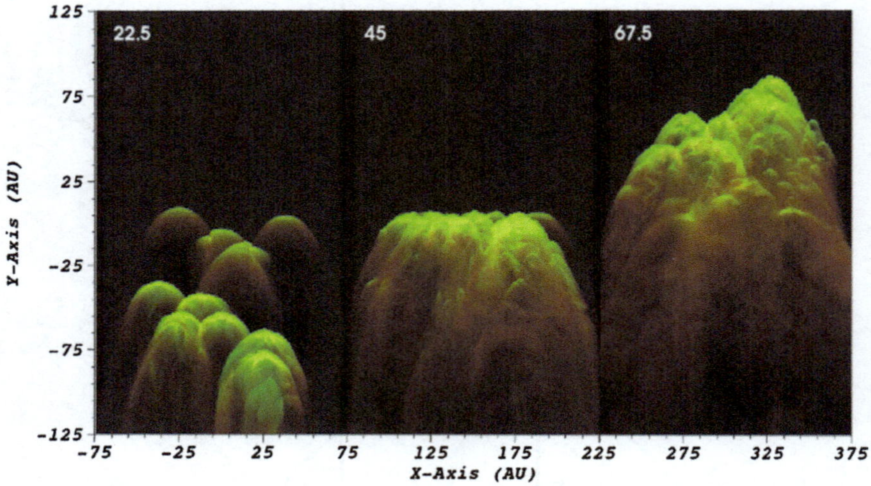

Figure 5.11 Numerical simulations of jets assumed to be composed of many small-scale clumps moving independently. Panels from left to right are at simulation times of 22.5, 45, and 67.5 years, and bow shocks from clumps with random velocities may combine to create a sheet of shocked material.
Reproduced by permission from the AAS from Hansen *et al.* (2016).[82]

support for the "buckshot" model; *i.e.* the jet beam is composed—at least in part—of small clumps with a range of velocities.

5.3.5.3 *Implications for Chemistry*

The two models described here may not be mutually exclusive. A variable velocity appears to be a necessary aspect of the buckshot model. Perhaps clumpiness may be incorporated into the variable velocity model. If ejection velocities are varying, then it seems plausible that densities are also varying. The main restriction on both models is that the velocity variation is not too great.

In terms of chemistry, it seems that there will be little to choose between these models. Both models are shock driven, but shock speeds cannot be very great. Otherwise, the shock becomes dissociative. The range of molecules likely to be established in each model is likely to be similar. However, the models differ in one significant respect; in the "buckshot" model, intersecting bow shocks play an important role, whereas these do not exist in the variable velocity model. These hotter and denser intersections should show some chemical differences. The line emission characteristics from the intersections, especially in the rotational excitation of H_2 and CO, should also be distinct. Thus, astrochemistry may be able to contribute to the understanding of the jet structure, but this work has yet to be done.

5.3.6 The Terminal Working Surface and Its Chemistry

The terminal working surface is where the collimated jet, with a velocity of $\sim$100–1000 km s^{-1}, impinges on the ambient molecular cloud, or—having escaped from the local star-forming region—impinges on more distant material. Jets up to $\sim$10 pc in length have been observed (see Section 5.2). As far as astrochemistry is concerned, very high velocity direct impacts are obviously dissociative for molecules and destructive for grains (see Section 2.3.4 for a discussion of dissociative shocks and the post-shock recovery) and the impact creates a characteristic nebulosity, the so-called Herbig–Haro (HH) object. Such objects have temperatures and densities similar to or rather larger than those of HII regions, and were first observed by optical emissions from ions and atoms independently by G. Herbig and G. Haro around 1950. There are no molecular tracers of such regions until re-formation of molecular hydrogen has occurred (see Section 2.3.4). The wings of some bow shocks are seen in H_2 emission.

The shape of the bow shock set up by the impact of a jet of finite size on the environment implies that the magnitude of the component of the flow velocity perpendicular to the shock front declines along the bow shock, so that at sufficient distance behind the shock head the post-shock conditions are no longer dissociative, and the bow shock can be traced in molecular emissions. In fact, the interaction of a finite jet with ambient gas is more complex than pre- and post-shocked gas, and—as shown by Lim *et al.*[84]—the impact sets up a number of distinct regions: (i) shocked jet gas; (ii) shocked molecular cloud gas; (iii) expanding shocked jet gas; (iv) expanded shocked molecular cloud gas; and (v) post bow shock environment gas (see Figure 5.12(a)). Lim *et al.*[84] developed a non-magnetic model in which hydrodynamics, thermal behaviour, and the chemistry of a hundred species interacting in over 1500 reactions are self-consistently determined using adaptive mesh techniques within all the distinct regions (i)–(v). Simulation results for several molecular species are shown in Figure 5.12(b). Potential molecular tracers of the interface regions in the form of rovibrationally excited molecular hydrogen, diatomic hydrides, and oxides of O, C, N, and S, and other conventional species such as HCO^+, H_2O, HCN, CS, and NH_3 were identified. The relative distribution of these tracers within the various post-shock regions is shown by Lim *et al.*[84] to depend sensitively on the physical conditions in the jet and in the environment.

Infrared counterparts to HH objects have been observed. They are detected mainly in lines of rotationally excited H_2 and CO. They have a bow shock shape, and are excited by much smaller velocities than their optical counterparts (Smith *et al.* 2003).[85]

5.3.6.1 HH Objects as Probes of Dark Clouds

The HH object created in the terminal working surface of a jet can be a major source of UV and optical radiation in molecular clouds, regions of high

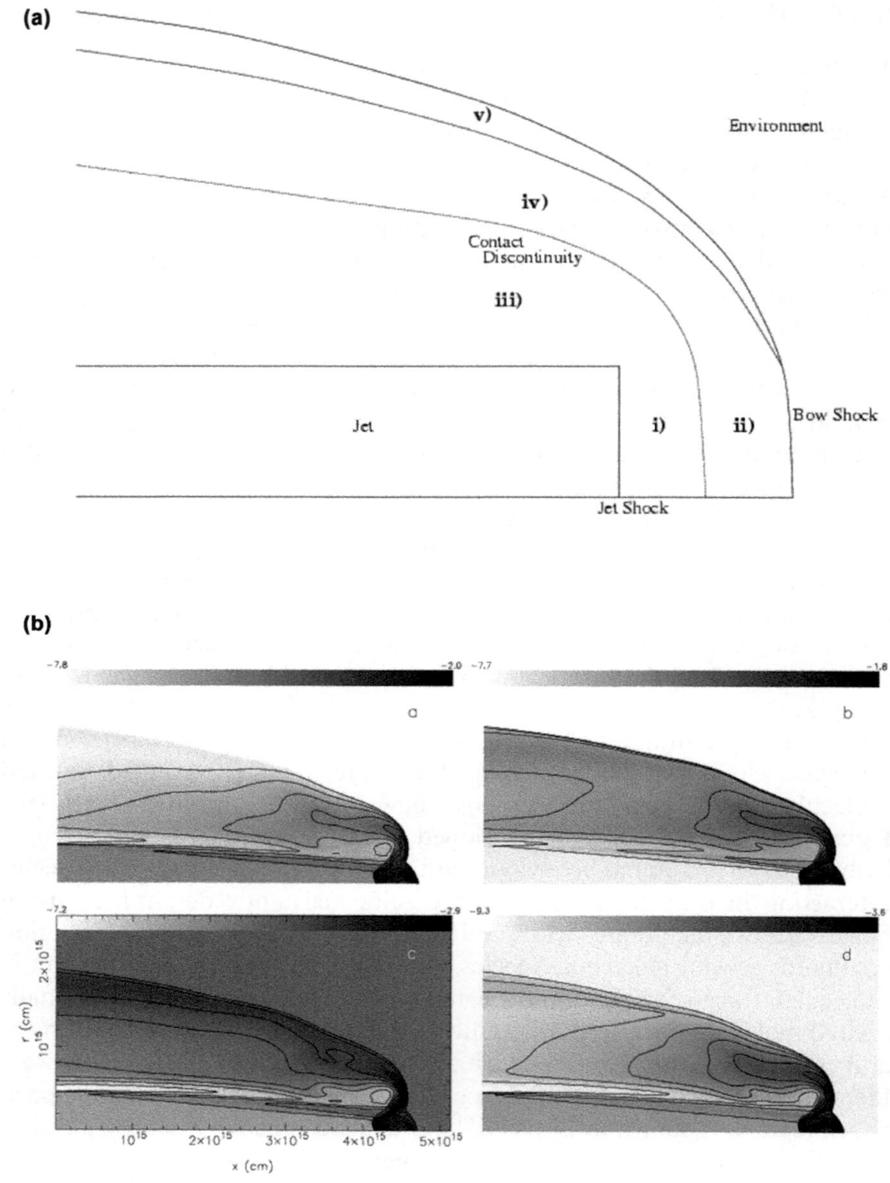

Figure 5.12 Numerical simulations of the impact of a fast jet on interstellar gas, and consequent chemistry. (a) A schematic diagram shows the various regions of interaction. (b) The results of the simulations are shown for abundances of CS, HCN, NO, and SO for a jet of unperturbed number density 100 $H\,cm^{-3}$, travelling at 50 $km\,s^{-1}$ into environment gas also of 100 cm^{-3}. The fractional abundances are shown (as $\log_{10}$) indicated in the greyscale bar above each part of the figure.
Lim *et al.*, A&A, **376**, 336 (2001), reproduced with permission © ESO.[84]

extinction that would otherwise be expected to be very dark. Where this occurs, then the chemistry in the cloud responds to this additional energy source and specific chemical tracers can be formed.

For example, Lefloch *et al.* (2005)[86] made a detailed study of the protostellar jet HH 2 and the parental molecular cloud in the mid-IR (wavelengths 5–17 μm) and also in millimetre lines of SO, CO, and ^{13}CO. They interpret the mid-IR emission at 11.3 μm as arising from polycyclic aromatic hydrocarbons (PAHs) and very small grains (VSGs) transiently heated by a far-UV radiation field with intensity that is about 20x the intensity of the mean interstellar radiation field. The PAHs and the VSGs are released by the HH shock, which also creates the far-UV radiation field.

Torrelles *et al.* (1992),[87] Girart *et al.* (1994)[88] and Girart *et al.* (1998)[89] made observational studies of a number of molecular cloud regions near HH objects. They found clumps within the molecular cloud in rovibrational lines of HCO^+, NH_3, and CS, with implied abundances enhanced above those found elsewhere in the molecular cloud. These clumps are about 0.1 pc in size. They are cool and quiescent (*i.e.* they have not been shocked) and have velocities that are typical of the molecular cloud rather than of the HH object. Girart *et al.*[88] suggested that these molecular enhancements might arise because of the photodestruction of icy mantles on dust grains within the molecular cloud, caused by the local irradiation by the UV field of the HH object. In effect, the HH object could be considered as a *probe* of the interior of molecular clouds; we use a fortuitously located HH object to irradiate a cloud with UV and obtain structural information about the molecular cloud (see the sketch in Figure 5.13).

Indeed, the clumps identified in this way are physically similar to clumps in molecular clouds that have been found from straightforward high-resolution direct observations (*e.g.* Morata *et al.* 2005).[90] Most of those clumps were found to be unbound, but a few were possibly gravitationally

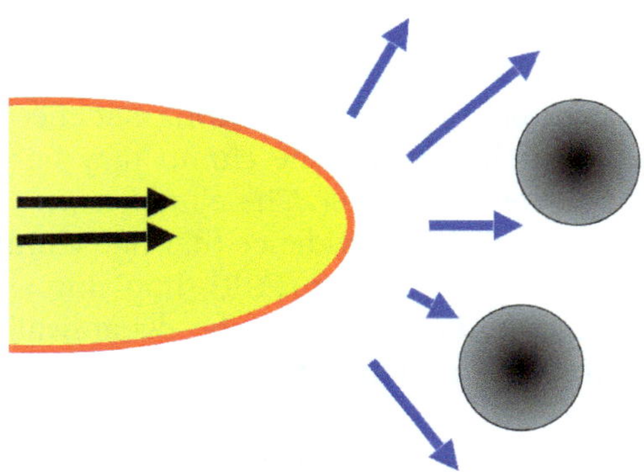

Figure 5.13 A diagram illustrating how a fast stellar jet (black arrows) interacting with a molecular cloud sets up a bow shock (in red), behind which the hot post-shock gas (yellow) is a source of UV (blue arrows) that may initiate chemical change within dense clumps (grey) of the molecular cloud.

unstable and may represent very early stages in potential star formation. Of course, it is possible that a nearby HH object not only irradiates a cloud clump, but also physically affects its evolution. Girart *et al.* (2001)[91] found evidence suggesting that the HH 80/81/80N outflow may have triggered the formation of a Class 0 object with bipolar winds. But in many situations, the interaction between the clump and the HH object appears to be simply irradiation of the clump.

Taylor and Williams (1996)[92] confirmed that the photochemistry induced by radiation from HH objects in molecular cloud clumps could be quite significant, as suggested by Girart *et al.*,[88] and Viti and Williams (1999)[93] investigated a similar but more extensive model. They showed that a rich chemistry, distinct from cold cloud chemistry and persisting for $\sim 10^4$ years, should result from such irradiation, and that a wide range of molecular tracers (other than the tracers HCO^+, NH_3, and CS observed by Girart *et al.*[88,89]) of photochemistry in otherwise dark clouds could be utilized. These include CH_3OH, H_2S, C_3H_4, H_2CO, SO, SO_2, H_2CS, and NS. The model included the gravitational collapse of interstellar gas from a diffuse state to typical molecular cloud clump densities and extinctions, with gas-phase reactions, ice deposition and surface chemistry on dust grains evolving during the collapse. The dark cloud clump is then abruptly irradiated by the HH object and the evolution of the resulting photochemistry is described. Viti and Williams[93] also concluded that these clumps were necessarily transient, with lifetimes of $\sim 10^6$ years. Thus, HH objects as probes are able to elucidate not only cloud structures but, in principle, may also begin to constrain the time dependence of cloud structures.

Detailed observational and theoretical studies (Girart *et al.* 2002;[94] Viti *et al.* 2003;[95] Girart *et al.* 2005[96]) have been made of the molecular gas associated with HH 2. A molecular line survey in the wavelength range 3–0.8 mm revealed emissions from 14 molecular tracers, confirming the prediction of a rich photochemistry (Viti and Williams 1999).[93] The molecular maps of this region show very complex morphological, kinematical and chemical structures (see Figure 5.14).

Several clump structures are found with strong evidence of photochemical evolution, confirming earlier work. One clump may be partially shielded from the UV source in the HH object. The clump nearer to the HH object may have higher density. There is evidence that one clump may also be affected by a photon dominated region (PDR), implying a fairly intense UV field. However, some material appears to be dynamically affected by the outflow from the protostar, although none of the conventional molecular shock tracers appear.

The main conclusion revealed by using HH objects as probes is that clumpiness seems to be present in many molecular clouds (see Viti *et al.* 2006[13]). The clumps are typically quiescent, but show enhanced abundances of certain species; these enhancements can be created by photochemistry initiated by the HH object. The evidence from the HH 2 study suggests that some of these clumps are affected dynamically by the HH jet.

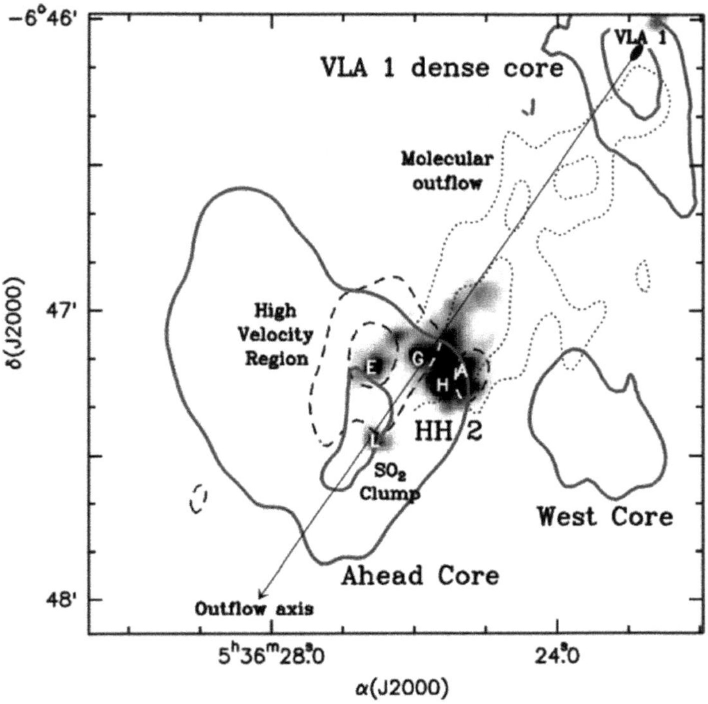

Figure 5.14 A sketch of the HH 2 region. The source of the jet is VLA 1, and the direction of the outflow and the HH 2 region caused by the outflow are indicated. The thick grey contours show the main molecular regions associated with HH 2; these are named the "ahead core" and the "west core". The dashed lines are HCO^+ contours that show a jet-like molecular outflow. The greyscale shows [SII] emission, with various knots within it identified.
Girart *et al.*, A&A, **439**, 601 (2005), reproduced with permission © ESO.[96]

5.3.6.2 HH Objects in Motion

HH objects are not static, but are in rapid motion with respect to the molecular clouds through which they penetrate. They travel a distance roughly equivalent to the diameter of a clump of the type observed by Girart *et al.*[88,89] and Morata *et al.*[90] in a time of the order of a thousand years. This is very roughly comparable to the time taken for the photochemistry to come into steady state, so, strictly, one should allow the UV source that is the HH object to move with respect to the clump. If so, then the orientation of the HH object with respect to the clump changes in time, and the photochemical response of the clump to the incident HH radiation affects different parts of that clump. Thus, the image of the irradiated clump in tracer molecules enhanced by photochemistry also changes in time; *i.e.*, the clump morphology as revealed by the distribution of the photochemical tracers in the clump appears to change (see Raga and Williams 2000[97]). The morphology also depends sensitively on the orientation of the line of sight with respect to the HH trajectory and the clump, and to the extinction in the clump. Emission regions from the photochemical tracers may appear as bars, arcs, or (as one might expect) clumps. Therefore, the photochemical tracers do not necessarily provide a complete map of an entire clump, and clump masses observed in this way may be significantly underestimated. These secular changes in clump description occur on timescales of a few hundred years.

If the distance between clumps is rather large, then the active photochemical state of one clump may cease before another clump is irradiated. If so, then an HH object that is irradiating a molecular cloud will be accompanied by no more than a single photochemically activated clump. Conversely, an HH object may irradiate several clumps if they are close enough together. It's like walking down a long corridor with occasional lights that are activated by timed switches. If the lights are close enough, then the corridor will be adequately lit, but if the lights are too sparsely distributed, one will be walking in the dark. From the limited statistics of clumps in molecular clouds irradiated by HH objects, it appears that the clumps may have a high filling factor, similar to that of the clumps identified separately by Morata *et al.*[90] From a sample of 22 clumps irradiated by HH objects, and using a calculation of the mean separation between such clumps made by Taylor and Williams,[92] Whyatt *et al.* (2010)[98] estimated that the area filling factor of molecular cloud clumps affected by HH objects is of the order of 10%.

Christie *et al.* (2011)[99] examined the photochemistry in a clump stimulated by the approach and departure of a nearby HH object. The time dependence of the radiation field has significant effects, because the radiation field not only creates tracers but may also destroy them, so the decline in the radiation field may help some tracers to survive. Some species have much higher abundances at all times for a "moving" HH source than for the "static" case; these include CH_3OH, SO, and SO_2. However, other species are initially found to have smaller abundances in the "moving" case than in the "static" case but are more abundant at (observationally significant) later times ($\sim 10^4$ years). These include SO_2, SO, NO^+, HCO^+, HCN, CN, CS, and OCS. The species with most notably enhanced abundances in the "moving" case are CH_3OH, NH_3, SO_2, SO, and H_2S.

5.4 Outflows

5.4.1 The Outflow/Core Interface

While collimated jets are perhaps the most dramatic component of low mass star formation, wide-angle bipolar outflows are probably the most important, even though their velocities (typically ~ 30 km s^{-1}) are very much lower than jet velocities ($\sim 10^2$–10^3 km s^{-1}). As discussed in Section 5.2, outflows are very visible signs of star formation. They inject mass, energy and momentum into the environment, creating large-scale cavities that grow in extent and width. They mix material from the star-forming core with cold molecular cloud material, and the mixing process may generate characteristic molecular tracers. As the opening angle of outflows increases, the infall onto the protostar may be suppressed and terminated. Thus, outflows may be responsible for determining the final stellar mass and the core : star efficiency of star formation.

An early study of the interaction between outflow and infall in a star-forming region was made by Velusamy and Langer (1998).[100] Their $^{12}CO(2–1)$ emission study of the "red" and "blue" wide-angle outflows from the low mass object IRS 1 in the Barnard 5 molecular cloud (B5) showed that the outflow opening angle near the source is increasing at a rate of about 0.006 degrees year^{-1}, while the infall measured in ^{13}CO (2–1) emission is confined to a disc perpendicular to the symmetry axis of the outflow lobes. Velusamy and Langer suggested that the opening of the outflow may eventually isolate the disc from further infall. If the rate of opening is sustained, this termination of infall would occur in a few $\times 10^4$ years. Cantó *et al.* (2008)[101] developed a thin shell analytic model for an outflow with an opening angle that increases with time. The form of the cavity depends on the structure of the environment and the mass loss rate. Cantó *et al.*[101] found that the form of the cavity in B5 IRS 1 is well fitted by their model, for a constant mass loss rate of 3×10^{-6} $M_\odot$ year^{-1} into a medium with density that falls with inverse square of distance from the protostar, an outflow velocity of 100 km s^{-1}, and a decollimation rate of 10^{-4} rad y^{-1} (see Figure 5.15).

A high resolution multi-molecular line study by Arce and Sargent (2006)[102] of nine low mass protostars at different evolutionary stages found similar results, with a clear trend in morphology. Class 0 sources have outflow opening angles close to the source of less than 55 degrees, while the more evolved Class I sources have opening angles greater than 75 degrees (see Figure 5.16(a)). Class II sources have even wider lobes or have no definite shape (see Figure 5.16(b) for a sketch).

5.4.2 Chemistry in the Mixing Zone

Arce and Sargent[102] detected ^{13}CO and HCO^+ emission from the cavity walls and from gas close to the protostar. The HCO^+ abundance is enhanced above that normally found in cold molecular clouds. However, it is consistent with a special chemistry in the turbulent mixing layer in the interface between the outflow and the ambient or infalling gas. In this interface, there is a mixture of dense cold gas from the molecular cloud with warmer gas from the outflow. The release of water molecules from icy mantles on dust grains and their reaction with C^+ ions creates HCO^+ molecular ions in abundances that can be as much as several orders of magnitude larger than those in cold quiescent molecular clouds (Rawlings *et al.* 2000;[103] Rawlings *et al.* 2004[104]). Note, however, that the interface chemistry is transient, and its observability depends on the interface renewing itself rapidly enough as

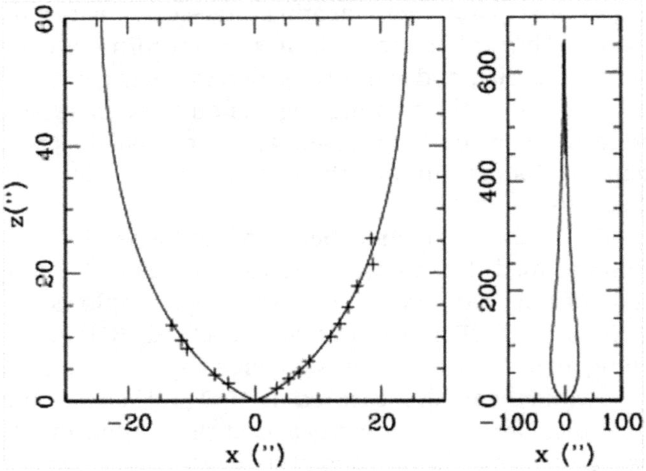

Figure 5.15 The solid line represents a fit obtained for the shape of the NE cavity of B5 IRS1 (crosses) as measured by Velusamy and Langer (1998),[100] with distances in arcseconds. The model assumes an inverse square density variation with distance from the star and a constant mass loss rate of 3×10^{-6} $M_\odot$ $year^{-1}$. The left hand figure shows the base of the cavity, while the right hand figure shows the full shape of the predicted cavity. Reproduced with permission from Cantó *et al.*, *Rev. Mex. Astron. Astrofis.*, 2008, **44**, 293.[101]

the cavity expands. This type of interface chemistry also promotes relatively high abundances of some other species, including H_2S, CS, H_2CS, SO, SO_2, and CH_3OH (Viti *et al.* 2002[105]).

The interaction of outflows with their cores has recently been the subject of intense study. For example, Arce *et al.*[9] and Zhang *et al.*[14] used ALMA Cycles 0 and 1 observations to study the HH 46/47 outflow, and Zapata *et al.*[4] studied the young star DG Tau B using the SMA. Arce *et al.*[9] estimated the outflow mass, momentum, and energy to be sufficient to disperse the parent core within the lifetime of the embedded phase of a low mass star, and the core-to-star efficiency as 0.25–0.33. The outflows in HH 46/47 and in DG Tau B are found to be slowly rotating, a situation interpreted as arising in a wind launched from large radii on the disc. The molecular outflow is found to be composed mainly of locally entrained core material, implying that core material joins the outflow as the cavity broadens. In the case of the red-shifted outflow (at least) the outflow cavity wall is composed of multiple shells entrained in a series of bow-shock events. Such locations provide the mixing that was the basis of the chemical models discussed in the previous paragraph.

Nearly all studies of the outflow/core interactions have been carried out for low mass protostars. However, as described in Section 5.2.3, Codella *et al.*[29] observed an extended dynamically distinct feature in the Cep A East region in which massive star formation is occurring. The chemistry and dynamics

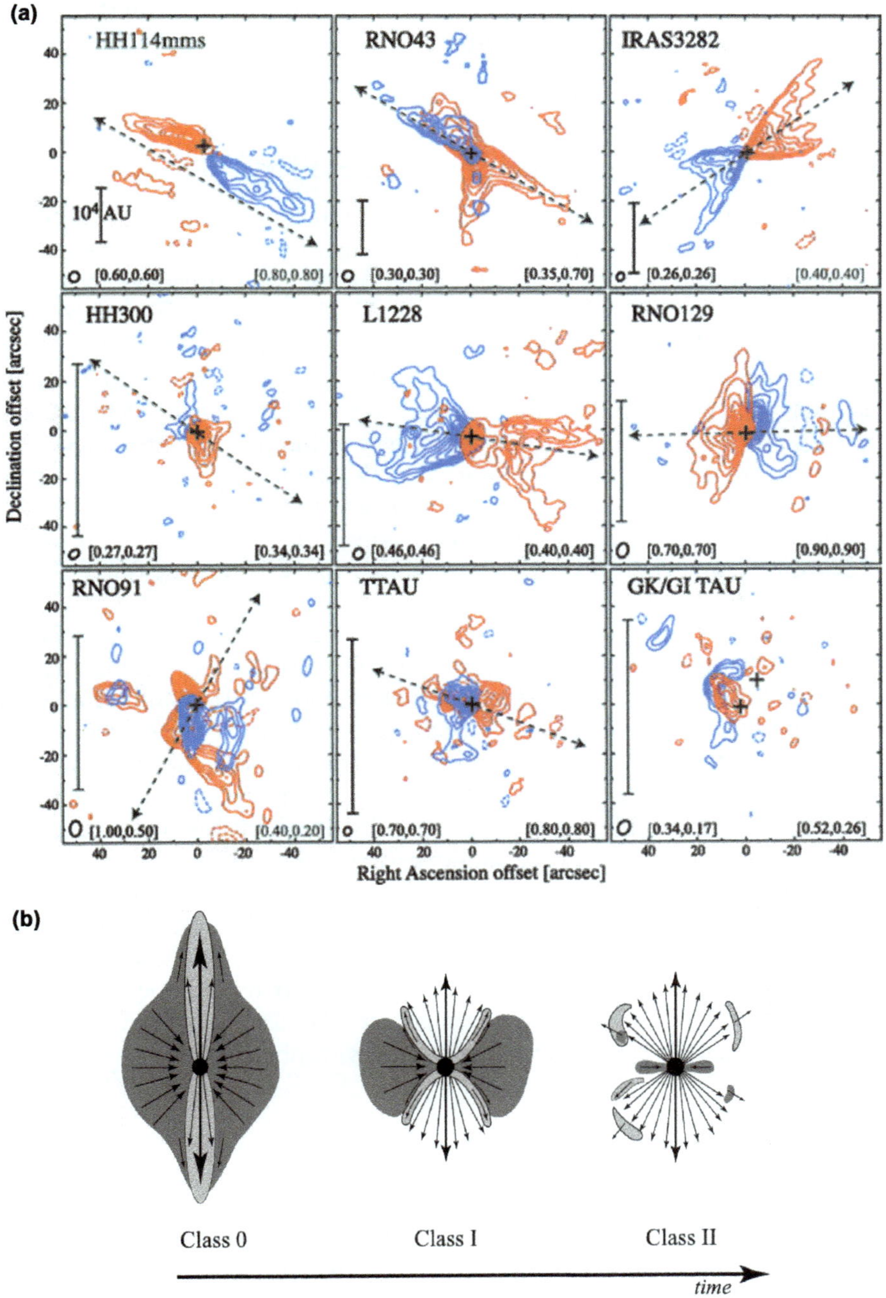

Figure 5.16 (a) ^{12}CO (1–0) outflows for Class 0 (top), Class I (middle), and Class II (bottom) sources. The dependence of the outflow opening angle on Class is clearly evident. (b) A schematic diagram of the evolution of the infall/outflow interaction.

Reproduced with permission from the AAS from Arce and Sargent (2006).[102]

of this feature are consistent with a core/wind mixing model, and the feature may be the first detection of interface chemistry in a region of high mass star formation.

5.4.3 Outflow Morphologies and Chemistry

Rollins *et al.* (2014)[106] note that in studies of low mass protostellar outflows there is an apparent observational preference for wide-angled cavities in carbon monoxide, within a few thousand au of the source (see, *e.g.*, Arce *et al.*,[9] Rawlings *et al.*,[104] Zhang *et al.*[14]). Rollins *et al.*[106] termed the morphologies that are wide-angled at the base as "brandy snifters", while the alternative (a small opening angle at the base, and the outflow always narrow)—of which very few are observed—was described as a "champagne flute" morphology. Why are there so few of the champagne flutes, given that in the picture first discussed by Velusamy and Langer[100] the evolution of the outflow cavity is, in all cases, from a champagne flute to a brandy snifter? For example, Li *et al.* (2013)[107] modelled cavity shape in terms of turbulent entrainment, and their results also suggested that champagne flutes should be observable.

Rollins *et al.*[106] developed a model of outflow chemistry and line emission based on the analytic model of cavity morphology and evolution of Cantó *et al.*[101] Rollins *et al.*[106] used the parameters adopted by Cantó *et al.*[101] to describe the B5 IRS 1 source, and assumed that an interface thickness at a particular radius was ten per cent of the radius, a value suggested by observations (Langer *et al.* 1996[108]). Material from the outflow (including ions H^+, He^+, C^+, and O^+) is mixed turbulently into the interface at a rate that depends crucially on the angle between the radius vector and the tangent to the cavity. The mixing is considered to be most effective when the wind from the protostar makes a large angle with the interface. In particular, the winds in a narrow "champagne flute" cavity cause a low level of mixing, while in a broad "brandy snifter" cavity, the mixing can be significant. Rollins *et al.*[106] showed that emissions from CO and HCO^+ in the older, broader cavities are qualitatively consistent with observations.

5.5 Conclusions

Stellar jets and outflows are some of the most dramatic—and unexpected—situations in astronomy. Who, half a century ago, could have predicted the existence of collimated hypersonic jets extending from newly formed stars into their environments for distances up to parsec scales? Who could have predicted that supersonic outflows would sweep out cavities around those stars, and that those cavities would be bounded by molecular-rich interfaces? We now know that these situations are the inevitable consequences of star formation by gravitational collapse in a partially ionized, slowly rotating medium. Given the violent nature of these events, it is perhaps also surprising that fragile simple molecules turn out to be very effective tracers of the underlying physics that determines the evolution of these regions.

The physical processes that generate jets and outflows are essentially described using an MHD approach. However, molecules play two very important and active roles in controlling these MHD processes. First, molecules are important coolants; they effectively remove gravitational potential energy from a system, allowing collapse to continue, while the energy is radiated away in low energy rovibrational transitions. Second, the chemistry that generates molecules is largely driven through ion-molecule reactions in which ionization is largely created by the action of cosmic rays. This ion–molecule chemistry determines the fractional ionization, and thereby the rate of ambipolar diffusion. Therefore, chemistry determines the ability of the magnetic field threading the gas to drag that gas with it, creating jets and outflows. Understanding the chemistry in these regions is therefore absolutely fundamental to a proper description of the formation of jets and outflows.

Molecules also play a passive role—as tracers of the physics of star formation and its consequences. This role is different in nature, but no less important. Models of star formation and its consequences are highly complex—perhaps some of the most complex models in astrophysics. These models are successful but still developing. As we have described in this chapter, molecules provide the observational tests required to assess the success of new ideas in the theoretical modelling, and in making progress in our understanding. However, while the observational and theoretical study of the relevant dynamical processes near newly formed stars has been making rapid progress, a great deal of work remains to be done on the associated astrochemistry.

References

1. A. Frank, T. P. Ray, S. Cabrit, P. Hartigan, H. P. Arce and 8 co-authors, *Protostars Planets*, 2014, **VI**, 451.
2. R. E. Pudritz, R. Ouyed, Ch. Fendt and A. Brandenburg, *Protostars Planets*, 2007, **V**, 277.
3. H. Shang, Z.-Y. Li and N. Hirano, *Protostars Planets*, 2007, **V**, 261.
4. L. A. Zapata, S. Lizano, L. F. Rodríguez, P. T. P. Ho, L. Loinard, M. Fernández-López and D. Tafoya, *Astrophys. J.*, 2015, **798**, 131.
5. H. Shang, S. Lizano, A. Glassgold and F. Shu, *Astrophys. J.*, 2004, **612**, 69.
6. B. Nisini, 2009, *Protostellar Jets in Context, by Kanaris Tsinganos, Tom Ray, Matthias Stute. Astrophysics and Space Science Proceedings Series*. Berlin: Springer, 2009, pp. 215–224.
7. J. Le Bourlot, G. Pineau des Forêts, D. R. Flower and S. Cabrit, *Mon. Not. R. Astron. Soc.*, 2002, **332**, 985.
8. R. D. Schwartz and T. P. Greene, *Astron. J.*, 2003, **126**, 339.
9. H. G. Arce, D. Mardones, S. A. Corder, G. Garay, A. Noriego-Crespo and A. C. Raga, *Astrophys. J.*, 2013, **774**, 39.

10. L. Podio, S. Mevdes, F. Bacciotti, J. Eislöffel and T. Ray, *Astron. Astrophys.*, 2009, **506**, 779.
11. J. M. Girart, L. F. Rodriguez, G. Anglada, R. Estalella, J. M. Torrelles and 5 co-authors, *Astrophys. J.*, 1994, **435**, L145.
12. S. D. Taylor and D. A. Williams, *Mon. Not. R. Astr. Soc.*, 1996, **282**, 1343.
13. S. Viti, J. M. Girart and J. Hatchell, *Astron. Astrophys.*, 2006, **449**, 1089.
14. Y. Zhang, H. A. Arce, D. Mardones, S. Cabrit, M. M. Dunham, G. Garay, A. Noriego-Crespo, S. S. R. Offner, A. C. Raga and S. A. Corder, 2016, arXiv:1602.02388.
15. J. M. C. Rawlings, M. P. Redman, E. Keto and D. A. Williams, *Mon. Not. R. Astr. Soc.*, 2004, **351**, 1054.
16. S. Viti, S. Natarajan and D. A. Williams, *Mon. Not. R. Astr. Soc.*, 2002, **336**, 797.
17. T. Umemoto, T. Iwata, Y. Fukui, H. Mikami, S. Yamamoto, O. Kameya and N. Hirano, *Astrophys. J. Lett.*, 1992, **392**, L83.
18. M. Benedettini, S. Viti, C. Codella, R. Bachiller, F. Gueth, M. T. Beltrán, A. Dutrey and S. Guilloteau, *Mon. Not. R. Astron. Soc.*, 2007, **381**, 1127.
19. F. Gueth, S. Guilloteau and R. Bachiller, *Astron. Astrophys.*, 1996, **307**, 891.
20. B. Lefloch, S. Cabrit, G. Busquet, C. Codella, C. Ceccarelli, J. Cernicharo, J. R. Pardo, M. Benedettini, D. C. Lis and B. Nisini, *Astrophys. J.*, 2012, **757**, 25.
21. J. Holdship, S. Viti, I. Jimenez-Serra, B. Lefloch and 7 co-authors, *Mon. Not. R. Astron. Soc.*, 2016, **463**, 802.
22. S. Viti, I. Jimenez-Serra, J. A. Yates, C. Codella and 4 co-authors, *Astrophys. J.*, 2011, **740**, 3.
23. G. Busquet, B. Lefloch, M. Benedettini and 9 co-authors, *Astron. Astrophys.*, 2014, **561**, 120.
24. D. Shepherd, *Massive star birth: A crossroads of Astrophysics, IAU Symposium Proceedings of the international Astronomical Union 227, held 16–20 May, Italy*, ed. R. Cesaroni, M. Felli, E. Churchwell, M. Walmsley, Cambridge: Cambridge University Press, 2005, pp. 237–246.
25. Q. Zhang, T. R. Hunter, H. Beuther and 5 co-authors, *Astrophys. J.*, 2007, **658**, 1152.
26. Á. Sánchez-Monge, A. López-Sepulcre, R. Cesaroni, C. M. Walmsley, C. Codella, M. T. Beltrán, M. Pestalozzi and S. Molinari, *Astron. Astrophys.*, 2013, **557**, 94.
27. S. Feng, H. Beuther, Q. Zhang, L. Qizhou, H. B. Liu, Z. Zhang, K. Wang and K. Qiu, *Astrophys. J.*, 2016, **828**, 100.
28. C. Codella, R. Bachiller, M. Benedettini, P. Caselli, S. Viti and V. Wakelam, *Mon. Not. R. Astron. Soc.*, 2005, **361**, 244.
29. C. Codella, S. Viti, D. A. Williams and R. Bachiller, *Astrophys. J. Lett.*, 2006, **644**, L41.
30. N. J. Turner, S. Fromang, C. Gammie, H. Klahr, G. Lesur, M. Wardle and X.-N. Bai, *Protostars Planets*, 2014, **VI**, 411.

31. C. Baruteau, A. Crida, S.-J. Paardecooper, F. Masset, J. Guilet, B. Bitsch, R. Nelson, W. Kley and J. Papaloizou, *Protostars Planets*, 2014, **VI**, 667.
32. C. Carrasco-González, L. F. Rodríguez, G. Anglada, J. Martí, J. M. Torrelles and M. Osorio, *Science*, 2010, **330**, 1209.
33. T. Ray, C. Dougados, F. Bacciotti, J. Eislöffel and A. Chrysostomou, *Protostars Planets*, 2007, **V**, 231.
34. S. Cabrit, C. Codella, F. Gueth, B. Nisini, A. Gusdorf, C. Dougados and F. Bacciotti, *Astron. Astrophys.*, 2007, **468**, L29.
35. O. Dionatos, B. Nisini, R. Garcia Lopez, T. Giannini, C. J. Davis, M. D. Smith, T. P. Ray and M. DeLuca, *Astrophys. J.*, 2009, **692**, 1.
36. O. Dionatos, B. Nisini, S. Cabrit, L. Kristensen and G. Pineau des Forêts, *Astron. Astrophys.*, 2010, **521**, A7.
37. C. J. Davis, K. W. Hodapp and L. Desroches, *Astron. Astrophys.*, 2001, **377**, 285.
38. F. Casse and R. Keppens, *Astrophys. J.*, 2002, **581**, 988.
39. F. Casse and R. Keppens, *Astrophys. J.*, 2004, **601**, 90.
40. C. Zanni, A. Ferrari, R. Rosner, G. Bodo and S. Massaglia, *Astron. Astrophys.*, 2007, **469**, 811.
41. P. Tzeracos, A. Ferrari, A. Mignone, C. Zanni, G. Bodo and S. Massaglia, *Mon. Not. R. Astron. Soc.*, 2009, **400**, 820.
42. G. C. Murphy, J. Ferreira and C. Zanni, *Astron. Astrophys.*, 2010, **512**, A82.
43. S. Sheikhnezami, C. Fendt, O. Porth, B. Vaidya and J. Ghanbari, *Astrophys. J.*, 2012, **757**, 65.
44. D. Stepanovs and C. Fendt, *Astrophys. J.*, 2014, **793**, 31.
45. D. Stepanovs, C. Fendt and S. Sheikhnezami, *Astrophys. J.*, 2014, **769**, 29.
46. D. Stepanovs and C. Fendt, 2016, arXiv:1604.0731.
47. D. Panoglou, S. Cabrit, G. Pineau des Forêts, P. J. V. Garcia, J. Ferreira and F. Casse, *Astron. Astrophys.*, 2012, **538**, A2.
48. W. Yvart, S. Cabrit, G. Pineau des Forêts and J. Ferreira, *Astron. Astrophys.*, 2016, **585**, A74.
49. R. D. Blandford and D. G. Payne, *Mon. Not. R. Astron, Soc.*, 1982, **199**, 883.
50. F. H. Shu, S. Lizano, D. Galli, M. J. Cai and S. Mohanty, *Astrophys. J. Lett.*, 2008, **682**, L121.
51. F. Casse and J. Ferreira, *Astron. Astrophys.*, 2000, **361**, 1178.
52. A. E. Glassgold, J. Najita and J. Igea, *Astrophys. J.*, 1997, **480**, 344.
53. H. Shang, A. E. Glassgold, F. H. Shu and S. Lizano, *Astrophys. J.*, 2002, **564**, 853.
54. A. Dalgarno, M. Yan and W. Liu, *Astrophys. J. Suppl.*, 1999, **125**, 237.
55. L. Podio, S. Medves, F. Bacciotti, J. Eislöffel and T. Ray, *Astron. Astrophys.*, 2009, **506**, 779.
56. C. Pinto and D. Galli, *Astron. Astrophys.*, 2008, **484**, 17.
57. R. W. Schunk, *Planet. Space Sci.*, 1975, **23**, 437.
58. J. M. C. Rawlings, D. A. Williams and J. Cantó, *Mon. Not. R. Astr. Soc.*, 1988, **230**, 695.

59. A. E. Glassgold, G. A. Mamon and P. J. Huggins, *Astrophys. J.*, 1989, **336**, L29.
60. A. E. Glassgold, J. Najita and J. Igea, *Astrophys. J.*, 2004, **615**, 972.
61. H. Nomura and T. J. Millar, *Astron. Astrophys.*, 2005, **438**, 923.
62. L. Podio, C. Codella, F. Gueth, S. Cabrit, R. Bachiller, A. Gusdorf, C.-F. Lee, B. Lefloch, S. Leurini, B. Nisini and M. Tafalla, *Astron. Astrophys.*, 2015, **581**, A85.
63. E. F. van Dishoeck, L. E. Kristensen, A. O. Benz, E. A. Bergin and 68 co-authors, *Pub. Astron. Soc. Pacific*, 2011, **123**, 138.
64. L. E. Kristensen, E. F. van Dishoeck, E. A. Bergin, R. Visser and 21 co-authors, *Astron. Astrophys.*, 2012, **542**, A8.
65. S. Matt and R. E. Pudritz, *Astrophys. J.*, 2008, **681**, 391.
66. C. Zanni and J. Ferreira, *Astron. Astrophys.*, 2013, **550**, A99.
67. M. C. White, G. V. Bicknell, R. S. Sutherland, R. Salmeron and P. J. McGregor, *Mon. Not. R. Astr. Soc.*, 2016, **455**, 2042.
68. C. J. Davis, B. Cervantes, B. Nisini, T. Giannini, M. Takami, E. Whelan, M. D. Smith, T. P. Ray, A. Chrysostomou and T. S. Pyo, *Astron. Astrophys.*, 2011, **528**, A3.
69. A. C. Raga, D. A. Williams and A. J. Lim, *Rev. Mex. Astron. Astrofís.*, 2005, **41**, 137.
70. B. Reipurth, S. Heathcote, Ka Chun Yu, J. Bally and L. F. Rodríguez, *Astrophys. J.*, 2000, **534**, 317.
71. T. Giannini, S. Antoniucci, B. Nisini, F. Bacciotti and L. Podio, *Astrophys. J.*, 2015, **814**, 52.
72. A. C. Raga, L. Binette, J. Cantó and N. Calvet, Astrophys. *J.*, 1990, **364**, 601.
73. J. Eislöffel and R. Mundt, *Astron. Astrophys.*, 1992, **263**, 292.
74. S. Heathcote and B. Reipurth, *Astron. J.*, 1992, **104**, 2193.
75. B. Reipurth and S. Heathcote, *Astron. Astrophys.*, 1992, **257**, 693.
76. B. Reipurth, L. F. Rodríguez, G. Anglada and J. Bally, *Astron. J.*, 2002, **123**, 362.
77. K.-S. Jhan and C.-F. Lee, *Astrophys. J.*, 2016, **816**, 32.
78. A. C. Raga, A. Noriega-Crespo, V. Lora, K. R. Stapelfeldt and S. J. Carey, *Astrophys. J. Lett.*, 2011, **730**, L17.
79. A. C. Raga, J. C. Rodríguez-Ramírez, J. Cantó and P. F. Velázquez, *Mon. Not. R. Astr. Soc.*, 2015, **454**, 412.
80. P. Hartigan, S. Heathcote, J. A. Morse, B. Reipurth and J. Bally, Astron. *J.*, 2005, **130**, 2197.
81. A. Frank, A. Ciardi, K. Yirak and S. Lebedev, *Rev. Mex. Astron. Astrofís.*, 2009, **36**, 193.
82. E. C. Hansen, A. Frank, P. Hartigan and S. V. Lebedev, *Astrophys. J.*, 2017, **837**, 143.
83. P. Hartigan, A. Frank, J. M. Foster, B. H. Wilde, M. Douglas, P. A. Rosen, R. F. Coker, B. E. Blue and J. F. Hansen, *Astrophys. J.*, 2011, **736**, 29.
84. A. J. Lim, J. M. C. Rawlings and D. A. Williams, *Astron. Astrophys.*, 2001, **376**, 336.

85. M. D. Smith, T. Khanzadyan and C. J. Davis, *Mon. Not. R. Astr. Soc.*, 2003, **339**, 524.
86. B. Lefloch, J. Cernicharo, S. Cabrit and D. Cesarsky, *Astron. Astrophys.*, 2005, **433**, 217.
87. J. M. Torrelles, L. F. Rodríguez, J. Cantó, G. Anglada, J. F. Gomez, S. Salvador and P. T. P. Ho, *Astrophys. J.*, 1992, **396**, 95.
88. J. M. Girart, L. F. Rodríguez, G. Anglada, R. Estalella and 6 co-authors, *Astrophys. J. Lett.*, 1994, **435**, L148.
89. J. M. Girart, R. Estalella and P. T. P. Ho, *Astrophys. J.*, 1998, **495**, 59.
90. O. Morata, J. M. Girart and R. Estalella, *Astron. Astrophys.*, 2005, **435**, 113.
91. J. M. Girart, R. Estalella, S. Viti, D. A. Williams and P. T. P. Ho, *Astrophys. J.*, 2001, **562**, L91.
92. S. D. Taylor and D. A. Williams, *Mon. Not. R. Astron. Soc.*, 1996, **282**, 1343.
93. S. Viti and D. A. Williams, *Mon. Not. R. Astr. Soc.*, 1999, **319**, 517.
94. J. M. Girart, S. Viti, D. A. Williams, R. Estalella and P. T. O. Ho, *Astron. Astrophys.*, 2002, **388**, 1004.
95. S. Viti, J. M. Girart, R. Garrod, D. A. Williams and R. Estalella, *Astron. Astrophys.*, 2003, **399**, 187.
96. J. M. Girart, S. Viti, R. Estalella and D. A. Williams, *Astron. Astrophys.*, 2005, **439**, 601.
97. A. C. Raga and D. A. Williams, *Astron. Astrophys.*, 2000, **358**, 701.
98. W. Whyatt, J. M. Girart, S. Viti, R. Estalella and D. A. Williams, *Astron. Astrophys.*, 2010, **510**, A74.
99. H. Christie, S. Viti, D. A. Williams, J. M. Girart and O. Morata, *Mon. Not. R. Astron. Soc.*, 2011, **416**, 288.
100. T. Velusamy and W. D. Langer, *Nature*, 1998, **392**, 685.
101. J. Cantó, A. C. Raga and D. A. Williams, *Rev. Mex. Astron. Astrofís.*, 2008, **44**, 293.
102. H. G. Arce and A. Sargent, *Astrophys. J.*, 2006, **646**, 1070.
103. J. M. C. Rawlings, S. D. Taylor and D. A. Williams, *Mon. Not. R. Astr. Soc.*, 2000, **313**, 461.
104. J. M. C. Rawlings, M. P. Redman and P. B. Carolan, *Mon. Not R. Astr. Soc.*, 2013, **435**, 289.
105. S. Viti, N. Sukina and D. A. Williams, *Mon. Not. R. Astron. Soc.*, 2002, **336**, 797.
106. R. P. Rollins, J. M. C. Rawlings, D. A. Williams and M. P. Redman, *Mon. Not. R. Astr. Soc.*, 2014, **443**, 3033.
107. G.-X. Li, K. Qiu, F. Wyrowski and K. Menten, *Astron. Astrophys.*, 2013, **559**, A23.
108. W. D. Langer, T. Velusamy and T. Xie, *Astrophys. J. Lett.*, 1996, **468**, L41.

CHAPTER 6

Outflows and Explosions of Evolved Stars

6.1 Introduction

Observations and theory provide strong evidence for the fact that the outflows from evolved stars are the overwhelmingly dominant sources of interstellar dust and, of course, chemical enrichment of the interstellar medium. AGB stars and supergiant winds are the main sources of s-process element enrichment, whilst violent cataclysmic variables, especially core-collapse supernovae, provide the enrichment of r-process elements. In addition, it has long been suggested that novae may be, in part, responsible for the variety of pre-solar grain inclusions in meteorites. The relative contributions of each class of object to the interstellar dust budget are not well known—and this is largely due to our poor understanding of the dust formation process. In this chapter we assess our understanding of the dynamics, astrochemistry and dust formation in outflows, and how these feed into the interstellar medium.

6.2 Stellar Evolution

In his own 1968 book on stellar evolution and nucleosynthesis D. D. Clayton claimed that four books had 'dominated' the history of research on stellar evolution.[1] The first, *Gaskugeln*, was published in 1907 and authored by R. Emden. It contains an exposition on the hydrostatic equilibria of self-gravitating spheres of gases having polytropic equations of state. In his 1926 work *The Internal Constitution of the Stars*, A. S. Eddington demonstrated how radiative transfer is a key mechanism for the transport of energy in the interiors of many stars. In *An Introduction to the Study of Stellar Structure*,

Dynamical Astrochemistry
By David A. Williams, Thomas W. Hartquist, Jonathan M. C. Rawlings, Cesare Cecchi-Pestellini and Serena Viti

Published by the Royal Society of Chemistry, www.rsc.org

published in 1939, S. Chandrasekhar formalized the results presented in the earlier books and summarized how the theory of the quantum statistical mechanics of degenerate matter can be applied to understanding the structures of white dwarfs. This book appeared shortly after H. A. Bethe and C. F. von Weizsäcker independently considered the carbon–nitrogen cycle as the source of fusion energy in stars,[2–4] and S. Chandrasekhar included a treatment of thermonuclear reactions as the source of stellar energy. As D. D. Clayton described, laboratory work on nuclear reactions during the 1940s and 1950s made possible the calculation of the rate of energy generation in a hot plasma and the construction of static, spherically symmetric, numerical models of stars in which the generation of energy by the relevant nuclear reactions was included. In 1958 M. Schwarzschild summarized these developments in *Structure and Evolution of the Stars*.

Subsequently the use of electronic computers enabled further advancement of the field. A major aim of research in stellar evolution is to gain an understanding of the distribution of stars within individual clusters on a luminosity–colour diagram, known as a Hertzsprung–Russell diagram. A model of a star with a specified initial mass and elemental composition allows one to calculate the corresponding position on such a diagram for the star as a function of its age.

6.2.1 The Hertzsprung–Russell Diagram

As noted by D. D. Clayton, in 1911 the Danish astronomer E. Hertzsprung plotted the apparent magnitudes of the stars in an individual stellar cluster against their colours. The apparent magnitude of a star is proportional to the logarithm of the flux of its radiation measured at Earth. Since all of the stars in an individual cluster are at approximately the same distance, the apparent magnitudes of the stars studied by E. Hertzsprung are very nearly proportional to their intrinsic magnitudes, which are proportional to the logarithms of their luminosities. Wien's law states that the wavelength at which the spectrum of a black body with a temperature T peaks is inversely proportional to T; consequently, the colour of a radiator is a measure of its temperature. Astronomers often use the difference between the logarithms of the fluxes in two broad spectral bands to define the colour.

D. D. Clayton also mentioned related work conducted in 1913 by the American H. N. Russell, who, in addition to making substantial contributions in astronomy, recognized that the spectral characteristics of some complex atoms arise due to Russell–Saunders coupling between electronic orbital angular momentum and electronic spin. H. N. Russell plotted the intrinsic magnitudes of some nearby stars against their colours.

When a diagram like those produced by Hertzsprung and Russell is constructed for a recently formed cluster, the stars lie on a diagonal curve; such stars are referred to as main-sequence stars. Figure 6.1 shows where main-sequence stars, as well as the other types of stars existing in evolved clusters, lie on a Hertzsprung–Russell, or H–R, diagram. In the figure, the axes are the

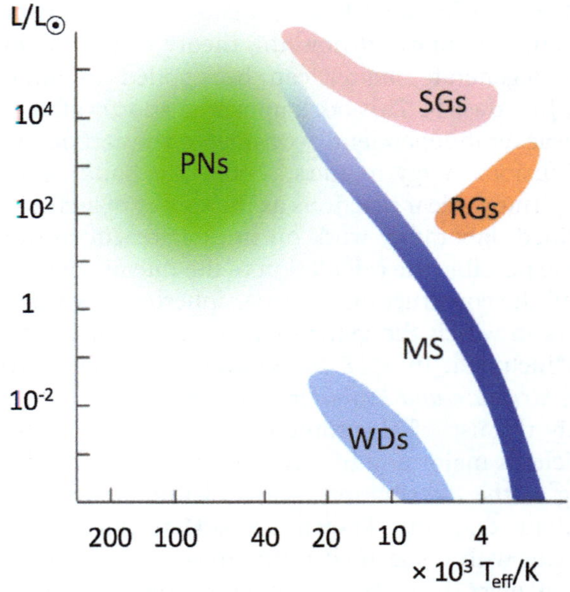

Figure 6.1 The locations of different classes of stars in the Hertzsprung–Russell diagram. Stars with hydrogen-burning cores lie on the main sequence (MS). Low-mass and intermediate mass stars become red giants (RGs) during the RGB and AGB phases described in Section 6.2.2. Higher mass stars become supergiants (SGs) after they evolve off the main sequence. AGB stars evolve into the central stars of planetary nebulae (PNs), which in turn eventually become white dwarfs (WDs).
Data for the figure were taken from Ryan and Norton.[7]

luminosity normalized to the solar luminosity, which is about 4×10^{33} erg s^{-1}, and the effective temperature of the photosphere of the star. Stars lie on the main sequence when they are born but move in an H–R diagram as they evolve. Consequently, such a diagram for one cluster of stars differs from that for another due largely to the different ages of the clusters but also to some extent due to other differences, such as in the metallicity, between clusters.

6.2.2 Evolutionary Tracks of Solar-mass and Intermediate-mass Stars

The numerical modelling of stars has elucidated how they move in the H–R diagram as they evolve. This subject is treated concisely but expertly in a single book chapter by D. Schönberner and T. Blöcker.[5] Another very informative book chapter is *Nucleosynthesis of Low and Intermediate-Mass Stars* by A. I. Karakas.[6] More lengthy treatments of stellar evolution are contained in the undergraduate textbook by S. G. Ryan and A. J. Norton[7] and the comprehensive two volume magnum opus by I. Iben, one of the great figures in the field.[8,9]

A star remains on the main sequence as long as hydrogen burning in its core continues. For stars with initial masses that are 0.8, 1, 2, 5, 15 and 25 times that of the Sun, the main sequence lifetimes are about 2×10^{10}, 9.2×10^{9}, 8.7×10^{8}, 7.8×10^{7}, 1.1×10^{7} and 7.5×10^{6} years, respectively.[6] The shorter main sequence lifetimes of more massive stars are due to their greater luminosities, which result from the temperature dependences of the nuclear reactions and the temperatures of the cores in larger stars being larger. At the end of its time on the main sequence a star has a helium core surrounded by a hydrogen burning shell, which in turn is enclosed by those parts of the star that are too cool for nuclear reactions to occur. In a star with a mass greater than 0.8 times the solar mass, eventually helium burning will occur in the core leading to the formation of a carbon core. In a low-mass star the helium ignition takes place under degenerate conditions, whereas in an intermediate-mass star the core is not degenerate at the onset of helium burning. High-mass, or massive, star are those in which the carbon cores are not degenerate when they start to burn and which evolve to become supernovae. Low-mass and intermediate-mass stars have masses in the ranges of 0.8 to about 2.25 and 2.25 to about 8 solar masses, respectively.[6]

As a low-mass or intermediate-mass star leaves the main sequence, it moves towards the red giant region of the H–R diagram and reaches the red giant branch (RGB). The hydrogen burning shell is the source of most of the star's luminosity, which increases. Consequently, the star expands to a radius several hundred times larger than the solar radius,[9] leading the photosphere to cool and the star to become redder. During the RGB phase convection dominates the transport of energy in the stellar envelope. The first dredge-up occurs. It causes material that underwent nuclear processing in the core during the main sequence phase to mix with material initially near the stellar surface.

Evolutionary tracks given in Figures 11.1.1, 11.1.32, 11.2.1 and 11.2.42 in one[8] of the volumes by I. Iben show that stars with initial masses of 1 and 5 solar masses take about 3×10^{9} and 8×10^{7} years, respectively, to evolve from the main sequence to the onset of core helium burning. The luminosity and effective temperature of a 1 solar mass star vary between 1.8 and 1600 solar luminosities, and 5500 K and 2700 K, respectively, during the evolution from the main sequence phase to the onset of the helium core burning. In contrast, the luminosity and effective temperature of a 5 solar mass star vary between 350 and 1700 solar luminosities, and 17 000 K and 4200 K, respectively.

The contraction and the associated heating of the helium core leads to its ignition and the termination of the RGB phase. Even during the helium core burning phase, most of the energy is generated in a hydrogen burning shell. In the core the triple alpha reaction fuses three helium nuclei to form a carbon nucleus. The production of carbon leads to its reaction with helium, yielding oxygen. After the onset of helium core burning the luminosity of the star containing 1 solar mass initially drops to about 40 solar luminosities as its effective temperature simultaneously increases to roughly what it was during the main sequence phase and then decreases again as the luminosity simultaneously increases. The luminosity of the star containing 5 solar masses initially decreases much less but its temperature also increases before decreasing again. The roughly horizontal movement of such stars on the H–R diagram leads to them being designated horizontal branch stars.

After about 1.2×10^8 and 1.8×10^7 years, respectively, the helium in the cores of the low-mass and intermediate-mass stars is exhausted.[9] Following the cessation of core burning, a star moves upward and to the right in the H–R diagram onto the asymptotic giant branch (AGB). During the AGB phase, helium burning becomes confined to a shell surrounding the core. That shell lies below the hydrogen burning shell. A second dredge-up occurs in an intermediate-mass star but not in a low-mass star as the luminosity increases, the envelope expands and the effective temperature drops. The dredge-up can lead to significant enhancements in the carbon and oxygen elemental abundances in the envelope of an intermediate-mass AGB star. In both the low-mass and the intermediate-mass stars, the helium shell burning is stable initially but becomes unstable, with the consequence that the AGB phase is divided into the early-AGB and the thermally pulsing AGB phases (characterized by alternate hydrogen-shell and helium-shell burning). The thermal pulses drive what is called the third dredge-up in both low-mass and intermediate-mass stars, even though low-mass stars do not experience the second dredge-up.

The maximum luminosity and minimum effective temperature of the low-mass AGB star are roughly comparable to those of its RGB counterpart.[6] In contrast, the maximum luminosity and minimum temperature of the intermediate-mass AGB star are about an order of magnitude larger and somewhat smaller, respectively, than those of its RGB counterpart.[6]

The evolution and lifetime of an AGB star depend sensitively on the mass loss rate.[5] In a model of a star with an initial mass of 3 solar masses, the mass loss rate during the AGB phase increases from about 10^{-9} solar masses per year to about 10^{-4} solar masses per year over a period of about 6×10^6 years.[5] During the final roughly 2×10^6 years, the thermal pulses cause the mass loss rate to vary by several orders of magnitude as the pulses occur. The AGB phase begins to cease when the envelope mass diminishes to about 0.01 solar masses due to burning and mass loss, at which time the mass loss rate begins to decrease by several orders of magnitude.[5]

The optical depth of the ejected material drops as it flows away from a star as it enters the post-AGB phase, and photodissociation and photoionization fronts driven by the stellar far ultraviolet radiation propagate through the ejecta. The hot surface of the remnant star becomes visible and the star moves horizontally in the H–R diagram as its effective temperature increases to over 10^5 K. The ionized ejecta form a protoplanetary nebula and then a planetary nebula. Figure 6.2 shows the Eskimo Nebula, which is a planetary nebula. The timescale over which the effective temperature of the remnant increases depends very sensitively on its mass; for remnant masses of 0.55, 0.6 and 0.94 solar masses the timescales are about 10^5, 4×10^3 and 50 years, respectively.[5] Hence, a star with an initial mass of 5 solar masses increases in temperature

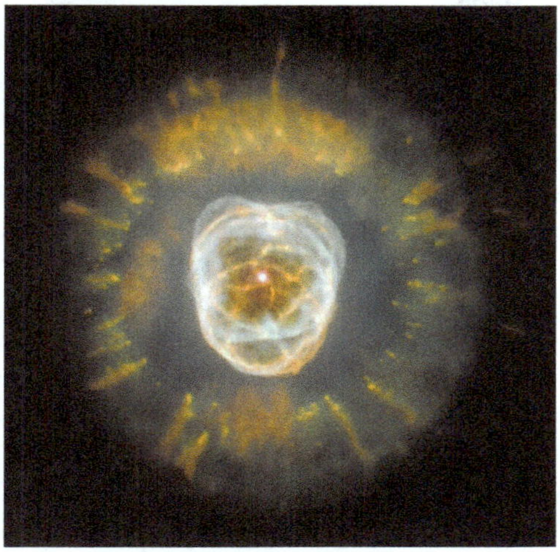

Figure 6.2 The Eskimo Nebula (from the Hubble Space Telescope). © NASA/Andrew Fruchter (STSci).

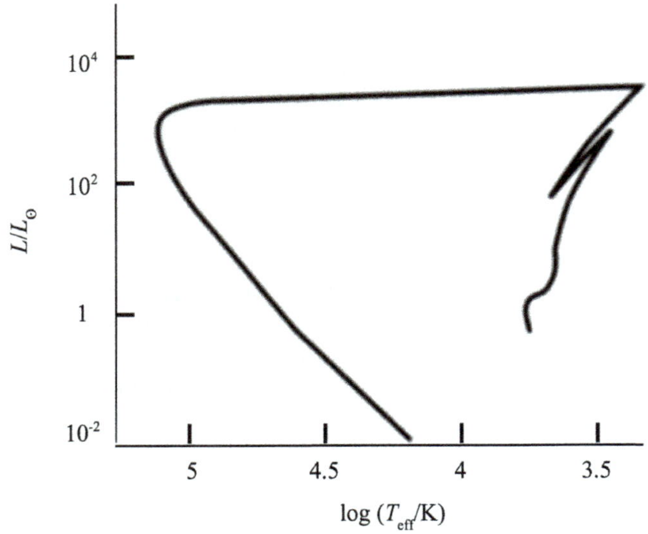

Figure 6.3 The evolutionary track of a star with an initial mass of one solar mass. Data for the figure were taken from Iben.[9]

on a timescale of many decades to several centuries after it leaves the AGB phase. In contrast, a star with an initial mass of 1 solar mass does so in roughly 10^4 years.[9] Then, over several hundred thousand to a million years,[5,9] the post-AGB star dims from its maximum luminosity of about 10^3 to 3×10^4 solar luminosities to less than several solar luminosities and becomes a white dwarf supported by electron degeneracy as dimming continues.

Figure 6.3 shows the evolutionary track in the H–R diagram of a star with an initial mass of one solar mass.

6.2.3 The Evolution of High-mass Stars

When a high-mass star leaves the main sequence, it evolves to become a supergiant. One with an initial mass of about 40 solar masses or more never becomes red, though less massive high-mass stars do. The carbon core produced by helium burning in a high-mass star undergoes non-degenerate carbon burning, which is followed by subsequent burning phases. Mass loss rates are as large as 10^{-5} to 10^{-4} solar masses per year.[5] The inner region of the star eventually consists of an iron core surrounded by a silicon shell, which in turn is surrounded by an oxygen shell. A neon shell, a carbon shell and a helium shell lie exterior to that. Iron has one of the largest binding energies per nucleon and does not burn, and once the core becomes iron it starts to cool, leading to collapse. When the central density approaches nuclear densities, the compressibility of the core decreases, and matter falling onto its surface bounces. The bounce, together with the pressure provided by the neutrinos emitted by the core, drives ejecta in a supernova.

6.3 Dynamics of Winds and Mass-loss

In an important early paper on stellar winds[10] C. S. Beals, whose contribution to the discovery of interstellar clouds was mentioned in Section 2.1, suggested that in 1904 J. K. E Halm became the first to hypothesize that the characteristic shape of lines formed in the star P Cygni is due to a continuous flow of material from the stellar surface.[11] Similar features in the spectra of other stars established the widespread existence of stellar winds. Though impulsive solar mass ejections had been recognized earlier, the presence of a continuous solar wind was first inferred in the 1950s by L. Biermann from a thorough analysis of the acceleration, ionization and excitation of cometary tails.[12] In 1960 E. N. Parker presented a theoretical description of steady, spherically symmetric, stellar winds.[13] Stellar pulsations and radiation pressure on dust are important mechanisms for driving mass loss from the asymptotic giant branch stars.

6.3.1 Simple Steady Hydrodynamic Wind

To make the analysis as simple as possible, we shall consider a steady, spherically symmetric, isothermal, radial hydrodynamic wind subjected to only the forces due to gravity and thermal pressure. Eqn (2.4) and (2.10) give

$$(1/r^2)\mathrm{d}(r^2\rho v_\mathrm{r})/\mathrm{d}r = 0, \tag{6.1}$$

and

$$(1/r^2)\mathrm{d}(r^2\rho v_r^2)/\mathrm{d}r + c_s^2\mathrm{d}\rho/\mathrm{d}r = -GM\rho/r^2, \tag{6.2}$$

where M is the mass of the star. Integration of eqn (6.1) shows that the mass loss rate

$$\mathrm{d}M/\mathrm{d}t = 4\pi r^2\rho v_\mathrm{r} \tag{6.3}$$

is constant.

The use of eqn (6.3) to eliminate ρ in eqn (6.2) allows one to find that

$$\mathrm{dln}(v_r)/\mathrm{dln}(r) = 2(c_s^2 - GM/2r)/(v_r^2 - c_s^2). \tag{6.4}$$

Eqn (6.4) implies that if the flow is subsonic and the sound speed is less than half the escape speed, the flow will accelerate. It will also accelerate if the flow is supersonic and the sound speed exceeds one half the escape speed. Eqn (6.4) has solutions that describe flows that are subsonic near the star, pass through sonic points at which the sound speed equals half the escape speed and are supersonic at distances beyond the sonic points. Of course, terms for forces in addition to those due to gravity and thermal pressure can be included, and those other forces would lead to a modification of the condition that the sound speed must equal half the escape speed at the sonic point.

The sonic point is referred to as a critical point. In spherically symmetric flow, the sonic point lies on a spherical surface. More generally, one considers sonic and critical surfaces. For an MHD wind, more than one critical surface is relevant due to the existence of more than one type of wave mode.[14]

6.3.2 Driving AGB Winds with Stellar Pulsations and Radiation Pressure on Dust

In optically thin winds with constant composition, ionization and excitation the force per unit mass due to radiation pressure is proportional to $1/r^2$, and the effect of radiation pressure leads to a particularly straightforward modification of eqn (6.4). However, in most winds at least one of composition, ionization and/or excitation varies with position. For example, in outflows from AGB stars dust formation occurs, and the radiation pressure on the dust is key for the production of AGB winds with speeds of about 10–20 km s^{-1} and mass loss rates as large as about 10^{-4} solar masses per year for thousands of years.[5,15–18]

AGB stars pulsate with periods of about 100 to 1000 days, but the theory of the instabilities in AGB stellar atmospheres driving those pulsations is considered to be inexact compared to that of the instabilities driving pulsations in warmer variable stars.[19] Much of the atmosphere of an AGB star is convective, and convection strongly influences the instability of the atmosphere.[19–21] Computational studies of the instability of AGB stars with atmospheres indicate that although convective transport of energy generally acts as a damping mechanism, the turbulent pressure associated with the convection usually acts to induce instability.[20] This contributes to the development of an understanding of the instability of an AGB stellar atmosphere that causes pulsations to be challenging. The computational models contain free parameters, which must be inferred by a combination of physical arguments and comparisons with observational results.[19] In any case, the pulsations in AGB stars play a key role in the generation of winds by driving shocks, which extend the stellar atmospheres sufficiently to allow cooling and dust formation.[15–18] Three-dimensional radiation hydrodynamics models, in which convection, convection-driven acoustic noise, fundamental-mode radial pulsations and atmospheric shocks are treated self-consistently, indicate that the convection, waves and shocks all contribute to the extension of the atmospheres.[21]

6.4 Dust Formation in AGB Outflows

Some atmospheres of AGB stars are carbon-rich with an elemental carbon to oxygen ratio greater than unity, while others are oxygen-rich with an elemental carbon to oxygen ratio less than unity. Throughout most of an atmosphere of the first type, CO contains most of the oxygen and the carbon not in CO forms complex organics, including soot. Extensive work on terrestrial combustion and flames has informed the development of kinetic models for the production of dust in the carbon-rich atmospheres.[16,22–24] Though much of the work on the formation of dust in oxygen-rich AGB stars has been based on classical nucleation theory, recently a kinetic network was used to model the formation of molecules and dust in the oxygen-rich atmosphere of IK Tau.[18] Many AGB stars can be assumed to be single and isolated, but environmental effects will influence dust formation in some.

6.4.1 Dust in Carbon-rich Outflows

The temperature in an atmosphere drops sufficiently at a large enough density for some simple molecules to survive under thermal equilibrium conditions. CO and acetylene C_2H_2 are the two most abundant molecules other than H_2. Much of the subsequent chemistry occurs under conditions at which thermal equilibrium is not obtained, though some three-body reactions can play roles.

The analysis of one model allowed the identification of the dominant path to the formation of benzyne, which is C_6H_4 and, like benzene, has a ring structure and simple polycyclic aromatic hydrocarbons (PAHs).[22] The sequence is initiated by the collision of C_2H_2 with another species yielding H_2CC. This is followed by

$$H_2CC + C_2H_2 \rightarrow C_4H_4, \tag{6.5}$$

and

$$C_4H_4 + H \rightarrow C_4H_3U + H_2, \tag{6.6}$$

where the U indicates that the species is a vinyl-type radical with an unpaired electron located on its terminal carbon atom. Then

$$C_4H_3U + C_2H_2 \rightarrow \text{benzyne} + H. \tag{6.7}$$

Further reactions involving H and C_2H_2 form PAHs from the benzyne. The PAHs then coagulate to form dimers in which the aromatics are linked by van der Waals forces in the next step to eventual soot production.[23] The production of dust reaches completion in the region where the number density and temperature are about 10^{10} cm^{-3} and 1500 K, respectively.[16]

6.4.2 Dust in Oxygen-rich Outflows

In a kinetic model used to study the formation of molecules and dust in the oxygen-rich atmosphere of IK Tau,[18] the dust clusters were considered to be the dimers of forsterite (Mg_2SiO_4) and alumina (Al_2O_3). The key reaction sequence affecting the abundances of forsterite and enstatite ($MgSiO_3$) begins with reactions involving SiO, which is formed deeper in the stellar envelope. It includes the reactions

$$SiO + H \rightarrow HSiO, \tag{6.8}$$

$$SiO + H_2 \rightarrow HSiO + H, \tag{6.9}$$

$$HSiO + H \rightarrow H_2SiO, \tag{6.10}$$

$$HSiO + H_2 \rightarrow H_2SiO + H, \tag{6.11}$$

$$H_2SiO + H_2SiO \rightarrow H_2Si_2O_2 + H_2, \tag{6.12}$$

$$H_2Si_2O_2 + H_2O \rightarrow H_2Si_2O_3 + H_2, \tag{6.13}$$

$$H_2Si_2O_3 + Mg \rightarrow MgSi_2O_3 + H_2, \tag{6.14}$$

$$MgSi_2O_3 + H_2O \rightarrow MgSi_2O_4 + H_2, \tag{6.15}$$

$$MgSi_2O_4 + Mg \rightarrow Mg_2Si_2O_4, \tag{6.16}$$

$$Mg_2Si_2O_4 + H_2O \rightarrow Mg_2Si_2O_5 + H_2, \tag{6.17}$$

$$Mg_2Si_2O_5 + H_2O \rightarrow Mg_2Si_2O_6 + H_2, \tag{6.18}$$

and the corresponding backwards reactions. As well as the products of this chemistry, dimers of many species also form. These include $(SiO)_2$, which is key in models in which dust production is due to cluster formation of SiO molecules.[25] Dimerization of HSiO and of H_2SiO may be particularly important for dust formation.[18]

Throughout much of the atmosphere the abundances of H and H_2 are within an order of magnitude of one another, which leads to the reaction of OH with H_2 and the reaction of H_2O with H, maintaining significant abundances of both OH and H_2O, particularly in recently shocked regions. The production of alumina is triggered by AlO production in shocked gas by

$$Al + OH \rightarrow AlO + H. \tag{6.19}$$

$(AlO)_2$ is formed by the termolecular recombination of AlO, and Al_2O_3 results from the oxidation of $(AlO)_2$ in reactions with H_2O. Silicate cluster oxidation due to reactions involving water is also important.[26] Termolecular recombination of Al_2O_3 produces the alumina dimer.

In the model of IK Tau,[18] the coalescence and coagulation of $(Mg_2SiO_4)_2$ and $(Al_2O_3)_2$ to form dust grains were followed by a formalism that is appropriate when scattering, coalescence and coagulation are due to collisions resulting from Brownian motions.

6.4.3 The Influence of an Active Galactic Nucleus

In the preceding sections of this chapter the effects of photoabsorption on the chemistry were neglected. However, if an AGB star is near a sufficiently strong source of dissociating and ionizing radiation, the chemistry leading to the formation of dust in other AGB stars is suppressed. This could be particularly relevant in the inner regions of some active galactic nuclei (AGNs), including quasars.[27] The strong radiation fields of some AGNs, powered by accretion onto black holes having masses that are in some cases significantly greater than 10^8 solar masses, are capable of suppressing dust formation in AGB stars within parsecs of the black holes. This leads to reduced mass loss rates and extended stellar lifetimes, during which the AGB stars experience nucleosynthesis histories differing from those of AGB stars in less extreme environments. If the gas in the broad emission line regions of AGNs is supplied primarily by such unusually long-lived AGB stars, the elemental composition of those line regions should differ from the compositions in interstellar matter that is not in an AGN environment.

6.5 Masers and Measurements of the Magnetic Fields in AGB Outflows

Together with optical spectral polarimetry,[28] observations of molecular emission lines provide information about the strengths of magnetic fields near the surfaces of AGB stars with oxygen-rich atmospheres and in their outflows,[28,29] and, thus, also about the potential role of magnetic fields in the dynamics of mass loss. The relevant molecular lines are emitted by SiO, H_2O and OH masers. The polarization of the lines is used to measure the strengths of the magnetic field in the masers. A SiO maser is within several stellar radii of the stellar surface at a distance of up to about 10 astronomical units (1 au $= 1.496\times10^{11}$ m) from the centre of the star, whereas a H_2O maser and an OH maser are at distances of up to a few hundred au and from about 10^3 to 10^4 au, respectively. A compilation for the results of a number of observations of all these three types of masers suggests that on average the magnetic field strength decreases approximately inversely proportionally with distance from the central star. Field strengths can be as large as several hundred micro-Tesla at the stellar surface. Whether a field having this strength at the surface, and such a variation in the envelope, will lead to magnetic effects influencing the shaping of a nebula after the star leaves the AGB phase is a subject of debate.[28,29] In any case, if, as implied by a maser pumping model, the number density and temperature in a SiO maser are about $5\times10^9\,\mathrm{cm}^{-3}$ and 1500 K, respectively,[30] and if the magnetic field strength is about 3.5×10^{-4} Tesla,[29] then the thermal pressure is over two orders of magnitude less than the magnetic pressure.

6.6 Novae

6.6.1 Introduction

Novae are cataclysmic variable stars that originate from close binary stellar systems in which a late-type main sequence star overflows its Roche lobe and transfers matter, *via* an accretion disc, onto a white dwarf.

The discussion below is limited to 'classical novae', for which many observations of molecules and dust exist. Other nova types, dwarf and recurrent novae, are not believed to be efficient in the production of molecules or dust. Recurrent novae, as the name suggests, are seen to undergo relatively minor outbursts on historic timescales, whereas dwarf novae are believed to originate from catastrophic disc instabilities (similar to FU Orionis type events). Neither is known to result in dust production. Where classical novae (which, by definition, have only ever been observed to have undergone one eruption event) are particularly interesting is that they are often seen to produce optically thick dust shells on observable timescales (of the order of 10–100 days) and are thus unique laboratories for the study of the production of astrophysical dust. Whilst their contribution to the total interstellar dust budget is unclear, they are probably an important source of astrophysically significant isotopes such as ^{22}Na and ^{26}Al.[31]

In a classical nova a thermonuclear runaway (TNR) occurs in the accreted matter, which is subject to a degenerate equation of state. As the degeneracy is lifted a dense outflowing wind is produced. This wind can be massive ($\sim2\times10^{-4}$ $M_\odot$ year^{-1}) and rapidly expanding ($\sim$300–1300 km s^{-1}). The source luminosity is large (of the order of 10^4 $L_\odot$) and the expelled ejecta are very dense (initially $\sim10^{12}$–10^{13} cm^{-3}). In addition, this wind is extremely chemically enriched both by the TNR itself (the initial ejecta temperature is $\sim$2–3×10^8 K) and from material dredged up from the white dwarf,[31] yielding abundance enhancements of C, N, O, Ne, Mg, Si, S and other elements by factors of 10–1000 or more, relative to cosmic abundances.

The physical and chemical properties of novae vary widely so that they are hard to classify, and the classification system itself is subject to constant revision. But, broadly speaking, and particularly in the context of molecule and dust-formation efficiency, novae fall into two categories:

(i) The less bright, more slowly evolving novae, originating from low mass systems ($M_{wd}<1.2$ $M_\odot$) in which the white dwarf is a carbon–oxygen type. These produce dense winds that are strongly enriched in the elements C, N and O. These novae are often seen to exhibit CO, CN and—rarely—C_2 emission in the early stages. A sub-class of these are the so-called FeII novae, which tend to be prolific and efficient producers of optically thick carbon dust shells.

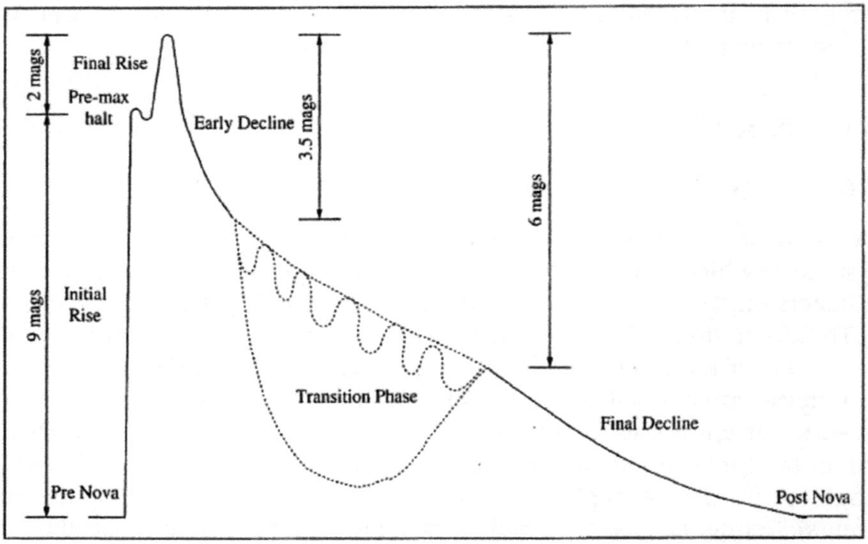

Figure 6.4 A schematic diagram of a nova light curve (visible brightness plotted as a function of time).
Reproduced with permission from *Classical Novae*, ed. M.F. Bode and A. Evans, Cambridge Astrophysics Series 43, Cambridge University Press, Cambridge, 2008, 2nd edition.[32]

(ii) The brighter, more rapidly evolving novae, originating from more massive systems ($M_{wd} > 1.2\ M_{\odot}$) in which the white dwarf is a carbon-oxygen type, which produce fast moving (but less-dense) winds that are strongly enriched in the elements He, Ne and Mg, and which either produce no dust at all or else optically thin shells of silicate dust, upon which are superimposed emission features for SiC and SiO.

However, even this categorization is a simplification as some novae exhibit features that are common to both of these categories—and some even show evidence for the existence of both silicate and carbon dusts.

6.6.2 A Physical Model of Novae

The physical and spectral development of novae is extremely complex and involves parcels of gas passing through a wide range of physical conditions. Outbursts are characterized by the production of a shell or series of shells of ejecta over a limited period of time, which probably possess non-spherical morphological attributes (such as equatorial rings and bipolar outflow lobes)—adding to the complexity of the spectral development. However, we can understand the basic processes by considering the generic light curve shown in Figure 6.4. In this, and the discussion that follows, we refer to times as the time post-outburst visual maximum, which we define to be $t = 0$.

The initial expansion ('fireball') phase results in the formation of a growing 'pseudo-photosphere' and a visual maximum occurs within a few days of the outburst. However, as the mass loss rate falls and the shell/s suffer geometrical dilution due to expansion, the pseudo-photosphere retreats to inner, hotter regions of the ejecta and the peak in the spectrum shifts from the visible to the UV (at near-constant luminosity). Shortly after, or near the peak in visual luminosity, some very simple molecules (such as CO and CN) are often detected.

The 'speed' of a nova is defined by the rate at which the visual light curve declines post-maximum and hence the rate at which the location of the pseudo-photosphere falls back towards the remnant. This, in turn, will depend on such factors as the density of the ejecta. A 'fast' nova will decline at a rate of $\sim$0.1 magnitudes per day (in the V band), whilst a 'slow' nova will decline at a rate of $\sim$0.01 mag day^{-1}. In fact there are several fairly well-constrained empirical relationships that connect the nova speed and fundamental properties. During this period of visual decline—and several tens of days after the initial eruption—many novae, most especially the 'slow' novae—exhibit a well-defined 'transition' phase during which the visual brightness can decline catastrophically (by several magnitudes), accompanied by brightening (re-distribution of luminosity) into the infrared. This is interpreted as being due to the formation of an optically thick dust shell in the ejecta. Approximately a third of all novae (usually the 'slow' novae) produce these optically thick dust shells.

Finally, some time *after* the dust formation epoch—during the so-called 'recovery', when the optical depth of the dust shell falls as it expands and the visual light curve brightens once again—more complex molecular signatures become apparent, indicative of hydrogenated carbon, or free PAH molecules. Throughout much of the evolution, the bolometric luminosity typically remains approximately constant.[33] The steady decline in the optical results from the blueward shift of the photosphere as it contracts to hotter regions of the outflow, and the transition follows from the redistribution of UV and visible light to the infrared.

As the epochs of molecule and dust formation occur on timescales of days to a few years, it is therefore apparent that novae provide unique opportunities to study molecule and dust formation on practical timescales.

6.6.3 The Molecule Formation Epochs

As described above, there are essentially two epochs during which molecules are detected in novae:

(i) in the early (pre-dust formation) phase, when small molecules, such as CO and CN are detected, often with very high excitation temperatures, and
(ii) in the late (post-dust formation) phase, when hydrocarbon/PAH features are detected.

The simple species seen in the early stages vanish as the nova evolves, which could be a result of (a) incorporation into more complex, as yet undetected, species, (b) destruction, or (c) excitation effects. Interestingly, no molecules of intermediate complexity have been observed in novae. Similarly, the appearance of the hydrocarbon features *after* dust formation is perhaps indicative of the presence of the breakdown products of the dust grains (maybe due to sputtering or photodisintegration) rather than being an intermediate species between the early-stage small molecules and the dust grains.

The molecular inventory of novae is not large—only CO, CN, SiO, SiO_2, SiC, PAH-features, several silicate and hydrocarbon features, and H_2 (at late times) have been reliably detected—but their presence is extremely common in certain types of novae.

6.6.4 Early Stage Chemistry

The presence of molecules in the early stages of the evolution of a nova may seem rather surprising, bearing in mind the very extreme physical conditions in the gas. That gas is extremely hot (~1000–5000 K), dense ($\sim 10^7$–10^{13} cm^{-3}) and located close to an intense source of ionizing (far-UV) radiation, and yet there are clear indications of the presence of simple molecules, such as CO and CN. The CN radical was detected as early as 1935 in nova DQ Her [34] but the largest body of observations relate to CO. The first indications came from photometric measurements that identified the presence of a '5 μm excess' with fundamental ($\Delta v = 1$) vibrational transitions (4.7 μm) in CO.[35] Subsequent high-resolution spectroscopic measurements have shown that the CO is often detected in both the fundamental and first overtone (2.3 μm, $\Delta v = 2$) transitions, and the band usually shows the presence of very highly vibrationally excited states. An example of CO observations, as seen in the ejecta of Nova V705 Cas, is given in Figure 6.5. This shows the presence of transitions up to $v = 6 \rightarrow 4$ which imply an extremely high excitation temperature of ~4500 K.[36]

In the conditions that exist in novae ejecta, a chemistry that is highly unlike that of the interstellar medium pertains; one that bears some similarity to that which is believed to exist in the early Universe. The ejecta are, of course, quite dust-free in these early epochs, so that only gas-phase processes can occur. Firstly it is helpful to identify the factors that help or hinder the chemistry in such extreme conditions. The high temperatures and densities are conducive to neutral–neutral reactions, many of which have significant activation barriers. Reactions involving anions (especially H^-, C^- and O^-), which require an abundance of free electrons and are also strongly temperature-dependent, are also enhanced. The conditions also allow for the possibility of reactions involving (vibrationally and/or electronically) excited reactants and, at the highest densities that pertain in the earliest stages of the outflows, three-body reactions are significant.

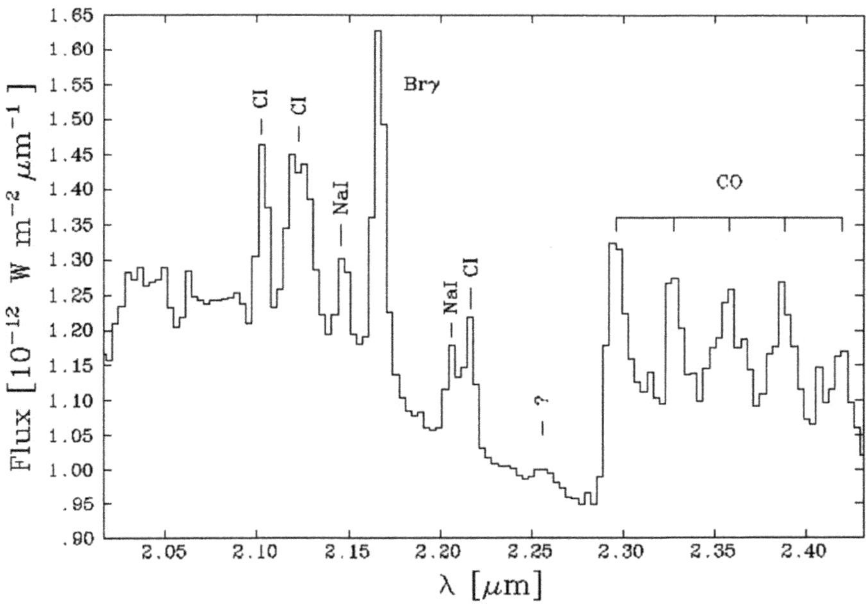

Figure 6.5 First overtone ($\Delta \nu = 2$) CO emission lines as seen in Nova V705 Cas. Reproduced from Figure 6.2 in Infrared spectroscopy of Nova Cassiopeia 1993 — I. The pre-dust phase, Evans *et al. Mon. Not. R. Astron. Soc.*, 1996, **282**,[36] by permission of Oxford University Press on behalf of the Royal Astronomical Society.

However, the reverse processes of collisional dissociation, which proceed by an n-body process that leads to the sequential excitation of the vibrational levels into the continuum, can be the dominant loss channel for many species, and are critically sensitive to both the temperature and the density. Potentially most damaging of all is the presence of an intense far-UV radiation field, which, if unshielded, would result in photodissociation/photoionization timescales of the order of 0.1–10 seconds. There are a variety of factors affecting the photoreaction rates; geometrical dilution, the shift in the spectrum to the UV as the photosphere contracts, the possible presence of shock-induced radiation fields, *etc.* The combined effect of these factors is for the ejecta to be overtaken by a series of ionization fronts. Molecular processes are only viable if the molecular species are part-shielded from the radiation field by the ionization continua.

Early studies[37] identified two important facts:

(i) To allow any form of chemistry to proceed with an efficiency that is required to explain the observed abundances of molecules such as CO and CN, the molecular region must be carbon neutral (CI). This fact follows quite simply from the necessity to shield key molecular species, such as H_2 and CO, from the radiation field. The absorption

bands that lead to photodissociation for these molecules are shielded by the carbon ionization continuum ($\lambda < 1102$ Å).
(ii) The chemical timescales, whether for two-body, three-body or photo-reactions, are many orders of magnitude smaller than the dynamical timescales (which are of the order of tens of days). Thus *chemical quasi-equilibrium* holds at all locations and at all but the later times (>100 days) so that the chemistry responds promptly to changes in the physical conditions.

To help understand the viability of molecule and dust formation we can also look at the development of a nova in the context of non-LTE model atmosphere calculations in which line-blanketing plays an important role;[38] an initially, highly optically thick 'fireball' phase is followed by a ballistic, freely expanding 'optically thick wind' phase (post-maximum) in which the outer layers are less optically thick. In this phase the line- and continuum-forming atmosphere becomes spatially extended. Significantly, in many novae, the UV spectrum can be strongly dominated by a forest of (mainly) FeII lines—the so-called 'iron curtain' that forms a quasi-continuous absorption potentially protecting molecules against photodestruction. Finally the opacity drops, the iron curtain is lifted and the nova enters a pre-nebular phase characterized by higher ionization/excitation. In this framework it is therefore apparent that the epochs of molecule and dust formation must occur in a fairly narrow 'window of opportunity' during this second, optically thick wind, phase. This is consistent with the observations that whenever low ionization potential metals are detected in neutral form (*e.g.* NaI, see Figure 6.5), dust formation is seen to ensue.[39]

If the conditions are conducive to molecule formation then, in the earliest stages of the ejecta evolution ($T < 3$–5 days), H_2 is mainly formed by three-body association and lost by collisional dissociation:

$$\mathrm{H + H + H \rightleftharpoons H_2 + H.} \tag{6.20}$$

The three-body formation rate for H_2 is, obviously, sensitive to the density, whilst collisional dissociation (which occurs as a multi-stage process, involving successive excitation up the vibrational ladder) is sensitive to the density and the temperature of the gas.

At somewhat lower densities (and higher ionization) the ion channels are significant:

$$\mathrm{H} + e^- \rightarrow \mathrm{H}^- + h\nu, \quad \mathrm{H}^- + \mathrm{H} \rightarrow \mathrm{H_2} + e^- \tag{6.21}$$

or, to a lesser extent

$$\mathrm{H + H^+ \rightarrow H_2^+} + h\nu, \quad \mathrm{H_2^+ + H \rightarrow H_2 + H^+} \tag{6.22}$$

In addition, reactions involving excited H may be significant,[40] particularly if Lyα emission is efficiently trapped:

$$\mathrm{H}_{(n=2)} + \mathrm{H/H^+} \rightarrow \mathrm{H_2^+} + e^-/h\nu \tag{6.23}$$

although, for these reactions, the intermediates, H^- and H_2^+, are both susceptible to the radiation field. These processes, whilst efficient in novae ejecta, would not be able to produce H_2 with fractional abundances larger than $\sim 10^{-8}$ if the H_2 were not protected from the intense radiation field, which—at the epoch at which molecules are seen—would imply a photodissociation timescale of the order of just ~1–10 s.

In the dense and hot, but essentially neutral, environment of the nova ejecta, the formation of simple molecules tends to be dominated by neutral–neutral condensation and association reactions, with lesser contributions from anion–neutral chemical reactions paths. This is very much the case for CO, which can be formed by direct radiative association:

$$C + O \rightarrow CO + h\nu \tag{6.24}$$

or *via* the OH and CH radicals

$$O + H_2/H \rightarrow OH, \quad C + H_2/H \rightarrow CH \tag{6.25}$$

which can also be formed from O^- and C^-:

$$O^-/C^- + H_2/H \rightarrow OH/CH \tag{6.26}$$

followed by

$$OH + C \rightarrow CO + H, \quad CH + O \rightarrow CO + H. \tag{6.27}$$

The CH and NH radicals also provide a lead-in to the formation of CN:

$$OH + N \rightarrow NO + H, \quad NO + C \rightarrow CN + O \tag{6.28}$$

or

$$CH + N \rightarrow CN + H \tag{6.29}$$

with some additional contributions from the $C^- + N$ reaction and direct radiative association of C with N, and loss *via* neutral exchanges with O and N atoms.

The C_2 molecule, which has potential significance in the dust formation process (see below), follows a similar chemistry, being formed from CO and CH:

$$CO + C \rightarrow C_2 + O, \quad CH + C \rightarrow C_2 + H \tag{6.30}$$

with lesser contributions from association reactions between C/C^- and C.

Subsequent studies[41] paid more careful attention to the photochemistry, employing accurate cross-sectional data, where available, and sophisticated model atmosphere calculations. These studies showed that the chemistry in the molecular zone is far less photon-dominated than previously thought, and is instead largely dominated by neutral–neutral reactions. The models confirmed that a region with a $H_2 : H$ ratio of $>10^{-4}$ and a gas temperature of <3500 K must exist within a few days of the outburst for molecules and dust to form. At higher temperatures, collisional dissociation reactions tend to inhibit efficient molecule formation. As with the simple chemistry, dust

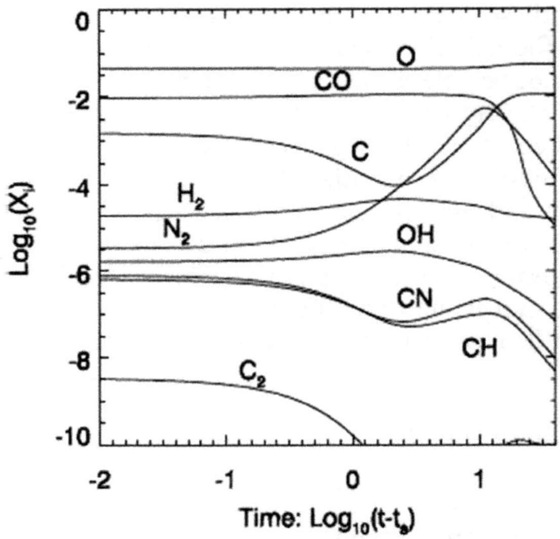

Figure 6.6 Results from a model of the (early-stage) time-dependence of simple molecules in a dust-forming nova. Reproduced from Figure 6.1 in The early chemical evolution of nova outflows, Pontefract and Rawlings, *Mon. Not. R. Astron. Soc.*, 2004, **347**,[41] by permission of Oxford University Press on behalf of the Royal Astronomical Society.

formation is very sensitive to the density. An example of the results obtained is shown is Figure 6.6, and demonstrates many of the key findings:

(i) The early-phase chemistry is essentially *not* photon-dominated and is largely determined by neutral–neutral reactions.
(ii) As per earlier studies, densities that are higher than the spherical average are needed to promote molecule formations. This is found to be particularly true for the key species C_2.
(iii) CO formation, whilst efficient, does *not* go to saturation for any combinations of the physical conditions, so that peak CO/O or CO/C fractions are only ~0.01–0.1.
(iv) The period of high CO abundance only lasts for ~10–15 days post-outburst, after which it declines rapidly.

These findings, particularly the last two, have been strongly supported by observations. Examples include the strong dust-producing nova Sagittarii 2015 no. 2 (V5668 Sgr),[39] V2615 Ophiuchi (2007)[42] and V2676 Ophiuchi,[43] which show evidence for the presence of low ionization states and CO first overtone emission, which only persists for of the order of ~10 days after maximum. The last of these, V2676 Ophiuchi, also showed evidence for C_2 in absorption. The amount of CO that is produced is not huge (typically in the range 4×10^{-10}–3×10^{-8} $M_\odot$), but the vibrational excitation temperature is always very high ($T\approx2500$–4500 K).

6.6.5 Carbon Dust Formation

More than one third of all novae produce copious amounts of optically thick ($\tau > 5$–10) carbon dust on a timescale of tens to hundreds of days. This potentially allows the processes of dust formation, growth and subsequent evolution to be observationally monitored. As such they allow us to test models of dust nucleation and growth.

The dust formation epoch has some interesting characteristics: (i) It is fast; entry into the transition stage occurs on a timescale of a few days. (ii) It is efficient throughout the shell; for those novae that are prolific dust producers the dust shell effectively blocks out the visible radiation from the central source and the source luminosity is re-radiated in the thermal infrared. This is an important point that must be considered in the context of the early-time chemistry discussed above; the dust formation must occur in such a way that it 'covers the sky' of the nova. It cannot be confined to clumps, rings, or bipolar lobes in the outflow. If, as suggested above, it occurs in thin shells of ejecta, then they must be subject to pressure confinement, or some other mechanism, to allow them to remain stable for timescales that are longer than the inferred sound-crossing time of a few tens of days. (iii) The formation process is itself efficient; observational and theoretical studies[44] have shown that in many cases the dust grains can grow to large (>1 μm) sizes and account for a significant fraction of the carbon budget in the ejecta.

As described in Section 6.4.1 above the dust-formation chemistries of AGB stars have relied heavily on (well-characterized) terrestrial flame chemistries, which are largely based on acetylene (C_2H_2) polymerization,[22] and the main channel for closure of the first aromatic ring is through the recombination of propargyl[16] (HC≡C–CH2–).

In the non-photon dominated environs of AGB stars, models predict that the carbon budget is essentially split between CO and C_2H_2. Despite the difficulty in detecting C_2H_2, it has been firmly detected in IRC + 10216. This mechanism is therefore potentially very viable. However, in the harsher, photon-dominated, environment of novae (and supernovae) C_2H_2 is liable to photodissociation. Using recent model atmosphere calculations for novae, supernovae and AGB stars it has been shown that C_2H_2 cannot survive in these harsher environments,[45] and that, therefore, it is most unlikely that a single dust formation mechanism operates in all astrophysical environments.

Theoretical studies of dust formation are conceptually difficult, as the process marks the transition between the microscopic (chemical)/kinetic approach and the macroscopic (condensed matter)/thermodynamic approach. The remit of astrochemical modelling is to try and understand the first of these, and to model the formation of dust grain nucleation sites. Once these have been formed, the subsequent (macroscopic) growth of (carbon) dust grains will be controlled by the requirement that the ejecta temperature drops below the condensation temperature (determined by the condition that the partial pressure of the accreting monomer is greater than the vapour pressure of the condensate), subject to any other constraints imposed by the level of ionization, radiation field strength, *etc.* The situation is further complicated by the fact that, at all times, the physical and chemical conditions are very far from LTE—the gas kinetic temperature is typically ~1000 K, whilst the radiation temperature is ~10–20 000 K. In addition, the intense UV radiation field continually acts to limit reaction complexity. Thus, whilst some smaller molecules may be shielded by ionization continua, the more complex organics will not. As a result, reactions between large molecules are inhibited:

$$C_aH_b + C_aH_b \rightarrow \text{products.} \tag{6.31}$$

Early studies[46] showed, on purely theoretical grounds, that ion–neutral radiative association reactions with the C_2 monomer (which is partially shielded against photodissociation by the carbon ionization continuum) could be effective in promoting molecular growth, *e.g.*:

$$C_aH_b^+ + C_2 \rightarrow C_{a+2}H_b^+ + h\nu \tag{6.32}$$

$$C_aH_b^+ + C_2 \rightarrow C_{a+2}H_{b-1}^+ + H \tag{6.33}$$

and

$$C_aH_b^+ + H_2 \rightarrow C_aH_{b+1}^+ + H \tag{6.34}$$

followed by dissociative recombination. This simple, and largely conjectural, ion–(simple) neutral network made no attempt to differentiate isomers and relies on some degree of hydrogenation. Thus an H_2 fractional abundance of $>10^{-4}$ is required to drive the chemistry, and this can be provided by the H^- channel described above—provided, once again, that the H_2 is shielded from the radiation field. Following the usually adopted paradigm, the presence of free oxygen atoms was assumed to strongly inhibit hydrocarbon formation (hydrocarbons 'burn' in hot oxygen-rich gas), and the models assumed that CO formation goes to saturation.

As other studies have shown, the formation of nucleation sites is probably the most difficult stage in the dust formation process; once the molecules are sufficiently large (typically ~8–9 carbon atoms), they become more stable and can act as nucleation sites for rapid and efficient monomeric grain growth.[47] The models predict the formation of carbon nucleation sites, but at a fairly low efficiency, so that the dust production is characterized by a smallish population of large grains. Results from an update of this model,[32] which include full incorporation of the model atmosphere calculations, as well as an empirically determined network of hydrocarbon/PAH chemistry,[16] is shown in Figure 6.7.

Again, the models imply that the presence of H_2 (even relatively low fractional abundances $\sim 10^{-3}$) is essential to drive much of the chemistry and that the high density requirement suggests that the region of dust (and perhaps molecule) formation must be dense shells of carbon–neutral gas, at

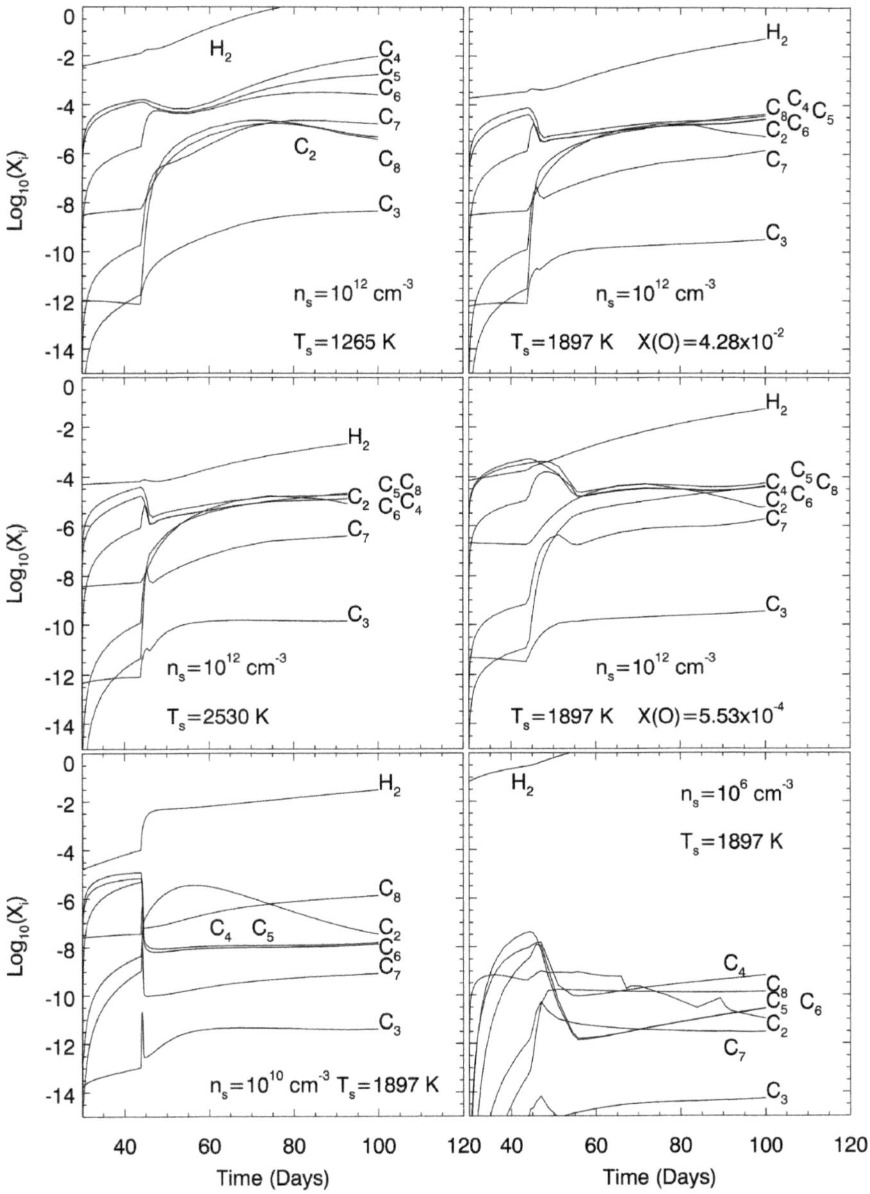

Figure 6.7 Modelled abundances as a function of time for C_n species in a carbon dust-producing nova, based on a generic hydrocarbon chemistry coupled to a PAH chemistry. Results are shown as a function of time for several values of the density, temperature and oxygen atom abundance.
Reproduced with permission from *Classical Novae*, ed. M.F. Bode and A. Evans, Cambridge Astrophysics Series 43, Cambridge University Press, Cambridge, 2008, 2nd edition.[32]

temperatures <5000 K. Significantly, the studies indicate that the formation of nucleation sites may be more efficient than previously believed. These calculations also include free oxygen atoms and, interestingly, although the presence of oxygen reduces the formation efficiencies, it clearly does *not* inhibit the formation of nucleation sites altogether.

6.6.6 Silicate Dust Formation

The majority of dust-producing novae (and especially those that produce optically thick dust shells) generate carbon (graphitic) dust, good examples being Aql 1995, Her 1991 and PW Vul 1984. Moreover, due to the paucity of spectral features, the dust is usually assumed to be graphitic in nature. However, novae also produce other dust types, *e.g.* SiC (Aql 1982, V842 Cen 1986), hydrocarbons (V842 Cen, V705 Cas 1993) or SiO_2 (V1370 Aql 1982, V705 Cas 1993). Some novae (notable examples being V842 Cen and QV Vul 1987) are also seen to produce multiple dust types. Clearly identified dust components include olivine (Mg_2SiO_4), *ortho*-pyroxene ($MgSiO_3$), silicon carbide (SiC), and, to a lesser extent, corundum (Al_2O_3) and cohenite (Fe_3C).

This variety and, in particular, those that produce multiple dust types, may seem to be difficult to explain in the context of the usual paradigm, in which CO formation goes to saturation leaving free oxygen or carbon atoms to form silicates or carbon dust in O-rich and C-rich ejecta, respectively. There are, however, some important factors to be considered:

(i) The ejecta may be stratified with strong abundance gradients, both radially and perhaps between the polar outflows and equatorial regions. The former of these is something that is predicted by recent nova models that show that a typical CNO-type white dwarf may have $C>O$ in its surface layers, and $C<O$ underneath.[48]

(ii) As commented on in Section 6.6.4, both the chemical models and observations indicate that CO formation does *not* go to saturation, and yet the formation of nucleation sites is still possible in these conditions.

In addition, if (ii) is correct, then models of dust formation may need to be augmented to include other mechanisms that bypass hydrogen-based chemistries—see below.

Since silicate dust formation in novae is not as efficient as carbon dust formation, it has not attracted the same amount of attention as the latter, and has yet to be properly investigated, although it probably proceeds along similar pathways as described for supernovae below.

6.6.7 Alternative Chemical Pathways to Carbon Dust Formation

As indicated above, it is possible that we have not yet identified all of the viable chemical pathways that may be applicable to the nova environment.

An example of this may be provided by the Boudouard mechanism,[49] which has yet to be fully investigated in astrophysical contexts. The mechanism is based on the redox reaction between two CO molecules that results in the disproportionation of CO into CO_2 and solid-state carbon:

$$CO + CO \rightleftharpoons CO_2 + C_{(s)}. \tag{6.35}$$

This reaction is exothermic, and is probably augmented by surface catalysis once the solid carbon starts to form, but the large formation entropy of CO means that at high temperatures the forward reaction is endergonic and the reverse reaction is favoured. However, if the gas cools sufficiently, then the forward reaction becomes viable. The process is, for example, a major cause of soot formation in blast furnaces and catalytic converters. It is not normally of relevance in astrophysical contexts as there is a large activation energy barrier that, in effect, can only be overcome if the reacting CO molecules are both highly vibrationally excited.

As we have seen above, such conditions may indeed be present in nova outflows, in which case the mechanism may be an important formation pathway for carbon dust. Obviously, a key diagnostic of the Boudouard mechanism would be the presence of significant amounts of CO_2. Observationally, CO_2 is difficult to detect and has so far been elusive, but improvements in sensitivities and resolutions may soon be able to determine whether or not this process is important.

6.6.8 The Subsequent Evolution of the Dust

As the ejecta expand, the density drops and dust formation processes are quenched. At the same time, the hardening radiation field, accompanied by the progression of ionization fronts through the outflow, increase the likelihood of grain erosion by ions. Additionally, X-rays produced by the nova will also result in grain destruction. Indeed, it is a well-observed fact that, in the recovery phase, the grain size falls. This response of the grains to the ever-harsher environment poses a serious question as to how much of the dust that is initially condensed survives into the interstellar medium, and this is still unquantified. The dust temperature and spectral energy distribution provide constraints on the erosion, but a corollary of this is that the erosion yields breakdown products, in the form of large molecules. Thus a number of infrared emission features are often detected,[32] most of which are associated with hydrocarbon stretch/bend modes: 3.29 μm/6.2 μm (aromatic C–H stretch), 3.4 μm (aliphatic C–H stretch and/or CH_2/CH_3 inclusions in silicates), 7.7–8.2 μm (blend of C–C stretching bands and/or $Si–CH_3$), 8.7 μm (aromatic C–H in-plane bend) and 11.25 μm (aromatic C–H out-of-plane bend). However, as already commented upon above, these features have not been detected *prior* to dust formation and are probably indicative of dust grain destruction by shocks and/or the passage of ionization fronts rather than bottom-up formation of complex organic molecules.

6.7 Supernovae

6.7.1 Supernova Types and Outflow Dynamics

The taxonomy of supernova types is rather complex, with several subdivisions. However, for the purpose of this book, we simply identify the distinction between type Ia (thermal runaway) supernovae, which are—in essence—extreme nova events in which the white dwarf secondary is completely disrupted, and types Ib, Ic and II (core collapse) supernovae.

Current studies have addressed molecule and dust formation in several supernovae, mostly type IIs —although there have been some (purely theoretical) studies of the early-Universe population III (massive supernovae; $M \approx 80$–190 $M_{\odot}$). SN1987a has been the subject of most investigations for several reasons: (i) it is close, located 50 kpc away in the Large Magellanic Cloud, (ii) there is a wealth of observational data at various wavelengths from the time of eruption to the present; it was the first supernova from which molecular emissions were detected and the first to show evidence for dust synthesis, and (iii) it is relatively young, so that the outflowing gas has not yet been overtaken by the reverse shock, nor has it mixed into circumstellar/interstellar material.

This is an important point; the reverse shock may raise the gas to very high temperatures and it is unlikely that molecules would survive the passage of the shock. Thus, molecular emission can only serve to diagnose the conditions in the outflow and perhaps be identified as intermediates to dust grain formation. The dust grains *may* survive the shock, but the molecules cannot be regarded as an input to the interstellar medium.

A type II core collapse supernova results when the central regions reach the iron peak and the fusion energy production rate falls dramatically. The core implodes and the outer helium core is traversed by the resulting explosion blast wave that deposits energy into the gas. As the wave encounters the base of the progenitor envelope a reverse shock wave is generated that propagates inwards and triggers Rayleigh–Taylor instabilities, and the breakup/mixing of the stratified layers into microscopically and/or macroscopically mixed clumps over a timescale of days. The helium core is disrupted but a degree of chemical stratification persists. In addition, Richtmyer–Meshkov instabilities may result in the penetration of the outer hydrogen-rich layers into the inner regions.

This is potentially very complex and the exact nature of the resulting structure is hotly debated; the layers may remain intact, break up into chemically similar clumps, or undergo mixing so that the clumps incorporate the chemical compositions from several layers.[50,51] Figure 6.8 gives a schematic illustration of the structure of a macroscopically mixed clumpy supernova.

Overall, ejecta from supernovae (SNe) are hydrogen poor (metal rich) and the compositional stratification, identified by the dominant components, is believed to be H, He/N, He/C, He/O/C, O/C, O/Ne/Mg, SiO and Si/S/Fe.

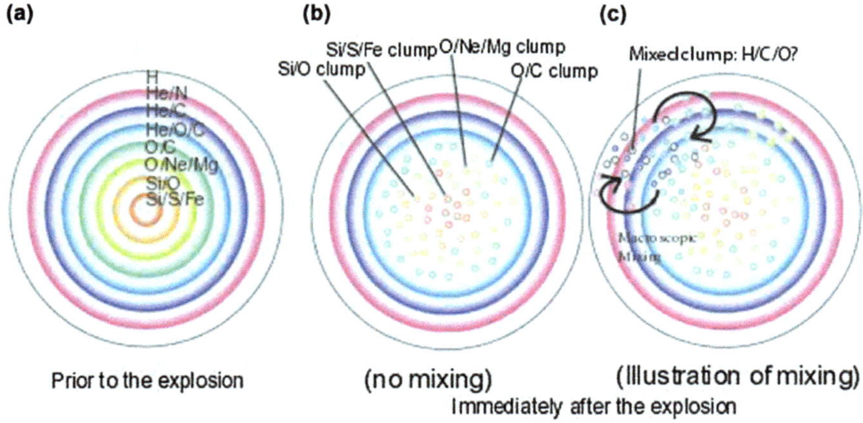

Figure 6.8 Schematic representation of the structure of a supernova, illustrating the possible role of macroscopic mixing. (a) Shows the pre-explosion chemically stratified layers, (b) shows the result of instabilities in forming chemically distinct clumps, and (c) illustrates the effect of mixing to/from the surface layers.
Reproduced with permission of Oxford University Press on behalf of the Royal Astronomical Society from Matsuura *et al.*, *Mon. Not. R. Astron. Soc.*, 2017, **469**, 3347.[50]

6.7.2 Molecule Formation

Compared to the other classes of object described in this chapter, the chemistry of supernovae ejecta is both very much more complex—in part due to the additional physical factors as described above, but also due to the elemental differentiations. Modelling efforts have risen to the challenge and, consequently, are at a much greater level of sophistication as compared to those used to study novae.[52–55]

As stated above, the greatest interest in molecule- and dust-formation studies is directed towards type II (specifically type II-P) supernovae, as there is now a substantial body of data to support the existence of dust formation in these objects. Stars with initial masses of ~8–30 $M_{\odot}$ usually become type II supernovae, and dust and molecules have been seen in these objects some hundreds of days after the eruption.

SN1987a was the first supernova in which both dust formation and molecular emission was observed and it still serves as a prototype/test-bed for chemical- and dust-formation theories. The molecules CO and SiO were seen within 120 days of the explosion and newly formed dust was detected by ~400 days.[56] Similar (warm) dust populations (as diagnosed by a mid-IR excess, coupled with a sharp decline in the optical emission—similar to novae) have been detected in several other supernovae. In addition, SiO vibrational emission has been detected and seen to wane, possibly due to the inclusion of SiO in condensing silicates.

The early-stage detections were fairly limited; with CO 2.3 μm and 4.6 μm emission observed from 112 days post-outburst onwards—implying an abundance equivalent to $CO:C \approx 10^{-3}$, and SiO $v = 1 \rightarrow 0$ (8.1 μm) emission being seen from $t = 160$ to, at least, 578 days post-outburst—implying a mass of $\sim 4 \times 10^{-6}$ $M_{\odot}$. Similar detections have been made in a few other (~8) young (<10 years old) SNe, but CO and SiO are the only molecules that have been detected, generally with low masses.

In SN1987a, after about 600 days the (mid-infrared) emissions gradually faded, most probably due to excitation effects. Then, after a hiatus of about 25 years, molecular and dust emissions strongly reappeared in millimetre and sub-millimetre wavelengths.[57,58] These reveal that the CO mass has risen sharply from the early stage value to $>0.01 M_{\odot}$.

The earliest modelling of SN1987a recognized that the conditions in the molecule and dust-forming regions of supernovae are most unlike other outflows; these were some of the earliest attempts at modelling molecule formation in hydrogen-poor astrophysically extreme environments.[59] Perhaps most significantly, the radiation field is largely intrinsic to the gas, being powered by the radioactive decay of ^{56}Co, ^{57}Co and other elements.

In subsequent studies[52–55] the contributions of the radioactive decays of ^{56}Ni, ^{57}Ni (*via* ^{56}Co and ^{57}Co respectively), ^{44}Ti and ^{22}Na to produce γ-rays, X-rays, electrons, positrons and neutrinos have been variously included. The subsequent degradation of the γ-rays leads to the production of hard X-rays by Compton scattering, which degrades further to UV photons through a cascade of inverse Compton interactions and ionization/recombination processes, which also produce a population of energetic electrons. These electrons can ionize/dissociate atoms and molecules (including forming chemically significant noble gas ions) and generate the radiation field (perhaps by the collisional excitation of the 2^1S metastable state of helium, resulting in two-photon continuum emission).

Results for a simple H, He, C, and O chemistry in a simplified two-layer stratification model[59] showed that only a handful of molecules could achieve appreciable abundances: CO, CO^+, C_2 and O_2. In the carbon/oxygen rich core zone hydrogen is almost completely absent and the formation channels are extremely simple, *e.g.* for CO:

$$C + O \rightarrow CO + h\nu \tag{6.36}$$

$$O + O \rightarrow O_2 + h\nu, \quad O_2 + C \rightarrow CO + O \tag{6.37}$$

$$C + C \rightarrow C_2 + h\nu, \quad C_2 + O \rightarrow CO + C \tag{6.38}$$

There are, however, several channels, so that the formation rate is robust and the abundance is essentially controlled by the destruction (photodissociation) rate. Thus, as the ^{56}Co decays, the CO abundance becomes significant after ~100 days and rises steadily thereafter—a fact that is significant in the context of recent observations of 'late' CO in SN1987a.

A subsequent (hypothetical) study of molecule formation in early-Universe, massive (20–270 $M_{\odot}$), zero metallicity progenitor population III supernovae[52] expanded the chemical network greatly, although concentrated on simple species, and predicted the presence of O_2, CO, SiS and SiO, suggesting that as much as ~44% of the mass of the ejecta could be molecular at late times. Similar CO formation channels were identified as above, but the reversible channel of

$$C + CO \leftrightarrow C_2 + O \tag{6.39}$$

was highlighted for being effective (at high temperatures) in producing C_2 and C_3 and may be very important in the formation of carbon chains and dust, even in oxygen-rich environments.

A more comprehensive model[53] included the chemistry of atoms, ions, molecules and small clusters (including chains, rings and small dimers, tetramers of silicates, *etc.*) utilizing a comprehensive gas-phase chemical reaction network. The results showed that CO, SiO, SiS, O_2 and SO are produced efficiently, although the SiO is expected to dimerize quickly and form grains—implying that SiS is expected to be the dominant sulfur-bearing component in the gas-phase.

Part of the reaction network is shown in Figure 6.9 and it bears significant resemblance to that discussed in Section 6.4.2 above, but with no hydrogen content. The initiating steps are the dimerization of SiO and growth (*via* reactions with O_2 and SO) to form the important building block for dust, Si_2O_3—see below. Concerning the simple molecules; the study showed the dominant CO formation network to be similar, as found in the earlier studies, and emphasized the importance of simple neutral–neutral and radiative association reactions in forming species such as SiO, C_2 and carbon chains C_n. A steady growth of CO mass from ~10^{-4} $M_\odot$ to 0.1 $M_\odot$ over the period 100–1500 days post-eruption was predicted.

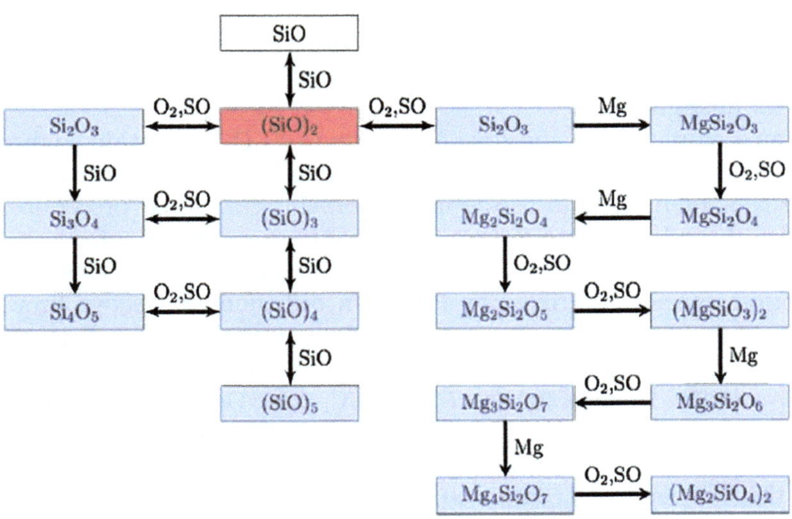

Figure 6.9 A possible formation network for enstatite dimers ($Mg_2Si_2O_6$) and forsterite dimers ($Mg_4Si_2O_8$).
Reproduced with permission from the AAS from Sarangi and Cherchneff (2013),[53] using data from ref. 60.

6.7.3 Dust Formation

For SN1987a, a warm dust component (of fairly low mass, $\sim 10^{-5}$–10^{-3} $M_{\odot}$) was seen at relatively early stages, but a number of recent observations have also revealed the presence of a large quantity ($\sim$0.06–0.8 $M_{\odot}$) of *cold* dust. A rich inventory of dust species is detected in old SNe, such as Cassiopeia A ($\sim$300 years old) in which alumina, carbon, enstatite, forsterite, magnesium protosilicates, silicon dioxide, silicon, iron, iron oxide and iron sulfide have been observed.[61]

The papers by Sarangi and Cherchneff (2013, 2015)[53,54] have applied molecular nucleation theory first of all to study cluster nucleation, then nucleation and grain growth. One of the key features of supernova outflows, as distinct from the other objects discussed in this chapter, is the presence of strong elemental abundance variations within the ejecta, leading to the formation of multiple dust types in any one source. The models combine/couple the early-stage chemistry to kinetic models of the formation of large molecules, clusters and then grain growth, and differentiate between the different layers/clumps within the ejecta. In contrast to novae, silicates are a very important component of the dust and their chemistry/dust formation has been modelled, paying careful attention to the various condensate types. The grain species considered are forsterite, alumina, carbon, magnesium, silicon carbide, silicon, iron and iron sulfide.

As was discussed for novae, classical nucleation theory—which is centred on the barrier-crossing problem of attaining a critical cluster size—is inappropriate to studies of supernova dust formation, in which conditions are very far from LTE and, instead molecular nucleation theory is more appropriate; the procedure adopted is to follow the gas-phase kinetics of the formation of the building blocks for the dust grains. Thus, the dimers of forsterite and enstatite as formed *via* the reaction network shown in Figure 6.9 then form 'clusters', which can then be treated as the nucleation sites for dust grains.

As an example of this, the growth of alumina (Al_2O_3) is considered through the dimerization of AlO, followed by oxidation. The tetramer of Al_2O_3, $(Al_2O_3)_4$, is then taken to be the nucleation cluster. Similar procedures are followed for other grain precursors. The molecular clusters grow by coagulation and (to a lesser extent) accretion of atoms and molecules. The coagulation itself is controlled by Brownian diffusion, convective Brownian enhancement, gravitational collection, turbulent motion and van der Waal's forces. This procedure is then followed in each the distinct chemical zones *e.g.* 1A: Si/S/Fe, 1B: Si/O, 2: O/Ne/Mg, 3: O/C, 4: He/O/C, 5: He/C, 6: He/N. In these models, the chemical layers retain their chemical integrity, so that, although the effects of clumping are considered, the clumps in any one layer have the same elemental composition. That is to say, the ejecta are assumed to be microscopically mixed, with no leakage between the zones. Both uniform and clumpy distributions were considered. In the case of carbon, chain growth was followed to C_{10}, and then ring-closure and growth by C_2 addition.

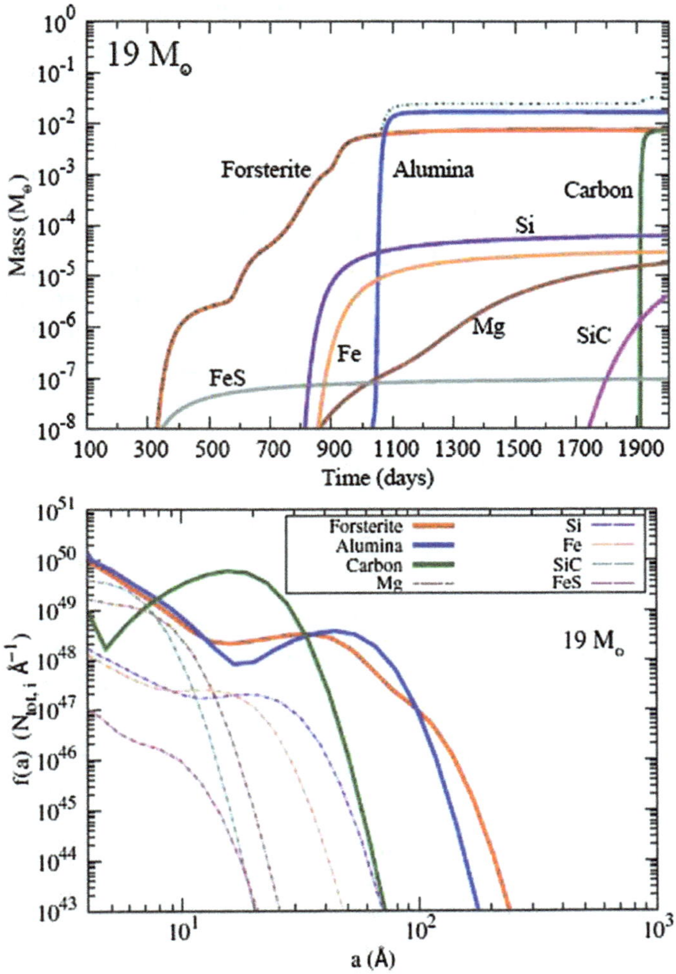

Figure 6.10 Modelled dust masses as a function of time, and dust size distributions at 2000 days for a 19 $M_{\odot}$ progenitor (mimicking SN1987a). Results are shown for the various dust types.
Sarangi and Cherchneff, A&A, 575, A95 (2015), reproduced with permission © ESO.[54]

An example of the result obtained is given in Figure 6.10, which considers a 19 $M_{\odot}$ progenitor (mimicking SN1987a) and a uniform (non-clumpy) outflow. The different dust components condense in the various chemical layers, so that first of all iron sulfide, pure iron and silicon dust condense in zone 1A (the Si/S zone), then in the oxygen core (zones 1B, 2 and 3); forsterite condenses in in zone 1B, forsterite, alumina and pure magnesium in zones 2–4, and finally carbon and silicon carbide in zone 5 (where C : O > 1). The efficiency of cluster coagulation is extremely high, so that ~99% of the clusters condense to form dust, and all models seem to predict a rapid period of grain growth between a period of ~200 and 600 days, but slower growth thereafter. Unsurprisingly, the dust condensation process is strongly sensitive to the gas density, so significant quantitative differences exist if the layers are assumed to be clumpy. In all cases the three main dust components (silicates, carbon and alumina) are produced. The predicted dust grain size distribution does not match the interstellar average but, since the grains will undergo shock processing, and their contribution to the overall budget is unknown, this is to be expected.

The dust formation models have been developed to ever-increasing complexity, and recent studies[55] have now included fourteen different species of dust grains. Other effects that have been considered include grain charge, accretion, sublimation from the dust grains, and grain surface 'weathering' by He^+, Ne^+ and Ar^+ ions.

6.7.4 Late Stage Molecules and Dust

CO, H_2O, OH^+, HCO^+ and SO have been detected in older SNe (>1000 years old), but probably originate from external molecular cloud material that is illuminated/excited by the SNe.

However, in the case of SN1987a, the current optical appearance, some 30 years after the explosion, can be described as a 'ring of hot spots', marking the location where the supernova blast wave collides with the (innermost) circumstellar ring (see Figure 6.11). Recently, a large reservoir (>0.01 $M_\odot$) of cold (<120 K) CO has been detected in SN1987a, 25 years after the explosion[57] and also a large mass (~0.5 $M_\odot$) of cold (~20 K) dust 23–25 years post-outburst.[58]

There is some debate as to whether this large amount of dust was formed early on, but then remained undetected for the last 25 years, or is simply the relic of some previous stage of the evolution of the system, or else is the product of steady, ongoing, dust formation. Existing astrochemical models tend to favour the former, and argue that it is hard to identify an efficient dust-formation mechanism at late times, when the ejecta is relatively less dense and cooler. However, simple accretion may explain the discrepancy.

Concerning the molecular gas, spectra indicate that the CO emission must be coming from a collection of clumps and Figure 6.11 shows that the location of the CO emission is *inside* the shockwave, and is therefore a product of the primordial supernova chemistry. Other observations have deduced CO and SiO masses of 0.02–1.0 $M_\odot$ and 4×10^{-5}–2×10^{-3} $M_\odot$, respectively (depending on excitation temperature assumptions). This implies that CO has continued to form during the 25 'dark' years. This is consistent with the predictions of both the early models and also those that invoke the $O_2 + C$ reaction as a source of CO in the O-rich layers and which predict a CO mass of ~0.1 $M_\odot$.

In addition to CO, the recent observations of SN1987a, 27 years after the outburst, have detected the presence of cold SiO, SO and, most surprisingly, HCO^+ (with a deduced mass of ~6×10^{-6} $M_\odot$), yet only an upper limit

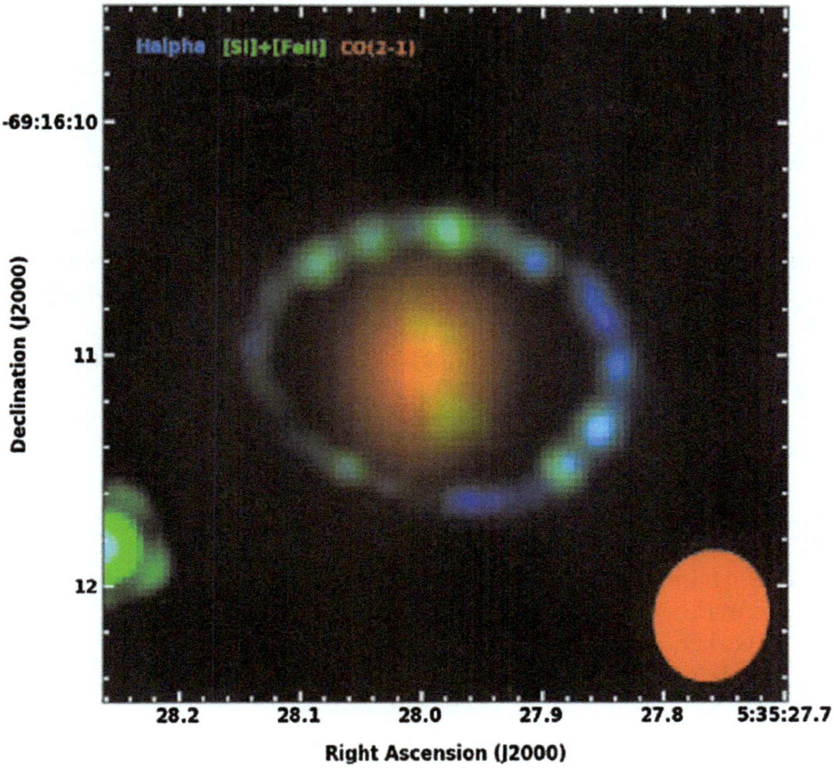

Figure 6.11 Late-time (ALMA) observations of CO in SN1987a. The ^{12}CO 2–1 (unresolved) line emission is shown in red. The blue and green/yellow show (Hubble Space Telescope) Hα and [Si I]+[Fe II] 1.644 μm emission. The ellipse in the corner is the ALMA synthesized beam. Reproduced by permission from the AAS from Kamenetzky *et al.* (2013).[57]

detection of SiS ($<6\times10^{-5}$ M$_{\odot}$).[50] HCO$^+$ has not been the subject of detailed modelling, as the C/O layers are believed to be hydrogen poor. However, it seems possible that the abundance of HCO$^+$ and the absence of SiS can be explained by the existence of macroscopic mixing (due to Richtmyer–Meshkov instabilities) shortly after the explosion; the presence of hydrogen from the outer layers in the inner C/O zone could allow HCO$^+$ formation via reactions of the type:

$$C + H_2 \rightarrow CH + H_2, \quad CH + O \rightarrow HCO^+ + e^- \tag{6.40}$$

whilst if oxygen atoms penetrate the Si/S zones, the formation of SiS (which is otherwise predicted to be the dominant reservoir of Si) could be suppressed. If so, then these species could act as powerful diagnostics of the supernova dynamics and the nature of the mixing in the ejecta.

References

1. D. D. Clayton, *Principles of Stellar Evolution and Nucleosynthesis*, McGraw-Hill Book Company, 1968.
2. C. F. von Weizsäcker, *Phys. Z.*, 1938, **39**, 633.
3. H. A. Bethe, *Phys. Rev.*, 1939, **55**, 436.
4. H. A. Bethe, *Science*, 1968, **161**, 541.
5. D. Schönberner and T. Blöcker, in *The Molecular Astrophysics of Stars and Galaxies*, ed. T. W. Hartquist and D. A. Williams, Claredon Press, 1998.
6. A. I. Karakas, in *Principles and Perspectives in Cosmochemistry*, ed. A. Goswami and B. E. Reddy, Springer, 2010.
7. S. G. Ryan and A. J. Norton, *Stellar Evolution and Nucleosynthesis*, Cambridge University Press, 2010.
8. I. Iben, *Stellar Evolution Physics –Vol. 1 Physical Processes in Stellar Interiors*, Cambridge University Press, 2013.
9. I. Iben, *Stellar Evolution Physics –Vol. 2 Advanced Evolution of Single Stars*, Cambridge University Press, 2013.
10. C. S. Beals, *Observatory*, 1934, **57**, 319.
11. J. K. E. Halm, *Proc. R. Soc. Edinburgh*, 1904, **25**, 513.
12. L. Biermann, *Observatory*, 1957, **77**, 109.
13. E. N. Parker, *Astrophys. J.*, 1960, **132**, 821.
14. E. J. Weber and L. Davis Jr., *Astrophys. J.*, 1967, **148**, 217.
15. G. H. Bowen, *Astrophys. J.*, 1988, **329**, 299.
16. I. Cherchneff, J. R. Barker and A. G. G. M. Tielens, *Astrophys. J.*, 1992, **401**, 269.
17. S. Höfner, *Astron. Soc. Pac. Conf. Ser.*, 2015, **497**, 333.
18. D. Gobrecht, I. Cherchneff, A. Sarangi, J. M. C. Plane and S. T. Bromley, *Astron. Astrophys.*, 2016, **585**, 6.
19. P. R. Wood, in *Proceedings of IAU Symposium No. 239 Convection in Astrophysics*, ed. F. Kupka, I. W. Roxburgh and K. L. Chan, Cambridge University Press, 2006.
20. D. Xiong and L. Deng, *Chin. Astron. Astrophys.*, 2013, **37**, 1.
21. B. Freytag, S. Liljegren and S. Höfner, *Astron. Astrophys.*, 2017, **600**, 137.
22. M. Frenkland and E. D. Feigelson, *Astrophys. J.*, 1989, **341**, 372.
23. I. Cherchneff, in *The Molecular Astrophysics of Stars and Galaxies*, ed. T. W. Hartquist and D. A. Williams, Claredon Press, 1988.
24. I. Cherchneff, *EAS Publ. Ser.*, 2013, **60**, 175.
25. H.-P. Gail, M. Scholz and A. Pucci, *Astron. Astrophys.*, 2017, **591**, 17.
26. M. R. Zachariah and W. Tsang, *J. Chem. Phys.*, 1995, **99**, 5308.
27. T. W. Hartquist, P. Bertoldi, R. H. Durisen, J. E. Dyson, R. J. R. Williams, J. M. C. Rawlings and D. A. Williams, in *The Molecular Astrophysics of Stars and Galaxies*, ed. T. W. Hartquist and D. A. Williams, Claredon Press, 1998.
28. A. Lèbre, M. Aurière, N. Fabas, D. Gillet, F. Herpin, R. Konstantinova, Antova and R. Petit, *Astron. Astrophys.*, 2014, **561**, 85.
29. F. Herpin, A. Baudry, C. Thum, D. Morris and H. Wiesemeyer, *Astron. Astrophys.*, 2006, **450**, 667.

30. R. C. Doel, M. D. Gray, E. M. L. Humphreys, M. F. Braithwaite and D. Field, *Astron. Astrophys.*, 1995, **302**, 797.
31. S. Starrfield, J. W. Truran, M. C. Wiescher and W. M. Sparks, *Mon. Not. R. Astron. Soc.*, 1998, **296**, 502.
32. *From Classical Novae*, ed. M. F. Bode and A. Evans, Cambridge Astrophysics Series, Cambridge University Press, Cambridge, 2nd edn, 2008, vol. 43.
33. A. R. Hyland and G. Neugebauer, *Astrophys. J.*, 1970, **160**, L177.
34. O. C. Wilson and P. W. Merrill, *PASP*, 1935, **47**, 53.
35. E. P. Ney and B. F. Hatfield, *Astrophys. J.*, 1978, **219**, L11.
36. A. Evans, T. R. Geballe, J. M. C. Rawlings and A. D. Scott, *Mon. Not. R. Astron. Soc.*, 1996, **282**, 1049.
37. J. M. C. Rawlings, *Mon. Not. R. Astron. Soc.*, 1988, **232**, 507.
38. P. H. Hauschildt, *et al.*, *Astrophys. J.*, 1997, **490**, 803.
39. D. P. K. Banerjee, M. K. Srivastava, N. M. Ashok and V. Venkataraman, *Mon. Not. R. Astron. Soc.*, 2016, **455L**, 109.
40. J. M. C. Rawlings, J. E. Drew and M. J. Barlow, *Mon. Not. R. Astron. Soc.*, 1993, **265**, 968.
41. M. Pontefract and J. M. C. Rawlings, *Mon. Not. R. Astron. Soc.*, 2004, **347**, 1294.
42. R. K. Das, D. P. K. Banerjee and N. M. Ashok, *Mon. Not. R. Astron. Soc.*, 2009, **398**, 375.
43. H. Kawakita, M. Fujii, M. Nagashima, T. Kajikawa, N. Kubo and A. Arai, *PAS Jpn.*, 2015, **67**, 17.
44. S. N. Shore, *et al.*, *Nature*, 1994, **369**, 539.
45. H. Dhanoa and J. M. C. Rawlings, *Mon. Not. R. Astron. Soc.*, 2014, **440**, 1786.
46. J. M. C. Rawlings and D. A. Williams, *Mon. Not. R. Astron. Soc.*, 1989, **240**, 729.
47. H. P. Gail and E. Sedlmayr, *Astron. Astrophys.*, 1987, **171**, 197.
48. J. José, G. Halabi and M. F. El Eid, *Astron. Astrophys.*, 2016, **593**, A54.
49. K. A. Essenhigh, *et al.*, *Chem. Phys.*, 2006, **330**, 506.
50. M. Matsuura, *et al.*, *Mon. Not. R. Astron.Soc.*, 2017, **469**, 3347.
51. A. Jerktrand, C. Fransson and C. Kozma, *Astron. Astrophys.*, 2011, **530**, A45.
52. I. Cherchneff and E. Dwek, *Astrophys. J.*, 2009, **703**, 642.
53. A. Sarangi and I. Cherchneff, *Astrophys. J.*, 2013, **776**, 107.
54. A. Sarangi and I. Cherchneff, *Astron. Astrophys.*, 2015, **575**, A95.
55. A. Sluder, M. Milosavljević and H. Montgomery, 2017, *Mon. Not. R. Astron. Soc.*, in press.
56. D. H. Wooden, *et al.*, *Astrophys. J. Suppl.*, 1993, **88**, 477.
57. J. Kamenetzky, *et al.*, *Astrophys. J. Lett.*, 2013, **773**, L34.
58. R. Indebetouw, M. Matsuura and E. Dwek, *Astrophys. J.*, 2014, **782**, L2.
59. J. M. C. Rawlings and D. A. Williams, *Mon. Not. R. Astron. Soc.*, 1989, **246**, 208.
60. F. Gouman and S. T. Bromley, *Mon. Not. R. Astron. Soc.*, 2012, **420**, 3344.
61. L. Dunne, *et al.*, *Mon. Not. R. Astron. Soc.*, 2009, **394**, 1307.

CHAPTER 7

Conclusions: Where Do We Go from Here?

7.1 What We Know Now About Dynamical Astrochemistry in the Milky Way

As described earlier in this book (Section 2.1), astrochemistry made its first tentative appearance in the detections of the molecular species CH, CH^+, and CN by their optical absorption lines along paths towards bright stars through interstellar clouds of rather low extinction (less than one visual magnitude). These detections—and the subsequent attempts to understand them by exploring the nature of chemical reactions in these regions of interstellar space—established the infant subject. It has now grown to an impressive maturity, as hundreds of thousands of research papers, this book and many other texts demonstrate. Even in the early astrochemical studies, it was evident that the detections contained important astrophysical information and that these studies provided useful probes of interstellar regions that were yet to be fully explored (and that were considered by many astronomers at the time to be merely a trivial and unimportant astronomical backwater—little did they know!). For example, it was clear by the 1960s that while the presence of CH and CN in interstellar diffuse clouds could be reasonably understood on the basis of simple steady-state astrochemical models, CH^+ might require an explanation based on gas dynamics, *via* a rapid, transient chemistry in post-shock gas with elevated temperatures. Thus, even at this early date, it seemed that there might be (at least) two types of interstellar regions: those in which chemical timescales were short compared to local physical timescales (in which the chemical steady state should provide an adequate description), and those regions in which the

Dynamical Astrochemistry
By David A. Williams, Thomas W. Hartquist, Jonathan M. C. Rawlings, Cesare Cecchi-Pestellini and Serena Viti

Published by the Royal Society of Chemistry, www.rsc.org

chemical processes had timescales that were comparable to, or longer than, those of the local physics. In the latter case, evidently chemical evolution in response to the changing conditions driven by local dynamical events needed to be considered. These dynamical regions have been the focus of this book.

The significance of astrochemistry as a major component of astronomy became clear from the late 1960s onwards, when—thanks to the availability of new detectors operating in the infrared, millimetre and sub-millimetre wavebands—a flood of molecular identifications arrived, and continues to this day. This increased the number of identified species in interstellar and circumstellar space from a mere handful (5) to an overflowing basketful (nearly 200), and detailed chemical modelling made clear that very many more species than those actually identified must also be present. Molecular identifications were made in a variety of astronomical locations, including dark clouds, star-forming regions, circumstellar envelopes and discs, jets and outflows, and in the ejecta of novae and supernovae. Dynamical astrochemistry in all of these regions of the Milky Way galaxy has been explored in this book, and we have shown that astrochemistry is in many cases the uniquely essential tool in these studies.

An early and very significant discovery (through millimetre-wave rotational emission lines of CO) was of the giant molecular clouds in the Milky Way and other galaxies. These giant clouds may contain a significant fraction of a galaxy's mass, and they always comprise almost all of the non-stellar baryonic matter in a galaxy. The observed association of molecular clouds with young bright stars confirmed that these clouds provide the raw material from which new stars are formed. The fundamental relevance of astrochemistry in tracing the conversion of gas to stars was thereby established. Astrochemistry deals with a previously unknown component of the Milky Way: the denser, cooler, and often more massive components of the interstellar medium; these components have an active existence and undergo dynamical changes.

While the basic information in molecular spectral lines is useful in determining gas densities and temperatures, detailed astrochemical modelling may reveal a much more complete description of the local chemistry and physics. As we have demonstrated in this book, such models are based on the response of the chemical networks to local physics, and may provide information not only on spatial variations in density and temperature in the emitting region, but also, for example, on the local electromagnetic and particle radiation fields, elemental abundances, and dust grain properties. In particular, such studies may reveal the state of evolution of studied regions, for they may show that the observed regions are in transition between one state and another. Therefore, astrochemistry—comprising observations of molecular lines and the subsequent detailed modelling—can provide insight into the dynamical behaviour of many aspects of interstellar and circumstellar gas in the Milky Way and other galaxies. These studies generate the compelling narrative of modern astronomy: how tenuous gas in interstellar clouds can be converted to stars, forming circumstellar and protoplanetary disks, how those newly formed stars are revealed through jets and outflows that the young stars generate, how these young stars influence their immediate environments through winds and radiation, how they populate their environment with dust and enrich the gas with heavy elements, and how they end their lives as white dwarf stars with AGB precursors, and as novae and supernovae explosions. The models have an almost insatiable appetite for chemical data on gas-phase, surface, and solid-state reactions. The title of Faraday Discussion 133 (2006), *Chemical Evolution of the Universe*, emphasizes both the advances in astronomical chemistry but also the role the subject plays in tracing and influencing the evolution of the Universe itself. The subject is now well-established, but some outstanding questions remain to be resolved. Some of these remaining issues are described in Section 7.2.

The focus of the book on dynamical chemistry in the Milky Way is not a limitation. In fact, the Milky Way's astrochemistry may be regarded as a template for other (at least nearby) galaxies. Studies of astrochemistry in the Milky Way have revealed the main drivers of chemistry to be the fluxes of cosmic particle and electromagnetic radiations, and reactions on dust grain surfaces and in their ices. The responses of the predicted chemical behaviour in models to changes in, for example, gas density, relative elemental and dust abundances, intense UV radiation fields, and very high fluxes of cosmic ray particles have been extensively explored and are well understood. The predictions of such models give insight into other situations, especially in external galaxies where the particle and photon fluxes, and relative elemental and dust abundances may be very different from those in the Milky Way. The Milky Way as a template means that extragalactic observations may be readily interpreted using chemical models derived from Milky Way studies. Some applications of dynamical astrochemistry external to the Milky Way are discussed briefly in Section 7.3.

7.2 Some Outstanding Questions Remaining for Dynamical Astrochemistry in the Milky Way

As this book illustrates, much progress has been made in the development of dynamical astrochemistry. However, many important and interesting questions remain to be answered. The rest of this section presents some examples of the need for further work, for all the areas covered by Chapters 2–6.

7.2.1 Shocks

Dust particles play important roles in astrochemistry: as recombination sites affecting the ionization structure, as sources of obscuration protecting molecular material from ionizing and dissociating radiation, as catalytic sites for reactions forming molecules that may be later injected into the gas phase, and as sites on which ices may be deposited and processed. The size distribution of dust grains determines how efficient grains are in these various roles. The ionization structure, radiative environment and chemical composition influence the magnetic field and temperature of astronomical sources, and these in turn govern the global dynamical evolution of environments. Dust grain destruction was treated in Chapters 2 and 3, but much more remains to be done before the size distribution of dust grains in a dynamic interstellar medium through which shocks pass can be reliably calculated from first principles.

7.2.2 Turbulence and Small-scale Structure

Tiny scale transient structure in diffuse interstellar clouds: small scale structure is inferred through the secular variation of atomic and molecular line intensities in the line of sight through diffuse regions towards a bright star. Intensities may vary on timescales as short as several years, implying that the structures may be as small as the solar system. The origin of such structures is not fully understood but may involve MHD waves of some kind.

Unidentified absorption bands (UIBs) are observed on many lines of sight through interstellar low density gas, and have been known for about a century. About 500 of these bands are known in the visual and near infrared. Five of these bands have now been firmly identified[1,2] as arising from the ionized fullerene C_{60}^{+}. Are fullerenes and their ions responsible for other UIBs? What is the origin of interstellar fullerenes, and what is their role in the interstellar medium?

The global cycling of matter from diffuse clouds into translucent and dark clouds containing star-forming regions, and the ejection of matter from star-forming regions and from both young and old stars back into the interstellar medium is understood only in qualitative terms.

7.2.3 Star Formation

Star formation is a very active field and it is clear from Chapter 4 that chemistry helps to diagnose the dynamics of the early stages of this process. However, many gaps in our knowledge remain. We have still to establish how the collapse depends on the initial conditions in the parent cloud. This dependence affects the formation of both low and high mass stars. The roles of magnetic fields and of turbulence are still being debated, and it is only very recently that both of these concepts have been included in models of star formation.

The study of high mass star formation is hampered by the limited observational data of the massive and dense parental clumps. There is a lively debate about massive star formation: does it result from the gravitational collapse of such a clump, or from the competitive accretion of a number of smaller clumps? The answer must lie in the way that cores and clumps fragment, and this in turn depends on the initial conditions of the parent cloud. Fortunately, interferometric studies are now providing vital information on the chemical and kinematic structure of such clouds.

7.2.4 Jets and Outflows

The launch of a jet from a protostar can be understood in general terms, but requires to be more firmly established. This may reveal the true nature of stellar jets, and the cause of the near-regular emission features along the jet. The variable velocity model of a uniform jet has been very successful, but does not account for all observed features of jets. Are jets clumpy, with varying clump velocities?

Interface regions between stellar winds and ambient material, the turbulent mixing that occurs there, and the resulting signature chemistry have been explored but with crude models and require more detailed modelling.

7.2.5 Evolved Stars, Novae and Supernovae

As noted in Section 6.3.2, radiation pressure on dust is important for generating the high mass loss rates that lead to the formation of planetary nebulae. Advances in the development of kinetic models of dust formation in the outflows have been achieved but much work remains to be done, particularly for dust formation in oxygen-rich AGB stars. An understanding of the initial phases of silicate dust formation requires a combination of *ab initio* electronic structure calculations and the application of approaches such as RRKM/master equation rate theory to estimate rate coefficients, before dust production networks are incorporated into models of the hydrodynamics of the outflows.

7.3 Applications of Dynamical Astrochemistry to Two Special Environments

7.3.1 Astrochemistry in the Early Universe

We described in Chapter 4 the physical and chemical processes involved in the formation of low mass and high mass stars from interstellar gas clouds in the Milky Way. We described in Chapter 6 how stars nearing the end of their lives inject—either steadily in a wind or violently in explosions from novae and supernovae—part of their mass into the interstellar medium, in the form of gas and dust. This cycling of material into and out of stars implies the enrichment of the interstellar medium in heavy elements such as carbon and oxygen, contained in atoms, molecules, and dust grains. But at some point in the evolution of the Universe, the very first stars—the so-called Population III stars—must have formed in a gas in which this enrichment had not occurred. What was the nature of that initial gas, and how did star formation occur in it? There were no heavy elements, and consequently no dust grains. We describe here, very briefly, the initial chemistry in the Universe and how star formation in that primordial gas was achieved.

According to current cosmology, the Universe began about 13.8 billion years ago. Observations suggest that mass (and its energy equivalent) in the Universe was, and still is, composed of several components: dark energy, dark matter, baryonic matter, and some other minor components, including photons and neutrinos. Dark energy has been invoked to account for the observed acceleration in the expansion of the Universe on the large scale. Dark matter is necessary to account for observed gravitational effects between and within galaxies that cannot be attributed to the mass of observed stars and gas. The third component, baryonic matter, is the familiar matter of which all interstellar gas, stars and planets (and their inhabitants), and galaxies are composed. The natures of dark energy and dark matter are completely unknown, although their effects are evidently real enough. In recent years, observations have established that the proportions of mass/energy attributed to the various components are dark energy 68.3%, dark matter 26.8%, and baryonic matter 4.9%. Neutrinos and photons make a negligible contribution. Baryonic matter is therefore a relatively minor component, but it is only through observations of it in the form of stars and galaxies that the role of the major components can be traced.

The initial state of the Universe was one of exceptionally high density and temperature, extreme conditions in which atomic nuclei could not survive and baryonic matter consisted entirely of sub-atomic particles; these were interacting with a background of very high energy electromagnetic radiation. However, as the Universe expanded and cooled atomic nuclei eventually became established in a background of black body radiation of the same temperature as the matter. The atomic nuclei formed by nucleosynthesis were—by the end of this phase—mainly hydrogen (75%) and helium (25%), with minor components of deuterium ($\sim 10^{-5}$ relative to H^+) and lithium ($\sim 10^{-10}$). The expansion and cooling continued and the peak of the radiation moved steadily to longer wavelengths and lower energy, so that photoionization declined and recombination of electrons on atomic nuclei allowed the first neutral atoms to exist. Helium atoms were the first to form (at a redshift ~ 2500) and later, as the hardness of the electromagnetic radiation declined further, atomic hydrogen was able to survive. When the matter was mainly neutral, the matter and radiation temperatures became decoupled because the radiation was no longer able to ionize hydrogen or helium; this occurred when the Universe was about 0.38 Myr old, at a redshift, z, of about 1100, and the temperature of both matter and radiation was about 4000 K. (The redshift, z, is defined such that $(1+z)$ is the ratio of the present value of the cosmic scale factor to the value at the time when the radiation reaching us now was emitted. The cosmic scale factor is a dimensionless function parameterizing the relative expansion of the Universe.) The temperature of the decoupled radiation continued to decline with the Universal expansion, and the radiation is detected at present as the so-called cosmic microwave background radiation, an almost perfect black body spectrum with peak wavelength in the millimetre range, representing a temperature of about 2.725 K. The radiation is a true relic of the Universe when it was less than 0.003% of its present age. The temperature and density of the baryonic matter also fell and remained low—these were the so-called dark ages—until at a redshift of $\sim$20 (when the age of the Universe was about 180 Myr) the very first stars in the Universe were formed. The radiation from these first stars then began to heat and ionize the gas, inhibiting further star formation. For reference, the current highest redshift of a detected galaxy is 11.09 (for GN-z11)[3] when the age of the Universe was about 400 Myr.

How were these first stars formed? As described in Chapter 4, star formation in the Milky Way proceeds essentially by matter falling into a gravitational potential well, created by baryonic matter itself; matter accumulates under its own weight. The release of gravitational potential energy in the form of heat occurs in radiation from molecular coolants emitted *via* rotational transitions at low temperature. The temperature in the collapsing region is held low, and the consequent pressure rise that would inhibit the collapse is suppressed. The necessary coolants are simple molecules such as CO, OH, H_2O and isotopic versions thereof. However, the situation in the early Universe was very different. No heavy elements were present. What coolants, molecular or otherwise, could have operated in primordial matter?

Chemistry during the dark ages ($z \approx 1100$–20) occurred in a range of densities (6700–0.046 cm^{-3}) and temperatures (4000–60 K), and the cosmic microwave background radiation field weakened and softened throughout this evolutionary period. In these dust-free conditions, molecular hydrogen can be formed through the radiative association

$$H + H^+ \rightarrow H_2^+ + h\nu, \tag{7.1}$$

followed by the charge exchange

$$H_2^+ + H \rightarrow H_2 + H^+. \tag{7.2}$$

Alternatively, the radiative attachment

$$H + e \rightarrow H^- + h\nu \tag{7.3}$$

may be followed by the associative detachment

$$H^- + H \rightarrow H_2 + e. \tag{7.4}$$

The first of these pairs of reactions dominated at $z \approx 400$, while the second pair was more important at $z \approx 100$, because H_2^+ is more tightly bound than H^-, the radiation field being stronger and harder at $z = 400$ than at $z = 100$.

Although hydrogen is dominant, helium may provide an alternative route to H_2 through the pair of reactions

$$He + H^+ \rightarrow HeH^+ + h\nu \tag{7.5}$$

$$HeH^+ + H \rightarrow H_2^+ + He \tag{7.6}$$

followed by charge exchange of H_2^+ with H, as before. Deuterium (at $\sim 10^{-5}$ relative to H) can form deuterium hydride from the atomic ion

$$H_2 + D^+ \rightarrow HD + H^+ \quad (7.7)$$

and HD may also contribute to cooling.

A detailed study (Galli and Palla 2013[4]) showed that, during the dark ages, H_2 (at a maximum fractional abundance of $\sim 10^{-6}$) and HD ($\sim 10^{-10}$) were probably the only effective coolants. Lithium and lithium hydride played no effective role. Thus, the coolants initially available during the formation of the first stars were very limited, had very low abundances, and were quite different from those that controlled the Milky Way star formation.

There is another significant difference between star formation in the Milky Way and that in the early Universe. According to current models of the Universal expansion, inhomogeneities developed in the dominant dark matter, creating centres of attraction into which the minor component of baryonic matter would have flowed. Therefore, these sites of attraction—called dark matter haloes—accumulated baryonic matter, and, if the cooling was rapid enough compared to the Universal expansion, could have maintained their collapse. If so, the formation of one or more stars from a halo may have occurred. Dark matter haloes are now regarded as the formation sites of the first stars in the Universe, the so-called Population III stars.

If the baryonic gas number density in the dark matter halo was not too great, say less than about $10^4\ cm^{-3}$, molecular hydrogen was an effective coolant down to a temperature of about 200 K. Above this density, collisions tended to quench upper state populations before radiation could be emitted. Nevertheless, cooling still occurred, although at a slower rate. At the critical density of gas at a temperature of 200 K, an accumulated mass of about 1000 $M_\odot$ becomes gravitationally unstable to thermal support, and therefore masses above this amount will collapse.

Unstable baryonic masses of this magnitude may have accumulated in dark matter haloes in the early Universe. If so, the number density of baryonic matter would have increased rapidly and when it exceeded about $10^{10}\ cm^{-3}$ hydrogen molecules would have formed very rapidly by three-body reactions:

$$H + H + H \rightarrow H_2 + H.$$

Thus, the temperature may have been maintained at the relatively low value of 200 K during the collapse until pressure support dominated. One or more Population III protostars may have formed from the large baryonic mass ($\sim 10^3\ M_\odot$) of gas in a dark matter halo. The large mass potentially available indicates that a Population III star would have been very massive and very bright, but would have had a short life, and would have ended as a supernova and populated its environment with the first heavy elements in the Universe. The UV radiation generated by the massive star would have tended to destroy molecular hydrogen in its environment, inhibiting further star formation until more dark matter haloes collapsed. Initially, the heavy matter enrichment from one of the first stars was tiny, but increased as star formation continued. Supernovae blast waves may also have affected the evolution of the gas. The accumulation of dark matter haloes within which star formation occurred is believed to have led to the formation of the first galaxies. Within them, the gradual increase of heavy elements from a very low level ($\sim 10^{-5}$ of solar fractional abundances) eventually allowed for the formation of early Population II stars, whose death throes in supernovae rapidly enhanced the heavy element abundances.

7.3.2 The Astrochemistry of External Galaxies

Of the nearly 200 molecular species identified to date in the interstellar and circumstellar regions of the Milky Way galaxy, about 60 have also been detected in external galaxies. Given the difficulties of angular resolution and of radiation intensity in making observations of molecular spectra of galaxies (typically at distances of megaparsecs from Earth), it is remarkable that so many molecular species have been confirmed to exist in these objects. The detections support the view that interstellar and circumstellar chemistry as discovered within the Milky Way is Universal and, consequently, that molecules may be useful probes of physical conditions and their evolution in these very distant objects, just as they are for various regions of the Milky Way. The products of chemistry in external galaxies are also responding to and influencing dynamics in those galaxies, just as they do in the Milky Way. Molecular detections in external galaxies also confirm that conventional astrochemistry occurred in galaxies early in the existence of the Universe. For example, observations[5] using the VLA detected CO (2–1) rotational emission in the hyperluminous IR galaxy SDSS J1148 + 5251. Together with earlier detections in other CO lines (including all five lines in the sequence 3–2 → 7–6), the observations imply a molecular gas reservoir in this galaxy of more than 10^{10} $M_{\odot}$, and of diameter of about 5 kpc centred on the active galactic nucleus. This object is at a redshift $z = 6.419$, which in conventional cosmology corresponds to an age of the Universe of 851 Myr (*i.e.*, we are viewing this galaxy when the Universe was about 6% of its present age).

Galaxies are present in astounding numbers (the current estimate is 2 trillion in the Universe[6]) and in a wide variety of forms. They have a huge range in mass and luminosity, from values in *dwarf galaxies* much smaller in mass and luminosity (say, at least 10^4 times smaller) than those of the Milky Way, to being perhaps an order of magnitude or so larger. Their shapes range from highly organized *disks* or *ellipticals* to quite *irregular* structures. The Milky Way galaxy is a *spiral*; spirals are highly flattened rotating disks of stars, gas and dust, and star formation occurs mainly in the spiral arms so these arms are where the very bright young stars are found; there is also a central bulge that contains older stars. Some spirals (*barred spirals*) have linear bars passing though the nucleus, with star formation observed to be occurring mainly at the intersection of the bars with the arms. Galaxies with large amounts of interstellar gas and dust may undergo intense bursts of star formation; such galaxies are called *starburst* galaxies. Starburst galaxies may be very bright in the infrared if the galaxies are particularly dusty, for then the UV from newly formed stars is absorbed by the dust and re-emitted at longer wavelengths. Starbursts may be regular or irregular in shape and like all galaxies they have a very wide range in mass and luminosity. Galaxies that have converted almost all of their interstellar matter into stars have completed their star-forming phase and show little gas and dust; they contain mainly older stars and are termed *early-type* galaxies. Most galaxies are believed to contain black holes in their nuclei. If the black hole in the nucleus of a galaxy is very massive, the accretion of stars, gas and dust onto it may be a very violent process in which huge amounts of radiation can be emitted. Such a nucleus is said to be *active.* A galaxy with an active nucleus at its centre (an *active galactic nucleus*, or *AGN*; see Section 6.4.3) can be exceptionally bright, especially in the IR, as in the case of the hyperluminous IR galaxy SDSS J1148 + 5251, mentioned in the previous paragraph. Galaxies may show the characteristics of more than one type. Given the enormous range of physical conditions in external galaxies, is it possible to use astrochemistry as a diagnostic? Remarkably, the answer is "yes". Since different types of environment can produce specific molecular species, it may be possible to use chemistry in distant galaxies to determine the classification of those galaxies. We indicate very briefly in the remaining paragraphs how chemistry may help to classify galaxy types.

As we have seen in earlier chapters, the chemistry in cold interstellar molecular clouds (the sites of star formation, where most of the dramatic dynamical events described in this book take place) does not occur to any appreciable extent unless one or more of several factors is operating.

(1) The temperature is raised, possibly in a shock or by turbulent dissipation, promoting neutral exchange reactions.
(2) The interstellar UV radiation field in regions in which the visual extinction is low creates atoms and ions to promote a low temperature exothermic ion–molecule chemistry.
(3) X-rays from AGNs can be powerful radiation sources affecting significant volumes (column densities up to $\sim 10^{24}$ $H_2\,cm^{-2}$) of an active galaxy, and the ionization produced by the X-rays may drive a complex chemistry.
(4) Ionization by cosmic rays affects almost all interstellar clouds, including very dark clouds, and drives an exothermic ion–molecule chemistry in them; cosmic rays also generate a weak UV radiation field that contributes to this chemistry.
(5) Surface reactions on bare grains may provide simple species (mainly hydrides).
(6) Cosmic rays and UV radiation fields may promote a complex solid-state chemistry within chemically simple ice mantles on dust grains in dark clouds; when these ices are desorbed (for example, in dense

regions near young stars, so-called hot cores), the newly formed complex molecules are released to the gas phase.

The main drivers of chemistry on a galactic scale are stellar UV, X-rays from AGNs, and the all-pervading flux of cosmic ray particles. These drivers may vary considerably among the many types of galaxy. The effects of these drivers are modulated by (a) the local gas number density, (b) the dust : gas ratio, and (c) the metallicity (*i.e.*, the abundances of carbon, nitrogen, oxygen and other elements relative to hydrogen).

Let's begin by considering the broad effects of these three main modulating parameters. First, we address the gas number density, (a). Observations of molecular emissions from external galaxies seem to come from regions of roughly similar densities to those of emitting regions in the Milky Way. Therefore, the role of gas number density in modulating the chemistry is roughly similar in external galaxies and in the Milky Way. Next, consider (b), the dust : gas ratios. In external galaxies, these ratios are generally similar to or rather smaller than those in the Milky Way. The effect of smaller ratios is to push deeper into molecular clouds the transition between regions strongly illuminated by UV starlight and regions in which UV is excluded. Molecules are abundant at points deeper than this transition. Beyond that point, gas phase chemistry proceeds roughly as in the Milky Way. A smaller dust : gas ratio may therefore reduce slightly the size of the molecular region. However, the gas phase chemistry is otherwise unaffected by the ratio. Reducing the dust abundance relative to gas also implies that there are fewer ice mantles but that these will be larger. Hence, processing of the ices by cosmic rays and UV as in (6) in the previous paragraph will give a similar hot core chemistry of complex molecules to that found in the Milky Way. However, the elemental abundances, (c), are known to vary from galaxy to galaxy, and changes in these abundances can significantly affect the molecular abundances, and in a non-intuitive way (see Bayet *et al.* 2008[7]). For example, in hot core models, while some species (*e.g.*, CO, H_2CO, CS, and SO_2) are approximately linear tracers of metallicity, others (*e.g.*, CH_3CN) are independent of metallicity, while others (SO, HCO^+) are apparently inverse tracers of metallicity. The Milky Way is, therefore, a poor template for the chemistry of hot cores in external galaxies that have different metallicities from those in the Milky Way. However, it may be possible to use the different behaviours of key molecules with respect to metallicity to determine relative elemental abundances in the molecular regions of external galaxies.

Having considered the effects of variations in gas density, dust : gas ratio, and metallicity, we'll now discuss very briefly the astrochemistry of galaxies that display intense UV and cosmic ray fluxes much larger than for the average of the Milky Way.

First, we consider the role of intense cosmic ray fluxes. Galaxy mergers and starbursts show regions of intense massive star formation, and these generate extremely high fluxes of cosmic rays; these may be ten thousand times (or more) higher than those in the Milky Way. Papadopoulos[8] has proposed that in these so-called "cosmic ray dominated regions", the conditions inside dark molecular clouds are modified to such an extent that star formation may be different to that in the Milky Way and, in particular, that the initial mass function (IMF) may be modified. (The IMF describes the distribution of stellar masses for a stellar population; normally a distribution extending from a relatively few massive stars to relatively many low mass stars). Modelling shows that high cosmic ray ionization rates may have a dramatic effect on the chemistry in star-forming regions. Consequently, cosmic rays affect the potential molecular coolants involved in star formation. For ionization rates about one thousand times larger than the Milky Way average, molecular hydrogen begins to be significantly dissociated, and the gas temperature also rises.[9] For even larger ionization rates, abundant hot atomic hydrogen erodes most interstellar molecules with ease, so that the normal gas phase chemistry in dark clouds is heavily suppressed. Thus, the cooling function normally supplied by simple molecules such as CO, OH, and H_2O is greatly reduced and star formation, at least for stars of modest mass, is predicted to be suppressed. Stars of high mass may still form, as in the early Universe (see Section 7.3.1).

In closing, we consider the effects of photo-processes. Galaxies with intense massive star formation, such as starbursts and mergers, also generate huge UV fields that can affect the chemistry within dark clouds, even if the cosmic ray fluxes are not increased. The affected regions, known as *photon dominated regions* are observed to be very powerful molecular line emitters in external galaxies. One may expect that strong UV will suppress all chemistry, and abundances of some molecules are reduced when high UV fields are imposed.[10] But some species seem to be insensitive to increases in the UV intensity up to a thousand times that of the Milky Way; these include CO, CN, OH, HNC, HCN, HCO^+, and H_3O^+, and some species actually show small increases if the UV intensity is increased (C_2, C_3). Evidently, destruction of one species may aid the formation of another; for example, destruction of CO releases free carbon atoms and ions that are quickly recycled back into CO and into minor species such as C_2 and C_3. Bayet *et al.*[10] have shown that the chemistries of starbursts (high UV intensity) and of starbursts with AGN (high UV intensity with high cosmic rays fluxes) may be distinguished by their molecular line emissions from active regions.

7.4 Final Remarks

Astrochemistry in the Milky Way is now a mature subject. It provides researchers with the capability to describe the physics and chemistry of dense, dark and massive regions of the interstellar medium, and to interpret the behaviour of specific regions where rapid evolution is occurring. This has been the main topic of this book. However, the subject is not complete, and in Section 7.2 we have listed examples of a few topics where further work is needed. Of course, the essential laboratory work that has validated our present understanding of astrochemistry and underpins the entire subject will continue to explore relevant gas-phase, surface and solid-state chemistry. In Section 7.3 we have shown that models of the chemistry in the Milky Way may be adapted to describe chemistry in the early Universe and in external galaxies. While the early Universe situation is essentially different because of the lack of heavy elements, it seems that models of Milky Way chemistry may be readily adapted to describe and understand the chemistry of external galaxies.

References

1. E. K. Campbell, M. Holz, D. Gerlich and J. P. Maier, *Nature*, 2015, **523**, 322.
2. G. A. H. Walker, D. A. Bohlender, J. P. Maier and E. K. Cambell, *Astrophys. J. Lett.*, 2015, **812**, L8.
3. P. A. Oesch, G. Brammer, P. G. van Dockum and 15 co-authors, *Astrophys. J.*, 2016, **819**, 129.
4. D. Galli and F. Palla, *Annu. Rev. Astron. Astrophys.*, 2013, **51**, 163.
5. I. I. Stefan, C. L. Carillia, J. Wagg and 7 co-authors, *Mon. Not. R. Astron. Soc.*, 2015, **451**, 1715.
6. C. J. Conselice, A. Wilkinson, K. Duncan and A. Mortlock, *Astrophys. J.*, 2016, **830**, 83.
7. E. Bayet, S. Viti, D. A. Williams and J. M. C. Rawlings, *Astrophys. J.*, 2008, **676**, 978.
8. P. P. Papadopoulos, *Astrophys. J.*, 2010, **720**, 226.
9. E. Bayet, D. A. Williams, T. W. Hartquist and S. Viti, *Mon. Not. R. Astron. Soc.*, 2011, **414**, 1583.
10. E. Bayet, S. Viti, D. A. Williams, J. M. C. Rawlings and T. Bell, *Astrophys. J.*, 2009, **696**, 1477.

Subject Index

Tables and charts listed in bold.